ALLE ZEIT WACH
1842

Nucleocytoplasmic Transport

Edited by
R. Peters and M. Trendelenburg

With 115 Figures and 12 Tables

Springer-Verlag
Berlin Heidelberg New York
London Paris Tokyo

Dr. Reiner Peters
Max-Planck-Institut für Biophysik
Kennedyallee 70
6000 Frankfurt 70, FRG

Dr. Michael Trendelenburg
Deutsches Krebsforschungszentrum
Im Neuenheimer Feld 280
6900 Heidelberg, FRG

The cover is based on an idea of J. Bernhardt and M. Scholz

ISBN-13:978-3-642-71567-9 e-ISBN-13:978-3-642-71565-5
DOI: 10.1007/978-3-642-71565-5

Library of Congress Cataloging-in-Publication Data. Nucleocytoplasmic transport. Includes index. 1. Cell nuclei. 2. Cytoplasm. 3. Biological transport. I. Peters, R. (Reiner), 1943-. II. Trendelenburg, M. (Michael) QH595.N84 1986 574.87'32 86-26008

Softcover reprint of the hardcover 1st edition 1986

2131-3130/54321

Preface

The exchange of information and matter between cell nucleus and cytoplasm is an intriguing aspect of the eukaryotic cell which has attracted the cell biologists' attention for many decades. Nevertheless, the elucidation of nucleocytoplasmic transport is still in a stage where questions can be more easily formulated than answered. Considering the explosive progress in molecular biology, the somewhat prodromal stage of nucleocytoplasmic transport studies may seem astonishing. However, the situation becomes immediately intelligible if technical aspects are taken into consideration. Nucleocytoplasmic transport is a function of the living cell and the development of suitable in-vitro systems proved to be difficult.

Recently, novel techniques for measuring molecular transport in single cells have been developed. Substantial progress has been also achieved by gene technology. In this situation a 'Workshop on Nucleocytoplasmic Transport' was held at the Deutsches Krebsforschungszentrum (German Cancer Research Center), Heidelberg, Federal Republic of Germany. A brief yet essentially complete picture of the field and its prospects was sought. The present book contains the proceedings of this workshop.

The book is organized in the following manner:
- Historical dimensions of the field are illuminated by Brachet.
- Techniques for single-cell studies are reported by Engelhardt et al., Garland & Birmingham, Krafft et al., Feldherr, Neuhaus & Schweiger, Riedel et al., Edström, and Trendelenburg et al..
- The structure and biochemistry of the nuclear envelope is discussed by Milligan, Maul & Schatten, Krohne & Benavente, and Dreyer et al..
- Nucleocytoplasmic transport of proteins is dealt with by Dingwall et al., Schulz & Peters, Roberts et al., Deppert & Staufenbiel, Colman & Davey, Richter & Jones, Hall, Dierich et al., and Gehring.
- Nucleocytoplasmic transport of ribonucleoproteins is described by Mattaj and by Skoglund et al., respectively.

Frankfurt and
Heidelberg

R. Peters
M. Trendelenburg

Contents

List of Contributors

You will find the addresses at the beginning of the respective contribution

A Few Reminiscences About Nucleocytoplasmic Transport

J. BRACHET
Department of Molecular Biology, University of Brussels, 67, Rue des Chevaux, 1640 Rhode St. Genèse, Belgium

To be or not to be, that was the question for the nuclear membrane when I attended, in 1927, my first course in Cytology. For a majority of cytologists, the nucleus was surrounded by a mere film resulting from a precipitation between nuclear and cytoplasmic colloids; the membrane seen in fixed and stained preparations under the microscope was nothing but an artifact; its very existerce was doubtful. This is what I was taught by our teacher, this is what E.B. Wilson said in his famous treatise "The Cell in Development and Heredity" (1925). There was so little interest in the nuclear membrane that Wilson devoted less than one page to the topic. Nobody was interested, 60 years ago, in exchanges of molecules between the nucleus and the cytoplasm; nobody believed that, if such exchanges existed, they might be important for the cell. This overcautious attitude toward the nuclear membrane was due to the powerful influence of Jacques Loeb: in "The Dynamics of Living Matter" (1906), he wrote on p. 191 : "the nucleus is surrounded by a solid film". And on p.39, he corrected Traub who had been speaking of a cell membrane : he wrote in a footnote : "we should now say, the surface film of the protoplasm".

In those days, colloids, coacervates, etc. explained everything ; macromolecules did not yet exist. DNA (then called thymonucleic acid) was restricted to the nuclei of animal cells; plant cells had instead RNA in their nuclei. Thymonucleic acid, with a molecular weight of only 1.387 dalton, could not possibly play a genetic role; however, it deserved some respect as a colloid and possibly a buffer controling the pH of the nucleus. The impact of colloid chemistry, in those days, has been very damaging to cell biology : everything one saw under the microscope was a fixation artifact. This laid Wilson to present in his book a grossly oversimplified diagram of a cell. The Golgi apparatus was clearly an artifact; skeptics expressed doubts about the reality of mitochondria and mitotic apparatuses.

Nucleocytoplasmic Transport
Edited by R. Peters and M. Trendelenburg

Very few people believed in cell organelles; they were right, but nobody believed them before the advent of phase contrast and electron microscopy.

However, as pointed out by Wilson (1925) and by our teacher in 1927, micromanipulation experiments by Kite and by Chambers spoke for the existence of a nuclear membrane; at least in some cells, they found that the so-called nuclear film is, in some cases, of a very tough and resistant nature. Pricking the germinal vesicle of an oocyte resulted in an outflow of nuclear sap into the cytoplasm followed by cytolysis. These experiments seemed convincing to the biochemist A.P. Mathews : in "Physiological Chemistry" (1927), he spoke openly of a nuclear membrane.

I was not particularly interested in the nuclear membrane until, around 1936, I isolated by accident the germinal vesicle of a large frog oocyte. I decided to test, with this material, the existing theories about the biochemical functions of the cell nucleus and I could show, in 1938, that, in contradiction to the general belief, the nucleus is neither the main site of cellular oxidations, nor a site of hydrolytic enzymes accumulation. During these studies, a few of the isolated germinal vesicles broke down accidentally and a broken nuclear membrane could unmistakably be seen under the microscope. This convinced me of its reality.

I needed some help for further reminiscences about nucleocytoplasmic exchanges and I found it in a book ("Biochemical Cytology", 1957) and a review article in "The Cell" volume 2 (1961) I wrote about this subject. Many references can be found by those interested in the History of Cell Biology in this book and this review. When I have read them again, before writing the present paper, I experienced very mixed feelings.

The nuclear membrane was treated in less than 8 pages (including 7 figures) in "Biochemical Cytology". These few pages dealt with the structure and the permeability of the nuclear membrane; they give a fair account of what was known about these two topics in 1957.

The pioneering work on the nuclear membrane <u>structure</u> is undoubtedly that of Callan and Tomlin (1950) who studied with the electron microscope the nuclear envelope of broken germinal vesicles isolated from amphibian oocytes. They discovered that the nuclear membrane is formed of two sheets: a continuous, homogeneous inner layer and an outer layer containing pores or annuli packed horizontally. A little later (1952-1954), a similar structure was described in amebae and in <u>Chironomus</u> larval salivary glands. By 1955, it was widely accepted that the nuclear membrane contains pores or holes; however, a minority of workers still believed that the pores are optical artifacts. Callan (1955) suggested that one of the functions of the porous outer layer might be to give more mechanical resistance to the continuous inner layer. On the other hand, M.L. Watson (1954, 1955), judging from the dimensions of the pores, concluded that the nuclear membrane is freely permeable to large molecules such as nucleic acids and proteins. For Gay, also in 1955, outfoldings of the nuclear membrane would contribute to the formation of the endoplasmic reticulum. My conclusion in 1957, about her work was that "it is a fascinating suggestion that exchanges of materials between the chromosomes and the RNA-rich cytoplasmic structures could occur in this manner; but (that) one could not avoid a vague suspicion that artifacts might play dangerous tricks when one comes to such very fine details".

H.G. Callan was also a pioneer in the study of nuclear membrane permeability (1948, 1952); he worked again with germinal vesicles isolated from aphibian oocytes and concluded that salts, sugars and polypeptides, but not proteins, readily cross the nuclear membrane barrier. Later, Holtfreter (1954) reported the penetration of hemoglobin into isolated germinal vesicles. But these oocyte nuclei are exceedingly fragile : slight damage to their slender extensions might have been responsible for Holtfreter's results. However, similar work done on somatic nuclei isolated from homogenates by centrifugation led to a confirmation of Holtfreter's conclusions: it was found that various enzymes leach out of isolated liver nuclei; conversely, addition of ribonuclease or deoxyribonuclease to isolated nuclei modifies the size and shape of the nucleoli. In his 1953 review of the subject, Anderson concluded that proteins can indeed cross the nuclear membrane.

I was not very happy with the experiments done on isolated nuclei because the permeability of their nuclear membrane might be altered as soon as they are isolated. I believed that it is essential, for a real understanding of nucleocytoplasmic exchanges, to find out whether the nuclear membrane is really permeable to proteins in living cells. I studied therefore, in 1955-1956, the penetration of ribonuclease into living amebae and onion roots. The truth is that the main purpose of these experiments was not to analyze nuclear membrane permeability : I wished to demonstrate that the integrity of RNA is required for protein synthesis in living cells. I had proposed, in 1942, independently of T. Caspersson, that RNA is somehow involved in protein synthesis; experiments by several groups of american biochemists (at a time where radioactive aminoacids were available in the USA, but unfortunately not in Belgium) had shown, in the early fifties, that the RNA-rich microsomes are the major site of protein synthesis; pretreatment of the microsomes with ribonuclease abolished their capacity to incorporate labelled amino acids into proteins. Is this also true for living cells ? Since it was known that proteins can penetrate into amebae and other cells by pinocytosis, it seemed important to find out whether addition of ribonuclease would inhibit protein synthesis in living cells. The result of these experiments was that addition of crystalline ribonuclease to onion root tips produced a 50% inhibition of aminoacid incorporation into the proteins within 1 hr, and a 90% inhibition after a 3 hr treatment; there was a parallel inhibition of the growth of the roots. Similar results were obtained, in my laboratory, with the ameba A. Proteus, frog and starfish oocytes, amphibian eggs, liver and ascites tumor cells, etc. The cytological analysis of ribonuclease-treated onion roots and amebae brings us back to our present topic, nuclear membrane permeability. I was immediately convinced that the enzyme penetrates into the nuclei of the still living cells : in amebae, the nucleoli were modified in shape and stainability as early as 15 min after addition of ribonuclease to the medium; later, the nucleus underwent swelling and the nucleoli completely lost their basophilia. In onion root tips treated in vivo with ribonuclease, staining of the nucleoli with basic dyes was suppressed, in all cells, after a 1 hr treatment. It was not clear, at the time, whether loss of staining in the nucleoli was due to formation of an enzyme-substrate complex or to actual RNA breakdown. But this did not alter the fact that ribonuclease, a protein of Mr 13.000, readily penetrated into the nuclei of intact cells: the nuclear membrane is thus permeable to proteins of that size. In order to prove completely this point, amebae were treated with radioactive (labelled with radioactive iodine) ribonuclease and submitted to autoradiography (Brachet, 1955): some of the tracks clearly originated from the nucleus.

Looking back to these now forgotten experiments (they had almost slipped out of my own mind !) one can conclude that they brought the first experimental evidence of a traffic of proteins from the cytoplasm into the nucleus : at least one protein, ribonuclease, can cross the nuclear membrane and penetrate into the nucleus of a living cell. However, the choice of ribonuclease was far from ideal for studies on nuclear membrane permeability : it could be argued that ribonuclease, after entering the cell, might digest some hypothetical

RNA-containing component of the nuclear membrane, allowing the enzyme to penetrate into the nucleus. One had to wait 15 years before Gurdon (Proc. roy. Soc. B 176, 303, 1970) solved elegantly the problem : injecting iodine-labelled proteins into *Xenopus* oocytes, he demonstrated that injected histones quickly accumulate into the nucleus while injected bovine serum albumin remains in the cytoplasm. Continuation of these experiments by Bonner and others led to the now familiar distinction between karyophilic and karyophobic proteins which are one of the major topics discussed in the present Symposium. Coming back to "Molecular Cytology", a long section was devoted to my merotomy experiments. I thought that cutting into two halves unicellular organisms and comparing RNA and protein synthesis in the nucleate and anucleate fragments might give some insight about the role played by the nucleus in macromolecule synthesis. Cytochemical and biochemical studies on sectioned amebae (*A. proteus*) showed that the nucleate halves kept their RNA content constant even after 12 days of fasting; in contrast, there was a steady and marked drop (60%) in the RNA content of the anucleate halves during the same period of time. The conclusion was that the nucleus maintains the cytoplasmic RNA level; the results were compatible with the idea that cytoplasmic RNA originates from nuclear RNA, implying that RNA synthesized in the nucleus crosses the nuclear membrane and moves into the cytoplasm. A more direct demonstration of the nuclear origin of cytoplasmic RNA was brought forward at the same time by Goldstein and Plaut (1955): they grafted a ^{32}P-labelled nucleus into a normal unlabelled amoeba and found, by autoradiography, that the cytoplasm of the recipient ameba became radioactive after 12 hours; in contrast, the originally unlabelled nucleus of this ameba did not acquire any significant amount of radioactivity. Digestion of the fixed amebae with ribonuclease established that all the autoradiographically labelled ^{32}P, in both the nucleus and the cytoplasm, was in the form of RNA. Although some aspects of these pioneering experiments remained open to discussion, they showed, for the first time, that RNA synthesized in the nucleus is transferred to the cytoplasm in a living cell. However, the experiment did not rule out the possibility that independent RNA synthesis occurs in the cytoplasm. We found that anucleate fragments of amebae indeed incorporate labelled adenine into RNA. Nothing was known about mitochondrial DNA and its transcription in those days and it was ignored that most strains of amebae harbor symbiontic organism. Their discovery by Prescott (1960) decided me to drop *A. proteus* merotomy experiments on *A. proteus.*

Our studies (Brachet et al., 1955) on RNA synthesis in Acetabularia led to still more confusing results. This green, very large unicellular alga is remarkable by the fact, discovered by J. Hämmerling in 1934, that anucleate fragments undergo extensive regeneration (formation of a large, species-specific cap or umbrella). We found, as expected, that RNA synthesis is intensive in the large vegetative nucleus of the alga (especially in its nucleolus). But quite unexpected results were obtained when we followed the RNA content of nucleate and anucleate fragments : there was considerable RNA synthesis during more than one week after removal of the nucleus. Paradoxically, its initial rate was even faster in the anucleate than in the nucleate halves. Thus removal of the nucleus speeded up RNA (and protein) synthesis, in agreement with the already known fact that caps form earlier in anucleate than in nucleate fragments of the alga. The presence of the nucleus thus exerts an inhibitory effect on cap formation, RNA and protein synthesis. The molecular mechanism for this negative control of the nucleus on cytoplasmic activities are not yet understood, but there is little doubt that nucleocytoplasmic transport is somehow involved.

Autonomous cytoplasmic RNA synthesis in anucleate algae was completely baffling in 1957 and I proposed no explanation at the time. Later work, in collaboration with H. Naora (1960),clarified the situation: there is, in anucleate fragments of Acetabularia, a net synthesis of chloroplastic RNA and a decrease in microsomal RNA. These findings allowed me to conclude in "The Cell" (Vol. II p. 806, 1961) that "synthesis of the RNA which is present in the chloroplasts (which may be considered as self-duplicating bodies) is largely independent of nuclear control; in this respect, chloroplasts behave very much like mitochondria in amebae. On the other hand, nonchloroplastic (microsomal + soluble) RNA lies under strict nuclear control".

I was rather satisfied with myself when I read recently this sentence which could still be written to-day. Satisfaction gave place to dismay when I came to the final scheme which summarized the proposed relationships between DNA, RNA and proteins in both "Biochemical Cytology" and "The Cell". This scheme does not deserve to be reproduced here, but its main features may be briefly commented.

As expected, I concluded that cytoplasmic RNA is involved in protein synthesis, and this remains true to-day. That nuclear RNA is the precursor of cytoplasmic RNA was not so obvious; this was my own belief, on the basis of the ameba work. But I had to admit that the biochemical data concerning the nuclear origin of cytoplasmic RNA were in a contradictory and confusing state : papers published in 1957 still concluded from experiments on homogenates of labelled cells that the results are not consistent with the assumption that nuclear RNA is the precursor of cytoplasmic RNA. Nevertheless, I finally concluded that a very large proportion of cytoplasmic RNA is of nuclear origin, but that synthesis of cytoplasmic RNA (probably chloroplastic and mitochondrial) also occurs in the absence of the nucleus.

So far so good ! But I was very reluctant in accepting the idea that nuclear RNA is synthesized under the influence of nuclear DNA: after discussing the rather meagre information then available, I concluded that a question mark should be left for the DNA → nuclear RNA step. Thus the reality of an exceedingly important event, transcription, was still in doubt in 1961.

Is DNA directly (i.e. without RNA intermediates) involved in protein synthesis ? My answer to that question was : "no", but I had no strong reason to support my belief: experimental evidence was still very conflicting. The idea that DNA makes directly proteins was still widespread but was steadily loosing ground.

My conclusions about the origin of the nuclear proteins look very strange to-day. Chromosomal proteins (histones and "residual" non-histone proteins) were thought to be synthesized directly by chromosomal DNA; I suggested that they are synthesized by a hypothetical chromosomal RNA bound to DNA. The nucleolar proteins must undoubtedly be synthesized by nucleolar RNA. The existence of a movement of neosynthesized cytoplasmic proteins into the nucleus was sadly overlooked, although I knew that proteins may cross the nuclear membrane and enter the nucleus. It has been proposed by T. Caspersson that the nucleus is the main center of protein synthesis; although I knew from my merotomy experiments that this could not be true for the great majority of proteins, I still believed that the nucleus is the center of nuclear protein synthesis. It did not occur to me (nor to anybody else) that nuclear proteins might be synthesized in the cytoplasm and transferred into the nucleus. In my scheme, arrows pointed from the nucleus toward the cytoplasm, indicating correctly the transfer of nuclear RNA to the cytoplasm; but no arrow pointed in the opposite direction for cytoplasmic proteins. This is why I read my old book and review with very mixed feelings.

We were groping in the dark, 25 years ago, because of our ignorance of too many essential facts : there were only two kinds of RNAs (microsomal and soluble); the molecular mechanisms of protein synthesis were not understood; RNA polymerases, mitochondrial and chloroplastic DNAs and many important other things were unknown. Nobody (except probably François Jacob) was interested in gene regulation, an exciting topic to-day. Gigantic progresses have been made, during the past quarter of a century, in our understanding of eukaryotic cells. We owe a great deal to the molecular biologists working on bacteria, phages and macromolecules, who had no reason to worry about nuclear membrane and nucleocytoplasmic transport. This allowed one of the greatest among them, Francis Crick (1958), to summarize in a few words (DNA makes RNA, and RNA makes protein) the essence of our still vague ideas about nucleocytoplasmic interactions.

References

Anderson, N. G. Science 117, 517

Brachet, J. (1955) Biochim. Biophys. Acta 18, 247

Brachet, J. (1955) in Nucleic Acids (Davidson, J. N. and Chargaff, E., eds.), vol. 2, pp. 771-841, Academic Press, New York

Brachet, J. (1957) in Biochemical Cytology, Academic Press, New York

Brachet, J. and Mirsky, A. E. (1961) in The Cell, vol 2, Academic Press, New York

Callan, H. G. (1948) Ricerca Sci. 18,3

Callan, H. G. (1952) Symp. Soc. Exp. Biol. 6, 243

Callan, H. G. (1955) Proc. VIII. Int. Congr. Cell Biol.

Callan, H. G. and Tomlin, S. G. (1950) Proc. Roy. Soc. B 137, 367

Casperson, T. O. (1950) in Cell Growth and Function, Norton, New York

Crick, F. H. S. (1958) Symp. Soc. Exptl. Biol. 12, 138

Gay, H. (1955) Proc. Natl. Acad. Sci. USA 41, 370

Goldstein, L. and Plaut, W. (1955) Proc. Natl. Acad. Sci. USA 41, 874

Gurdon, J. B. (1970) Proc. Roy. Soc. B 176, 303

Holtfreter, J. (1954) Exp. Cell Res. 7, 95

Loeb, J. (1906) in Dynamik der Lebenserscheinungen, Barth Verlag, Leipzig

Mathews, A. P. (1925) in Physiological Chemistry, Bailliere, Tindall and Cox, London

Naora, H. et al. (1960) J. Gen. Physiol. 43, 1083

Prescott, D. M. (1960) Exptl. Cell Res. 19, 29

Watson, M. L. (1954) Biochim. Biophys. Acta 15, 475

Watson, M. L. (1955) J. Biophys. Biochem. Cytol. 1, 257

Wilson, E. B. (1925) in The Cell in Development and Heredity, Macmillan, New York

Sensitive Techniques to Measure the Elastic Properties of Cell Membranes

H. Engelhardt, K. Fricke, H.P. Duwe, and E. Sackmann
Physics Department E22 (Biophysics Group), Technical University of Munich,
James-Franck-Straße, D-8046 Garching, FRG

Introduction

As plasma membranes of cells can hardly be treated in their whole complexity, it is useful to reduce them in terms of a three layer system consisting of (see Fig 1):

1) The glycocalix (control of the interaction with the environment or with other cells).
2) The lipid-protein bilayer (location of biochemical membrane processes).
3) The cytoskeleton (coupling with cytoplasm).

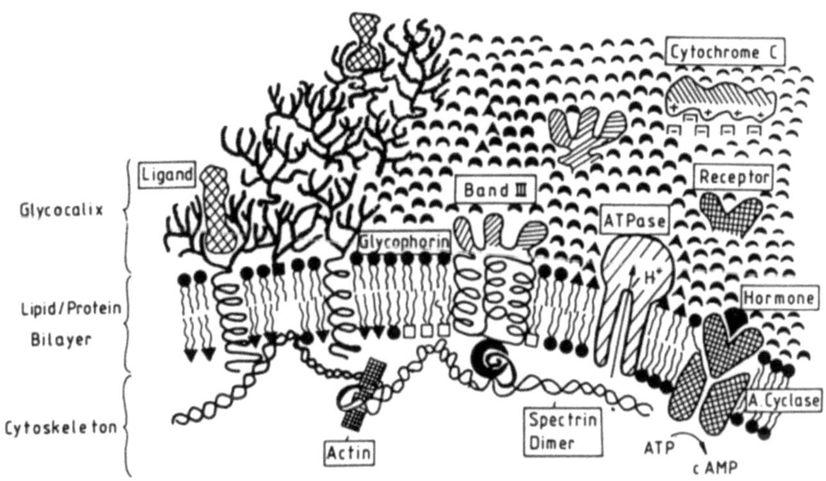

Fig 1: Cartoon of the present view of the plasma membrane.

Nucleocytoplasmic Transport
Edited by R. Peters and M. Trendelenburg

The elastic properties of a plasma membrane are determined by the structure of these three components and their mutual coupling. Therefore we can get an insight into the microstructure and the coupling of the membrane components by measuring the elasticity of a cell (Evans and Skalak 1980, Fricke and Sackmann 1984, Deuling and Helfrich 1976, Engelhardt et al 1984, Sackmann et al 1984).

In this contribution sensitive techniques are presented where small forces are involved. A membrane can be deformed in different ways (Fig 2). We can stretch it, shear it or bend it. For small forces acting on a cell usually only bending deformations will occur. For medium and strong forces, shearing and finally stretching will be the answer of the cell. A pure lipid membrane for example will not resist shearing deformations as long as it is in the fluid state. But if there is a network like the glycocalix or the cytoskeleton coupled to the lipid bilayer, this membrane is of course expected to resist shear forces. Obviously different techniques are necessary in order to obtain information about the different types of deformations. On the other hand different kinds of deformations allow us to distinguish the contributions of different components of the plasma membrane.

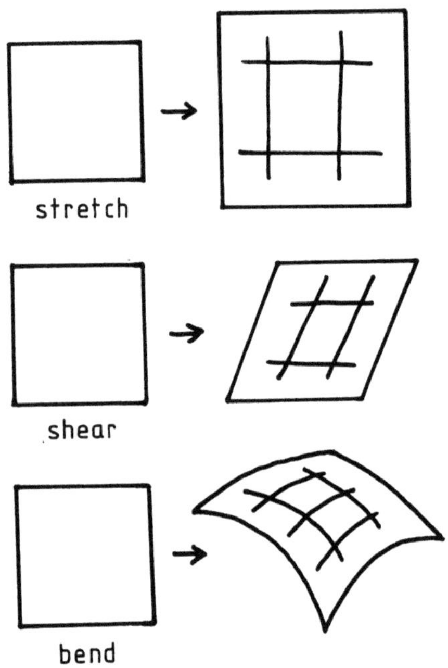

Fig 2: Schematic representation of different types of membrane deformations.

The flicker spectroscopy and the reflection interference contrast microscopy are used to study lipid-protein-bilayer properties of erythrocytes. Phasecontrast microscopy allows the study of pure lipid bilayers (vesicles). The electric field jump technique applies deformations in the order of 1 μm to the cells. The shear forces that occur in this case give information about biopolymer networks like the cytoskeleton and the glycocalix and their coupling.

1.) Flicker Spectroscopy

Everyone who ever looked through a microscope saw the Brownian motion of small particles, which is driven by thermal fluctuations. An erythrocyte in the phasecontrast microscope shows in addition fluctuations of the intensity of the transmitted light which are caused by thermally driven undulations of the plasma membrane. This effect is called flickering. The surface undulations can be seen even more clearly at large flaccid vesicles which exhibit pronounced shape fluctuations as is shown in Fig 8. The thermal undulations of the cell membranes allow us to evaluate their elasticity. Indeed the frequency analysis of the cell flickering, the flicker spectroscopy, is a very sensitive nondisturbing dynamic technique by which it is possible to detect even very small changes in membrane elasticity (Fricke et al 1986).

In the simplest case we analyse the oscillations in the middle of

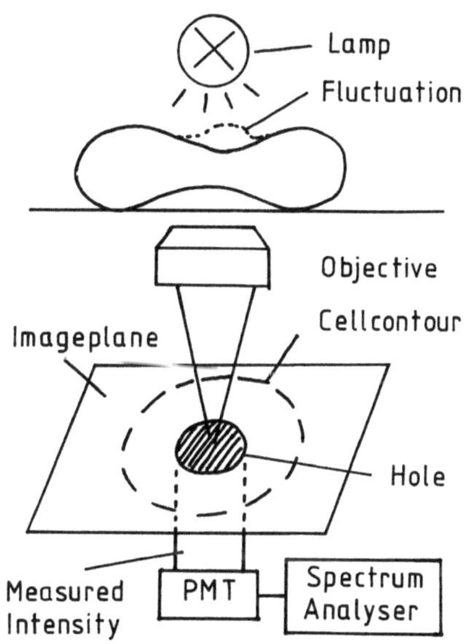

Fig 3: Flicker spectroscopy: Thickness fluctuations of the cell cause intensity fluctuation of the transmitted light. This is measured by a photomultiplier and analyzed with a spectrum analyzer.

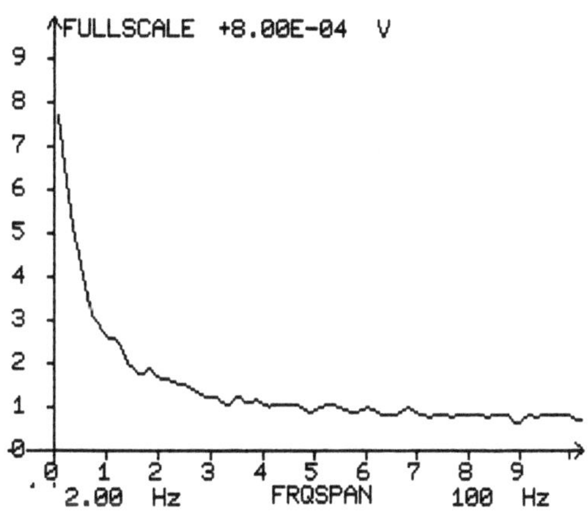

Fig 4: The flicker spectrum shows the amplitudes of the thermally excited undulations as a function of the frequency.

the cell. This is done by positioning a diaphragm with a diameter of 1/3 of that of the cell in the image plane of the phase contrast microscope (Fig 3). We measure the intensity with a photomultiplier tube. The frequency analysis is performed by a spectrum analyser. A typical spectrum obtained by this technique is shown in Fig 4. An example of the sensitivity of this technique is given in Fig 5. Here we treated human erythrocytes for example with 1/2 volume percent of ethanol. The amplitudes are reduced by about a factor of two. When we perfuse the cell again with normal medium, the amplitude recovers and reaches nearly the old value. Only a small irreversibility is observed (Fig. 5). This procedure can be repeated several times with the same cell. This example demonstrates that very subtile changes of the membrane structure caused by chemicals, drugs or diseases can be detected by the flicker spectroscopy.

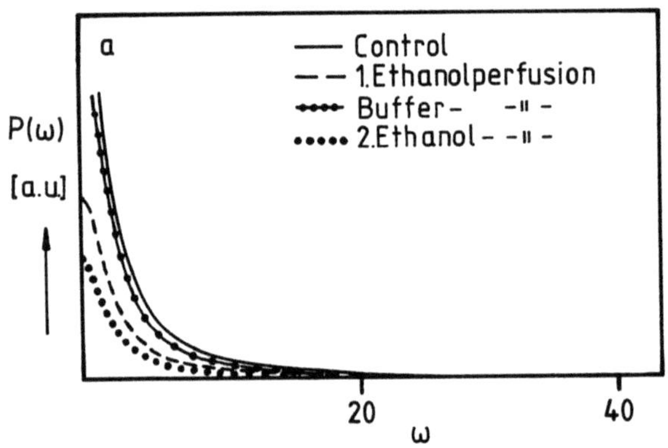

Fig 5: The amplitudes of the flicker spectrum are almost reversibly changed by perfusing the cell with normal buffer solution and buffer containing 0.5 % ethanol.

2.) Reflection Interference Contrast Microscopy

In an other modified technique we study the surface undulations by reflection interference contrast microscopy. The experimental device and the principles of the technique are shown in Fig. 6. By this technique Newtonian rings are formed by the interference of the light that is reflected from the surface of the microscope slide with that from the cell membrane. Membrane deformations can be measured by evaluating the lateral shift of the interference rings. With the help of the image processing system we can determine the contours as a function of time. In this way we are able to reconstruct the 3 dimensional shape of the cell surface and determine amplitudes of the undulations as small as 0.1 micron. This sophisticated method yields more information especially about the wavelength distribution of thermally exited undulations.

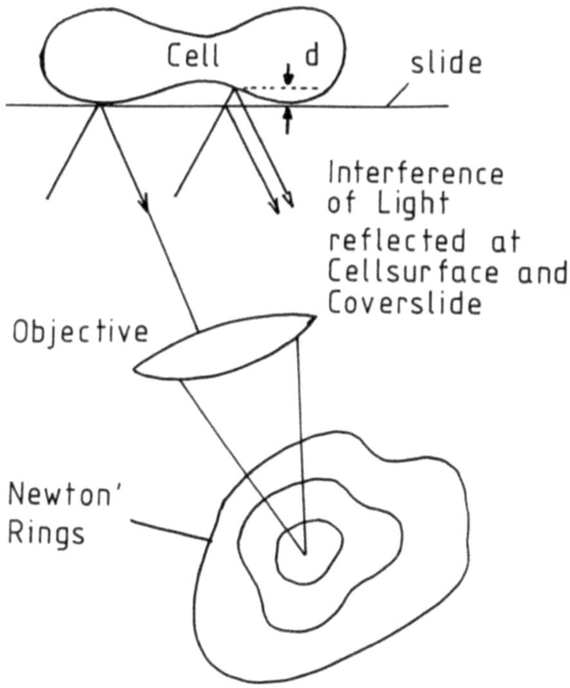

Fig 6: Newtonian rings are formed by the interference of the light reflected from the microscope slide with that from the cell membrane.

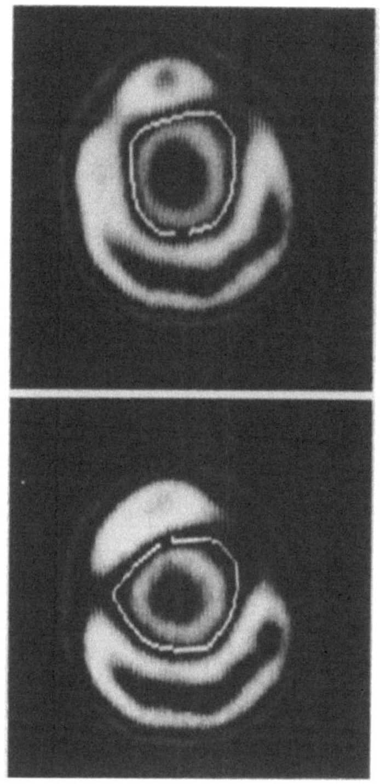

Fig 7: Interference contrast micrographs of an erythrocyte at different times (Δt 1 sec).

3.) Phase Contrast Microscopy

In order to study the elastic properties of pure lipid bilayers we use artificial membranes that is lipid vesicles (Engelhardt et al 1985). As vesicles are transparent the shape fluctuations can easily be seen by phase contrast microscopy (Fig 8) and are analysed as follows. As a first step we store images of the vesicel at different times. Then we tell the computer to search the coordinates of the vesicel contour. In the next step we analyse the deviations of the contour from a circle by Fourier analysis. Finally we have to convert the two dimensional fluctuations of the contour into three dimensional shape fluctuations. This procedure has been discribed previously (Schneider et al 1984, Engelhardt et al 1985).

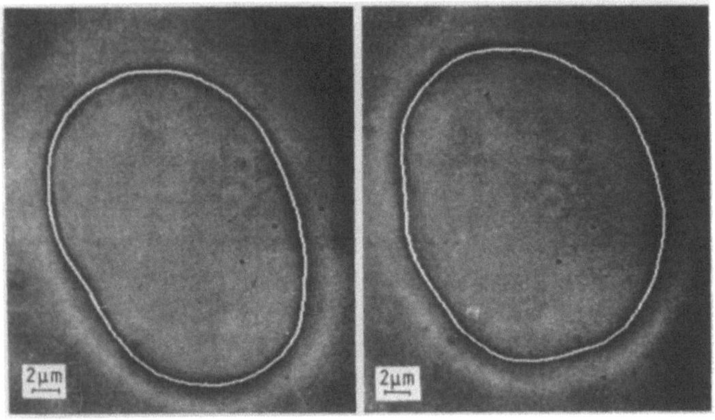

Fig 8: Photomicrographs of the contour of a vesicle. The time difference is 1 sec. The bright lines show the coordinates detected by the image processing system.

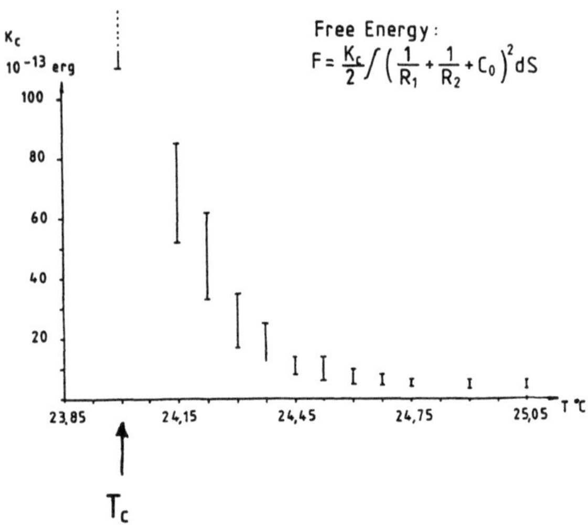

Fig.9: shows the curvature elastic modulus obtained for DMPC + 5% cholesterol near the main transition temperature. The curvature elastic modulus increases from a value of 4 * 10E-13 erg within 1/2 a degree celsius to a value of more than 100 * 10E-13 erg as one reaches the main transition.

4) The Electric Field Jump Technique

In the fourth technique which is a weak disturbation method we use high frequency electric fields to deform cells (Engelhardt et al 1984). Appreciable shear forces occure with this technique.
The effect of high frequency electric fields on cells situated between two electrodes with sharp edges is shown in Fig 10.

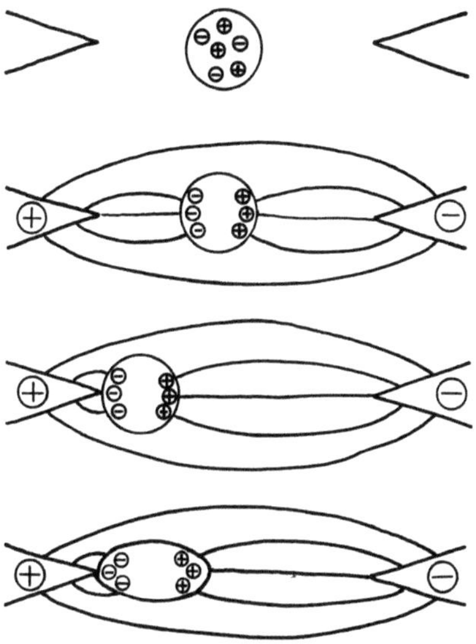

Fig 10: For a simplified description we assume the suspension medium to contain no ions. This is justified approximatly for our experimental conditions. If we apply a weak electric field the ions within the cell start to move, following the electric field lines until they bump against the insulating membrane. Therefore the cell is polarized and moves like an electric dipol to the point of highest field strength that is to the edge of one of the electrodes. In this way the cell is fixed at the electrode. If we now increase the field strength, the ions are pushing stronger against the cell wall and deforme the cell. As this effect does not depend on the polarity a net force acts on the cell as long as the change in polarity of the field occurs slow compared to the motion of the ions.

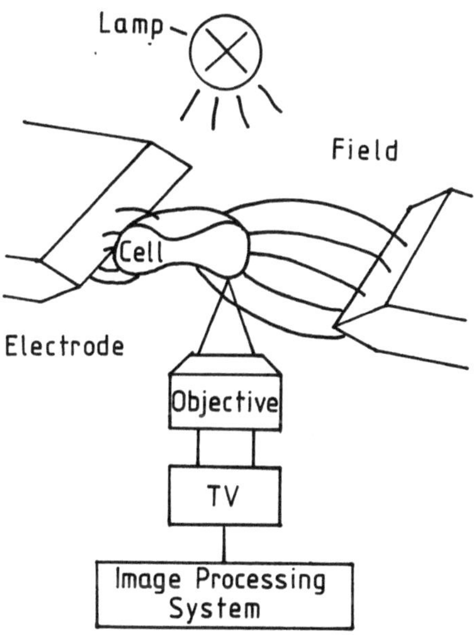

Fig 11: An erythrocyte is fixed and deformed by an electric high frequency field (3 MHz) between two electrodes with sharp edges. For data evaluation the microscope is connected to an image processing system.

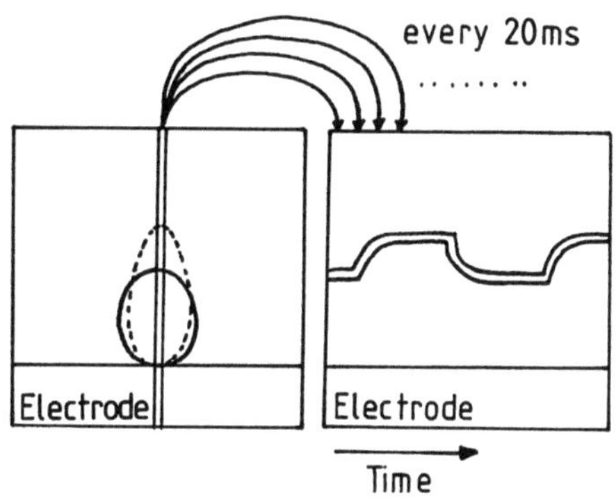

Fig 12: While switching on and off the force at a low frequency (0.1 Hz) the image processing system stores every 20 ms the center line of the cell image. This creates a deformation protocol of the cell. Then the computer searches for the steepest descend of the intensity of the cell contour. With this technique the deformation can be determined to an accuracy of 20 nm.

If we measure the elongation of the cell as a function of applied force created by the field, we obtain quantitative information about the elastic properties of the membrane especially about the shear elasticity which is determined by the cytoskeleton and the glycocalix. The experimental setup is shown in Fig 11 and measuring procedure is depicted in Fig 12.

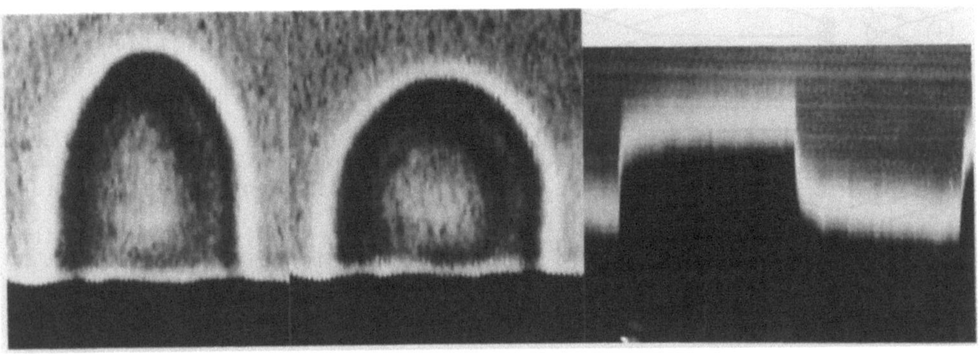

Fig 13: Photo micrograph of an erythrocyte deformed and not deformed. The right side shows the deformation protocol obtained by the procedure described in Fig 12.

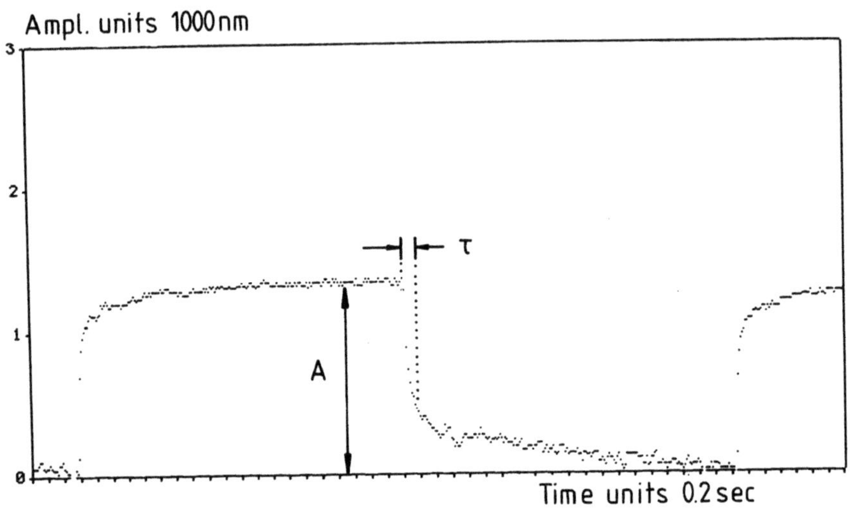

Fig 14: The deformation amplitude A gives the shear modulus of the cell membrane and the response time τ yields the viscosity of the membrane.

The elastic constants of the cell are obtained from the amplitudes of the deformation. The response time of the deformation yields information about the viscous or dynamic properties of the cell envelope. In addition, the plasticity of the cell may be studied by evaluation of the very slow (and irreversible) deformations. The latter property contains information about the cytoskeleton.

The presented technique may also be applied to other cells than erythrocytes. Fig 15 for instance shows the results for a hybridoma T-cell.

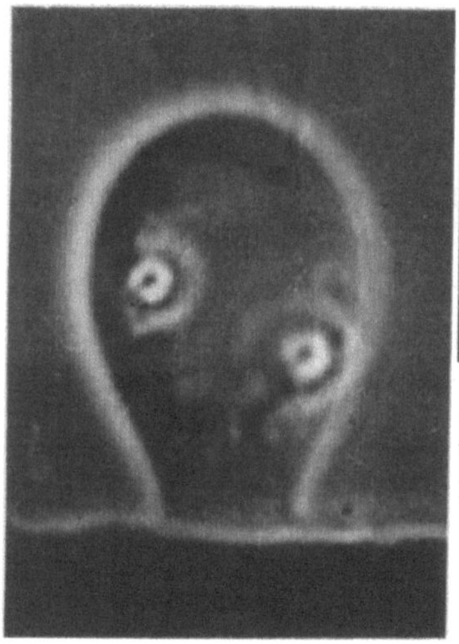

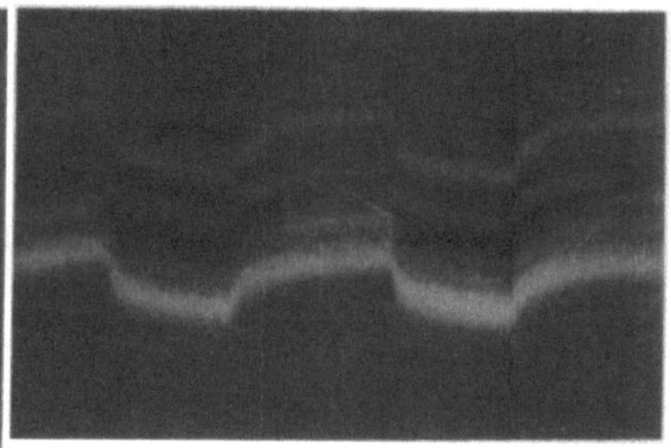

Fig 15: Left: deformed Hybridoma T-cell. Right: Deformation protocol of the Hybridoma T-cell.

Summary:

By combining different techniques it is possible to measure the different contributions to the membrane elasticity separately. This is essential in order to detect structural changes of the membrane in a selective way !

1) Non disturbing techniques, such as the flicker spectroscopy yield information about the curvature elasticity.

2) Weak disturbation techniques like the electric field jump method allow to measure shear elastic properties which provide information about the cytoskeleton and possibly about the glycocalix.

3) Techniques using large forces such as the micropipette method (not described here (Evans and Skalak 1980)) allow to measure the lateral compressibility of membranes.

(The forces and deformations that occur at Cell Poking tecniques ly in the range of 2) and 3).)
In order to understand the elastic properties of cell membranes, nuclear envelopes or other cellular membranes it is certainly necessary to study model membranes in a systematic way. Experiments with vesicles provide information about the lipid protein bilayer of membranes. By combining polymerizable and conventional phospholipids it is possible to build models of a lipid bilayer with a two dimensional cytoskeleton coupled to it (Gaub et al 1984).
We hope we could show you that sensitive techniques for measuring elastic properties of cells provide a valuable tool to gain insight into the complicated structure of cell plasma membranes. Such measurements are also of considerable practical interest since they allow to observe effects of drugs on membrane properties. In future these techniques could even have considerable impact in the diagnostic field.

References

1) Evans E.A and Skalak R. (1980) 'Mechanics and Thermodynamics of Biomembranes' CRC Press, Boca Raton USA, 1980.

2) Gaub H., Büschl R., Ringsdorf H. and Sackmann E. Biophysical J. 45 (1984) 725.

3) Fricke K. and Sackmann E. BBA 803 (1984) 142-152.

4) Deuling H.j. and Helfrich W. Biophysical J. 16 (1976) 861.

5) Sackmann, E. Engelhardt H., Fricke K. and Gaub H. Colloids and surfaces 10 (1984) 321-335.

6) Fricke K. Wirthenson K. Laxhuber R. and Sackmann E. European Biophysical Journal in press.

7) Schneider M.B., Jenkins J.T., and Webb W.W. J. Physique (Paris) 45 (1984) 1457.

8) Engelhardt H., Duwe H.P. and Sackmann E. J. Physique lettre 46 (1985) 395-400.

9) Engelhardt H., Gaub H. and Sackmann E. Nature 307 (1984) 378-380.

Modulation Methods and Slow Molecular Rotations

P. B. GARLAND[1] and J. J. BIRMINGHAM
Department of Biochemistry, Dundee University, Dundee DD1 4HN, Scotland, United Kingdom
[1]Present Address: Unilever Research, Colworth Laboratory, Sharnbrook, Bedford MK44 1LQ, United Kingdom

INTRODUCTION

The application of optical methods used with triplet - or other long-lived probes for measuring slow (i.e. μsec-msec) Brownian rotational diffusion of membrane proteins has been fully reviewed in the past few years (Cherry, 1979; Jovin et al., 1981; Garland & Johnson, 1985). We have previously emphasised (Garland & Birmingham, 1985) the fact that studies of the rotational diffusion of proteins in membranes or cells have, in many respects, lagged behind studies of lateral diffusion. Nevertheless, measurements of lateral diffusion and of rotational diffusion are complementary to each other, and in many instances it would be highly desirable to know D_R (the rotational diffusion coefficient) as well as the more readily obtained D_L (lateral diffusion coefficient). The same is true for measurements of that fraction of a macromolecular population that does not rotate or translate on the time-scale of the experiment. For example, a finding that D_L for a membrane protein measured over a distance of a micron or more is exceedingly slow (<10^{-13} cm^2 s^{-1}) does not indicate whether the protein is trapped by anchorage to some immobile structure (in which case D_R would be low) or merely confined by a limiting boundary to some sub-micron location within which D_L over short distances is high (<10^{-9} cm^2 s^{-1}). The extra information provided by D_R measurements could discriminate between these two possibilities.

The aim of our present work is to provide convenient optical methods suitable for measuring D_R of membrane proteins or other macromolecules in the μsec-msec time domain on microscopic samples such as single cells. We made considerable progress in that direction with the introduction of the so-called fluorescence depletion method, achieving a sensitivity of as little as 2 x 10^4 molecules of Band 3

Nucleocytoplasmic Transport
Edited by R. Peters and M. Trendelenburg

protein (the anion translocator) in a small area of erythrocyte ghost membrane (Johnson & Garland. 1981). Although there is still considerable scope for improving that method, it remains a pulse method which is likely to remain of limited time resolution in the 1-20 usec time domain for two reasons: (a) various artefacts that follow the photoselection pulse; (b) the need to keep the photoselection pulse duration long enough to spread the delivered energy over several usec if unacceptable membrane heating is to be avoided, at least for some probes of low triplet yield (Johnson & Garland, 1982). In this paper we describe some recent theoretical and instrumental developments in the application of phase modulation methods for measuring slow rotational diffusions.

COMPARISON OF PULSE AND PHASE-MODULATION METHODS

Both pulse and phase-modulation methods for the optical measurement of slow rotational diffusion require that (a) a suitable extrinsic and photoexcitable probe be attached to the protein of interest; (b) the probe has an excited state lifetime not very much less than the rotational relaxation time to be measured and (c) that the photoexcited state can be measured. An example of such a probe is erythrosin which is readily photoexcited to the phosphorescent state (Garland & Moore, 1979).

In the pulse method an intense but brief flash of plane polarised light excites some of the probe to a photoselected (and therefore anisotropic) triplet population. The resultant phosphorescence is also anisotropic, but becomes less so as the anisotropic population of triplet molecules randomizes by rotational diffusion (Cherry, 1979). In the phase method the excitation is continuous but has its intensity sinusoidally modulated with time. The modulation frequency is in the region of the reciprocal of the lifetime under study: e.g. 10-20 MHz for fluorescence, 0.1-50 KHz for phosphorescence. The emitted light is also intensity modulated, but delayed in phase and demodulated with respect to the excitation (Figure 1). For excitation modulated at an angular frequency w the relationship between the phase shift $\emptyset$ and lifetime τ is given by:

$$\emptyset = \tan^{-1} (w\tau) \qquad (1)$$

The demodulation ratio is the ratio of the modulations (i.e. the ratio of the AC amplitude to the average signal) of the excitation and signal waveforms, and is given by:

$$M = (1 + (w\tau)^2)^{-0.5} \qquad (2)$$

More generally, for multiple lifetimes where the fractional intensity of the ith component is f_i (and $\sum_i f_i = 1$), then:

$$\emptyset = \tan^{-1}(S/G), \; M = (S^2 + G^2)^{-0.5} \qquad (3)$$

where $S = \sum_i f_i M_i \sin \emptyset_i$ (4)

and $G = \sum_i f_i M_i \cos \emptyset_i$ (5)

If the excitation light is linearly polarized and the resulting signal (fluorescence, phosphorescence, triplet absorption, ground state depletion) is detected through polarizers orientated either parallel ($\parallel$) or perpendicular ($\perp$) to the excitation phase, then the differential signals of interest in obtaining rotation rates are:

$$\Delta = \emptyset_{\perp} - \emptyset_{\parallel}, \; Y = M_{\parallel}/M_{\perp} \qquad (6)$$

The differential polarized phase angle Δ is the phase angle difference between the signal waveforms detected through either perpendicular or parallel polarizers. The modulation ratio Y is the ratio of the modulation amplitudes of the two signal waveforms, parallel and perpendicular. Derivations of rotational correlation times from such signals are described by Lakowicz et al., (1985).

Phase-modulation methods have been used hitherto exclusively for fluorescence lifetime and depolarization studies, in the MHz frequency region upwards. The use of electro-optically modulated CW lasers has enabled data analysis to be carried out in the frequency domain, with the consequence that the resolution and convenience of phase-modulation methods now compares favourably with pulse methods and single-photon counting (Gratton et al., 1984a,b; Lakowicz et al., 1985; Beecham et al., 1983).

Pulse and phase-modulation methods are briefly compared in Table 1. The first four features listed are in favour of the phase-modulation method. The fifth feature, complexity of data analysis, becomes unimportant with the availability of dedicated and increasingly powerful micro- or mini-computers. The sixth feature, the need for several different frequency sets for the phase modulation method, can be overcome by automated measurements of signal phase and amplitude, and programmable frequency sweep. The seventh feature, two detection channels ($\parallel$ and $\perp$) is invariably a problem due to the need for accurate and noise-free differential measurements, irrespective of whether these detector channels are separated in time (i.e. one detector, change analysing polarizer) or in space (two detectors).

Feature	Pulse Method	Phase Modulation Method
Photo-excitation	High energy pulse	Continuous
Time resolution	Narrow pulse	High frequency
Photomultiplier gate	May be needed	Never needed
Pulse-induced errors	Common	No pulse
Data analysis	Less complex	More complex
Experimental conditions	One set	Several frequency sets
Detector channels	Two	Two

Table 1. Comparison of Pulse and Phase-Modulation Methods

A LOW FREQUENCY PHASE-MODULATION APPARATUS

We have constructed a variable frequency (0.1-50 KHz) phase modulation apparatus for measurement of long (usec-msec) lifetimes and slow rotation (Garland & Birmingham, 1985). It was not necessary to adapt the design features of high frequency machines (Lakowicz & Maliwal, 1983). Our instrument uses a programmable low frequency synthesiser that drives an acousto-optic modulator attached to the output of a 75 mW CW Argon Ion Laser. Photomultiplier signals are stored directly in a fast signal averager and transferred to disc for subsequent computation of phase angles, modulations, lifetimes and rotational relaxation times. Figure 1 shows experimental data obtained with this apparatus for the phosphorescence of erythrosin-labelled bovine serum albumin. Both the phase shift of the emission and the relative demodulation are clearly seen.

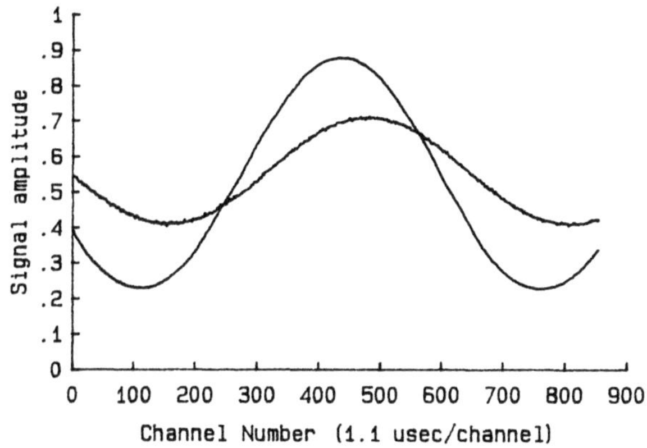

Figure 1. Phase modulation measurements of erythrosin-labelled BSA. Excitation of phosphorescence was at 514 nm, measurement at 650 nm upwards. The modulation frequency was 1.4 KHz. The signal amplitudes have been averaged from 1024 sweeps and normalized to the same mean value. The concentration of BSA (bovine serum albumin) was 10 uM in approx. 99% glycerol at 20°, and the laser power about 5mW. The trace with the larger modulation amplitude corresponds to the laser intensity, whereas the phase-delayed (24.6°) trace of laser modulation amplitude corresponds to phosphorescence. Polarization planes were set for the magic angle. Analysis of the frequency dependence yielded components of amplitude and lifetime 0.995 and 0.56 usec, and 0.0026 and 105 usec, and 0.002 and 405 usec.

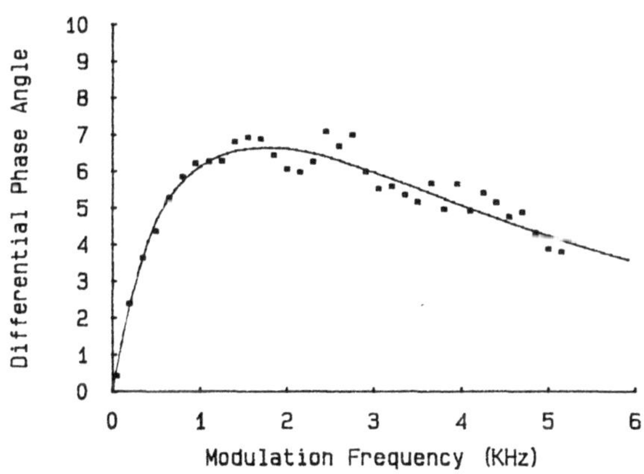

Figure 2. Frequency dependence of the phase differential for erythrosin-labelled BSA. Sample conditions as in Figure 1. The phase differential corresponds to the perpendicular channel minus the parallel channel. The non-linear least squares best fit, shown as a continuous curve, gave two rotational components with anisotropies and correlation times of 0.138 and 13 usec, and 0.009 and 118 usec. The anisotropies correspond to the zero time values in a pulse experiment.

A further example of experimental data is given in Figure 2, for the frequency dependence of the differential phase angle, for bovine serum albumin covalently labelled with erythrosin isothiocyanate and dissolved in 99% (w/v) glycerol. Non-linear least squares fit for the frequency dependence of the phase angle gave three lifetime components for the emission (α_1 = 0.99, τ_1= 0.56 usec; α_2 = 0.0026, τ_2 = 105 usec; α_3 = 0.002, τ_3 = 413 usec) of which the fast large component arises from the red-tail of prompt fluorescence. The frequency dependent differential phase angle plot was fitted with two rotational correlation times of 13 and 118 usec, with anisotropies 0.138 and 0.009 respectively. We believe that noise in the differential phase angle measurement arises from three main sources. Firstly, from overall suppression of all phase angles by the contaminating prompt fluorescence component measured in the same spectral region as phosphorescence (Garland & Moore, 1979; Garland & Birmingham, 1985). Secondly, from mismatch between the two photomultiplier/detection channels. Thirdly, from drift in conditions during data collection (full automation had not been achieved). Of these, the first can be got around in principle by use of other probe detection methods, such as ground state depletion; the second by having only one detector, in which case the plane of polarization of either the excitation beam or the detector polarizer must be rotated with time; and the third by rapid automated data acquisition and frequency sweeping.

A SINGLE DETECTOR SYSTEM

Recently, Barisas (1985) described a phase-modulation variation of the pulsed fluorescence depletion method of Johnson & Garland (1981). The most novel feature of this phase-modulation method was modulation of not only the intensity of excitation (with an acousto-optic modulator) but also its polarization plane (with a Pockels cell). Only one detector was required. Fast and continuous rotation of polarization planes is difficult. Conventional Pockels cell usage switches between orthogonal polarizations at $\pi/2$ phase shift, but provides elliptical polarization in between. Continuous rotation of a polarization plane by the Faraday effect is difficult in the visible region, where even the best glasses have low Verdet constants and require correspondingly high magnetic fields. Mechanical rotation of a half-wave plate has obvious drawbacks apart from limited maximum frequency. Nevertheless, a dual modulation system still has the great

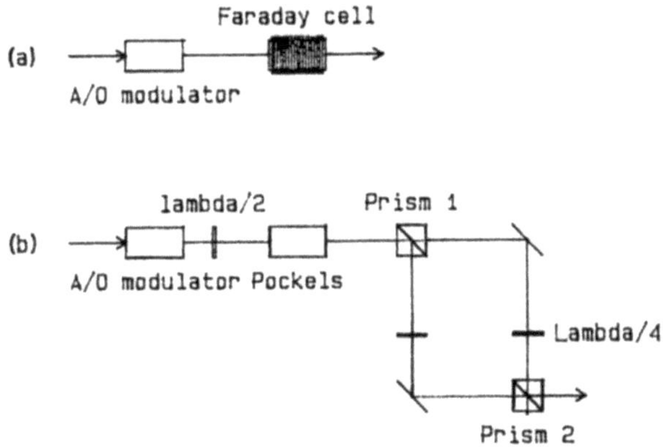

Figure 3. Continuous rotation of the polarization plane of a laser beam, with independent intensity modulation. Two schemes are shown. Both use an acousto-optic modulator to vary the beam intensity. Scheme (a) uses a Faraday cell to rotate the polarization plane. In practice a continuous 0-360° rotation is not readily achieved with visible wavelengths, but oscillation of a few degrees around a mean position could be explored. Scheme (b) uses an electro-optic modulator (Pockels cell) to introduce a phase shift between the two orthogonal components of a laser beam with its polarization plane at 45° to the electro-optic modulator axes. Prism 1 is a polarizing beam splitter and separates the two plane polarization states of equal intensities: these are then converted by appropriately set quarter wave plates into right- and left-handed circularly polarized states. Prism 2 is non-polarizing. The principle of the scheme is that when left- and right- circularly polarized light beams of equal intensity are added together, the resultant beam is polarized in a plane determined by the phase difference of the two initial beams. The phase shift is electro-optically modulated with a saw-tooth drive to the modulator.

attraction of not requiring two detection channels, provided that the data analysis is feasible. We have derived general equations for the frequency content of a single detection signal for the case where the excitation intensity is sinusoidally modulated at one frequency (w_f) and the plane polarization of excitation is continuously rotated through 360° at another frequency (w_p) with a fixed phase relationship to the first. The emission signal is viewed through a polarizer either parallel or perpendicular to the unrotated position of the excitation beam. The outcome is that there are four frequencies in the emission signal: w_f, $2w_p$, $2w_p \pm 1w_f$. At w_f only lifetime information (i.e. phase and amplitude) is present, whereas a mixture of lifetime and rotational information is present at the others. In effect, one experiment combines both a "magic angle" experiment and a differential phase experiment. A more formal description is given in the Appendix. We were encouraged by this theoretical outcome to explore ways of achieving continuous rotation of polarization of a laser beam. Figure 3. shows an optical set-up that we are currently exploring.

Acknowledgements: This work was supported by the Medical Research Council. We thank Dr. George Barisas for helpful discussions, and Dr. Martin Smith for suggesting the optical configuration of Figure 3b.

REFERENCES

Barisas, G. (1985) Abstracts, Workshop on Nucleocytoplasmic Transport, Heidelberg, September 1985 (R. Peters and M. Trendelenberg, eds.)

Beecham, J.M., Knutson, J.R., Ross, J.B.A., Turner, B.W. and Brand, L. (1983) "Global resolution of heterogeneous decay by phase-modulation". Biochemistry, 22, 6054-6058.

Cherry, R.J.(1979) "Rotational and lateral diffusion of membrane proteins". Biochim. Biophys. Acta, 559, 289-327.

Garland, P.B. and Birmingham, J.J. (1985) "Triplet probes and the rotational diffusion of membrane proteins" in "Fluorescence in the Biological Sciences" (Taylor, D.L., Lanni, F. and Waggoner, A.L., eds.). Alan Liss, Inc., New York (in Press).

Garland, P.B. and Johnson, P. (1985) "Rotational diffusion of membrane proteins: optical methods" in "The Enzymes of Biological Membranes" (A.N. Martonosi, ed.) Vol. 1, 421-439. Plenum Publishing Corporation, New York.

Garland, P.B. and Moore, C.H. (1979) "Phosphorescence of protein-bound eosin and erythrosin. A possible probe for measurements of slow rotational mobility". Biochem. J., 183, 561-572.

Gratton, E., Jameson, D.M. and Hall, R.D. (1984a) "Multifrequency phase and modulation fluorometry". Ann. Rev. Biophys. Bioeng., 13, 105-124.

Gratton, E., Limkeman, M., Lakowicz, J.R., Maliwal, B.P., Cherek, M. and Laczko, G. (1984b) "Resolution of mixtures of fluorophores using variable-frequency phase and modulation data". Biophys. J., 46, 479-486.

Johnson, P. and Garland, P.B. (1981) "Depolarization of fluorescence depletion: A microscopic method for measuring rotational diffusion of membrane proteins on the surface of a single cell". FEBS Lett., 132, 252-256.

Johnson, P. and Garland, P.B. (1982) "Fluorescent triplet probes for measuring the rotational diffusion of membrane proteins". Biochem. J., 203, 313-321.

Jovin, T.M., Bartholdi, M., Vaz, W.L.C. and Austin, R.H. (1981) "Rotational diffusion of biological macromolecules by time-resolved delayed luminescence (phosphorescence, fluorescence) anisotropy". Ann. N.Y. Acad. Sci., 366, 176-196.

Lakowicz, J.R., Cherek, H. and Maliwal, B.P. (1985) "Time-resolved fluorescence anisotropies of diphenylhexatrienes and perylene in solvents and bilayers obtained from multifrequency phase-modulation fluorimetry". Biochemistry, 24, 376-383.

APPENDIX : Dual Modulation Theory

To illustrate the potential advantages of simultaneously modulating the intensity and polarization plane of the incident light, we consider the simple case of a spherical rotor in an isotropic medium radiatively decaying with rate constant Γ (= $1/\tau$). If c_A (r,t) denotes the concentration of absorbtion dipoles in space and time then the dipole distribution will evolve according to

$$\frac{\partial c_A}{\partial t} = D_r \nabla^2 c_A - \Gamma c_A + F(r,t) \qquad (1)$$

where D_r is the rotational diffusion coefficient, ∇^2 the Laplacian operator and F the nonhomogeneous pumping function incorporating the intensity and polarization modulation functions. The solution of such a non-homogeneous parabolic partial differential equation can be expressed as an expansion in terms of the eigenfunctions of the associated homogeneous equation

$$\frac{\partial c_H}{\partial t} = D_r \nabla^2 c_H - \Gamma c_H \qquad (2)$$

For the isotropic model considered c_H will depend only on the azimuthal angle θ of our absorbtion dipole with respect to the standard laboratory photoselection axis (conventionally vertical), (2) being the ideal pulse evolution counterpart of the forced equation (1).

$$c_H(\theta,t) = \sum_{i=0}^{\infty} a_i P_i(\mathrm{Cos}\theta) e^{-(\Gamma + i(i+1)D_r)t} \qquad (3)$$

describes the general solution of equation (2), where a_i denote arbitrary constants and where P_i are the eigenfunctions required, the ith Legendre polynomial with argument Cos θ. The general solution to (1) will then be of the form

$$c_A(\theta,t) = \sum_{i=0}^{\infty} a_i(t) P_i(\mathrm{Cos}\theta) \qquad (4)$$

where $a_i(t)$ are arbitrary time dependent coefficients to be matched to the characteristics of the pumping function F. For dual modulation we take F of the form

$$F(\theta,t) = Af(t)\mathrm{Cos}^2(\theta - w_p t - \lambda_p) \qquad (5)$$

where f(t) is the intensity modulation function and A a lumped photoselection efficiency factor. We take f(t) to be sinusoidal,

$$f(t) = f_o + f_1 \mathrm{Sin}(w_f t + \lambda_f) \qquad (6)$$

w_f denotes the angular intensity modulation frequency, w_p the angular frequency for synchronous rotation of the incident polarization plane and λ_p, λ_f initial phase factors if present. The photoselection plane, inducing the usual cosine-squared distribution, is constantly re-orienting at a linear rate w_p. Thus $F(\theta,t)$ represents the probability of excitation to an orientation θ at any time t. We also expand $F(\theta,t)$ in terms of $P_i(\mathrm{Cos}\theta)$ with expansion coefficients $C_i(t)$

$$F(\theta,t) = \sum_{i=0}^{\infty} C_i(t) P_i(\mathrm{Cos}\theta) \qquad (7)$$

where $C_i(t) = \int_0^{\pi} \frac{2i+1}{2} F(\theta,t) P_i(\mathrm{Cos}\theta)\mathrm{Sin}\theta d\theta$

Substitution of equations 4-7 into equation 1 leads to the coupled set of nonhomogeneous first order ordinary differential equations (8) for the required expansion coefficients $a_i(t)$

$$\frac{da_i}{dt} + (\Gamma + i(i+1)D_r)a_i = C_i(t) \qquad (8)$$

The emission signal is determined not by the absorbtion dipole distribution $c_A(t)$ but by the emission dipole distribution $c_E(t)$ given by (9)

$$c_E(\theta,t) = \sum_{i=0}^{\infty} a_i(t) P_i(\mathrm{Cos}\lambda) P_i(\mathrm{Cos}\theta) \qquad (9)$$

where λ is the angle between the absorbtion and emission dipoles. The signal intensity viewed at $\theta=0$ to the reference laboratory vertical axis, is given by $I_{||}(t)$ (in normalised terms)

$$I_{||}(t) = \frac{1}{4\pi} \int_0^{2\pi} \int_0^{\pi} c_E(\theta,t) \mathrm{Cos}^2\theta \mathrm{Sin}\theta d\phi d\theta \qquad (10)$$

This equation coupled with the well known properties of the Legendre polynomials implies that only two terms in the series (9) will contribute to the viewed intensity (10) giving

$$I_{||}(t) = \frac{1}{3} [a_o(t) + \frac{2}{5} P_2(\mathrm{Cos}\lambda) a_2(t)] \qquad (11)$$

where we recognise the factor multiplying into $a_2(t)$ as the zero time anisotropy of the probe excited in a pulse experiment (r_o). The $a_o(t)$ and $a_2(t)$ are obtained upon solution of the ordinary system (8). The $C_o(t)$ and $C_2(t)$ functions obtained by substitution of (5) into (7) and integrating are:

$$C_o(t) = \frac{A}{2}\{f_o+f_1\text{Sin}(w_f t+\lambda_f)-\frac{1}{3}f_o\text{Cos}[2(w_p t+\lambda_p)]$$

$$-\frac{1}{6}f_1\text{Sin}[(w_f+2w_p)t+\lambda_f+2\lambda_p]$$

$$+\frac{1}{6}f_1\text{Sin}[(2w_p-w_f)t+2\lambda_p-\lambda_f]\}$$

$$C_2(t) = \frac{A}{3}\{2f_o\text{Cos}[2(w_p t+\lambda_p)]+f_1\text{Sin}[(w_f+2w_p)t+\lambda_f+2\lambda_p]$$

$$-f_1\text{Sin}[(2w_p-w_f)t+2\lambda_p-\lambda_f]\} \qquad (12)$$

We note that $C_2(t)$ does not have a pure w_f AC term. Therefore $a_2(t)$ will not contain this frequency component, meaning that the w_f-component of the signal $I_{||}(t)$ can only arise from $a_o(t)$ function. Because the coupled set (8) is independent of D_r for i=0, this means that the w_f-component of the signal can provide information only on Γ but cannot contain any rotational information (r_o,D_r). It is clear that the signal $I_{||}(t)$ will consist of a superposition of a DC term and four AC terms, ie. $w=w_f$, $2w_p$, $2w_p \pm w_f$. Writing $I_{||}(t)$ in terms of amplitude and phase characteristics (B_w,$\emptyset_w$),

$$I_{||}(t) = B_o + \sum_w B_w\text{Sin}(wt+\emptyset_w) \qquad (13)$$

we can calculate B_w, $\emptyset_w$ for each of the component frequencies. A similar frequency content describes the excitation input waveform. A dual modulation experiment would measure the relative phase ($\Delta\emptyset_w$) and demodulation of the single emission channel waveform compared to a monitor input channel at each of the component frequencies w. The expected relative phase differences are tabulated as Tan($\Delta\emptyset_w$) in Table 2.

AC component	w	$\mathrm{Tan}(\Delta\phi_w)$
1.	w_f	$-w/\Gamma$
2.	$2w_p$	$[4d_o r_o(\Gamma+6D_r)-d_2\Gamma]/[w(4d_o r_o-d_2)]$
3.	$2w_p-w_f$	$[w(4d_o r_o-d_2)]/[d_2\Gamma-4d_o r_o(\Gamma+6D_r)]$
4.	$2w_p+w_f$	$[w(4d_o r_o-d_2)]/[d_2\Gamma-4d_o r_o(\Gamma+6D_r)]$

where $d_o = \Gamma^2 + w^2 = d_o(w,\Gamma)$
$d_2 = (\Gamma + 6D_r)^2 + w^2 = d_2\ (w,\Gamma,D_r)$

Table 2. Extraction of Radiative Lifetimes and Rotational Correlation Times

We see that the relative phase delay at $w=w_f$ is just

$$\mathrm{Tan}^{-1}(w_f\tau)$$

where τ is the probe lifetime, which corresponds to a classical magic-angle lifetime determination. The estimation of the rotational parameters (r_o,D_r) then follows easily by using the τ-value from $\Delta\phi(w=w_f)$ to dissect any two of the remaining $\Delta\phi_w$'s. All these froquencies are of course present in the single emission waveform, the monitor sufficing to determine instrumental λ_p, λ_f if present. This decoupling phenomenon will hold however regardless of the complexity of the radiative decay law: there can never be any rotational information on the $w=w_f$ component of a dual modulation signal. Multi-frequency sweeping in conjunction with dual modulation will suffice to extract a set of radiative lifetimes and a set of rotational correlation times.

Fluorescence Photoactivation and Dissipation (FPD)

G. A. Krafft, R. T. Cummings, J. P. Dizio, R. H. Furukawa, L. J. Brvenik,
W. R. Sutton and B. R. Ware
Department of Chemistry, Syracuse University, Syracuse, NY 13210, USA

I. Introduction

Fluorescence methods have emerged as indispensable analytical and tracer techniques for a diverse range of biochemical, biophysical and physical studies that depend upon highly sensitive measurements of physical environment. This emergence has been hastened by advances in low-level light detection and by the development of coherent, variable-wavelength, and polarized excitation sources. Moreover, the spectrum of biological and physical applications of fluorescence methods has been expanded broadly by the development and availability of fluorescent molecules with specifically tailored characteristics. The attachment of these fluorescent molecules to molecular species has become a powerful approach for investigating the structure and dynamics of complex systems.

Fluorescence microscopy provides high-contrast images of the spatial distribution of specifically labelled species. Time resolved fluorescent images may reveal the dynamics of spatial rearrangements of labelled species after introduction into a complex system such as a cell. However, once the labelled species have attained a steady state or equilibrium distribution, the fluorescent images become invariant, irrespective of whether individual species are still in motion. The technique of fluorescence photobleaching recovery (FPR) extends the power of time-resolved fluorescent imaging by creating a tracer inequivalence among otherwise equivalent labelled species (Peters et al, 1974; Edidin et al, 1976; Jacobson et al, 1976; Axelrod et al, 1976). This inequivalence is accomplished by photobleaching a region of the sample using a brief pulse of high-intensity light in a spatially defined region of the

Nucleocytoplasmic Transport
Edited by R. Peters and M. Trendelenburg

sample. If all labelled species are free to move about in a random fashion, the bleached region will eventually disappear. The time constant for the disappearance of the bleached region will be determined by the translational diffusion coefficient of the labelled species and the square of the characteristic dimension of the bleached region. If some fraction of the bleached species is not free to move, then that fraction of the bleaching will persist, permitting determination of the fraction of immobile species. Since the first introduction of the FPR technique, the primary application has been the study of the diffusion of specifically labelled species in biological membranes. The FPR technique has also been applied successfully to the study of motion in synthetic bilayers, films, solutions, colloids, and cell cytoplasm (for reviews see Cherry, 1979; Peters, 1981; Ware, 1984).

Although the FPR technique is simple in concept, considerable experimental refinement has been required to achieve acceptable accuracy for specific applications. One intrinsic disadvantage of the FPR approach has not previously been addressed. FPR inherently involves the measurement of a subtractive signal in the presence of a background of bright fluorescence. This property limits the contrast achievable and the signal-to-noise ratio to be expected for a given system, and it also renders difficult or impossible the long-term monitoring of those species in the region of photobleaching as they diffuse throughout the specimen. We are seeking to address this problem by the introduction of a methodology which is essentially the positive analog of FPR. This new approach, which we call Fluorescence Photoactivation and Dissipation (FPD), has the potential for several inherent advantages over conventional FPR methodology. The essential requirement for an FPD measurement is the attachment of a molecular label that is initially non-fluorescent but which can be converted to a stable, fluorescent molecule when illuminated by a short pulse of intense light. Such a molecule is called a photoactivable fluorophore (PAF)

The photochemical generation of stable fluorescent molecules from non-fluorescent precursors has been noted in the literature for many years. One of the earliest observations of this type of process was described by Lewis et al (1940), who studied the photoisomerization of cis-stilbene (non-fluorescent) to trans-stilbene (blue fluorescence). Since then examples of photo-generated fluorescence via cis-trans isomerization have been reported for several classes of compounds, including indigo dyes (Rogers et al, 1957), β-styryl compounds (Feigl et al, 1955; Wheelock, 1959; Lippert and Luder, 1962), and tetraphenylcyclooctatetraene (White et al, 1969; Sharafy and Muszkat, 1971). These cis-trans isomerizations are typically reversible under photochemical conditions, such that an equilibrium mixture of fluorescent and non-fluorescent isomers is generated.

Other photochemical processes also lead to stable fluorescent molecules. Cis-stilbenes participate in oxidative photocyclization reactions and are thus converted to fluorescent phenanthrene derivatives (Mallory and Mallory, 1984). This process is the basis of a clinical assay for blood serum levels of the breast cancer drug tamoxifen, and its metabolite 4-hydroxytamoxifen, which form highly fluorescent 10-phenylphenanthrene derivatives (Mendenhall et al, 1978). Irradiation of aryl azides and subsequent C-H insertion generates secondary aryl amines, which often are fluorescent (e. g., aminonaphthalene). This process was employed in a photoaffinity labelling experiment of nerve cell sodium channels in which photolysis of an aryl azide analog of tetrodotoxin resulted in simultaneous covalent attachment and generation of a fluorescent label (Angelides, 1981). Additional reactions known to generate fluorescent products include the photo-Fries rearrangement of naphthyl esters to fluorescent hydroxynaphthyl lactones (Bellers, 1971), photo-oxidation of non-fluorescent dihydroaromatic systems to fluorescent aromatic molecules (e.g. dihydrofluorescein → fluorescein) (Uchida, 1963), and the photoisomerization of diarylisoxazoles to fluorescent diaryloxazoles (Ullman and Singh, 1966; Singh and Ullman, 1967). Huffman et al (1970) studied several 3-aroyl-2-(2-furyl)-chromones which participate in a photo-induced rearrangement to generate fluorescent 1-arylfuro[3,4b]chromone-3-acroleins. This reaction and several others that generate fluorescent products have been reviewed by Zweig(1973), who discussed the potential applications of these reactions in organic imaging processes and as mechanistic probes of photochemical processes.

In this paper we discuss design considerations and synthetic approaches for new types of photoactivable fluorophores and we report the use of these PAF molecules as specific markers for the study of diffusion by fluorescent tracer transport. We also present a description of the FPD apparatus and provide examples of FPD measurements.

II. DESIGN AND SYNTHESIS OF PHOTOACTIVABLE FLUOROPHORES (PAFs)

A photoactivable fluorophore (PAF) is a non-fluorescent molecule that can be converted to a stable fluorescent molecule when irradiated by light. The design of PAFs requires consideration of optimal characteristics of the PAF molecules, of the ultimate fluorophore molecules, and of the photoactivation process which effects the conversion. A number of ideal characteristics of PAF molecules are summarized in Figure 1.

PAF
-- Non-fluorescent
-- Chemically and thermally stable
-- Activated by light not strongly absorbed by ultimate fluorophore
-- Capable of covalent attachment to target molecules

PHOTOACTIVATION PROCESS
-- Efficient (high quantum efficiency)
-- Induced by relatively low energy light
-- Selective (no other processes induced)
-- Clean (no damaging by-products generated)

ULTIMATE FLUOROPHORE
-- Highly fluorescent
-- Excited by non-activating wavelengths
-- Chemically, thermally and photochemically stable

Figure 1. Ideal characteristics of photoactivable fluorophores (PAFs).

In addition to consideration of these characteristics, the design of PAF molecules also entails consideration of desired spectral and photoactivation parameters, PAF polarity, size limitations and functional group specificity for covalent attachment, as dictated by particular applications of the PAFs. In general, PAF design should result in a molecule that will be transformed into an ultimate fluorophore similar in structure and spectral characteristics to conventional probes already employed in fluorescence studies.

In order to design a particular organic molecule as an ideal PAF, it is useful to focus on a target fluorophore structure that, by a conceptual retrosynthetic transform, would be converted to a structurally modified non-fluorescent PAF. Two types of retrosynthetic transforms are capable of rendering the ultimate fluorophore target non-fluorescent. Attachment of a quenching auxiliary to the ultimate fluorophore molecule by a connecting chain would result in a non-fluorescent PAF molecule; alternatively, removal of the fluorescent chromophore by disruption or deconjugation of an extended pi system, or by removal or alteration of an integral functional group would provide a suitable PAF structure.

The photoactivation process for a PAF molecule necessarily requires a light induced reaction which reverses the retrosynthetic transform by which the target fluorophore was rendered non-fluorescent. In the case of a fluorophore linked to a quenching auxiliary, photoactivation would involve cleavage or alteration of the quencher. In other cases, activation processes would involve photochemical bond formation, rearrangement, or functional group manipulation to restore a fluorescing chromophore.

In a schematic sense, three general types of PAFs can be envisioned, as illustrated in Figure 2. A Type I PAF consists of a photoremovable quenching auxiliary, Q, attached to a fluorescent molecule F. A Type II PAF consists of two parts, A and B, which react with each other (bond formation) or rearrange when irradiated to form the ultimate fluorophore F. A Type III PAF might be considered a hybrid of the first two, consisting of X and Y, and a photo-removable protecting group P which prevents X and Y from reacting or rearranging via a thermal/chemical process that otherwise would occur spontaneously.

Figure 3 presents an example of each type of PAF. The Type I PAF consists of a substituted nitrobenzyl quenching chromophore coupled via a carbamate linkage to a fluorescent aromatic amine. Irradiation induces an intramolecular redox reaction of the nitro-

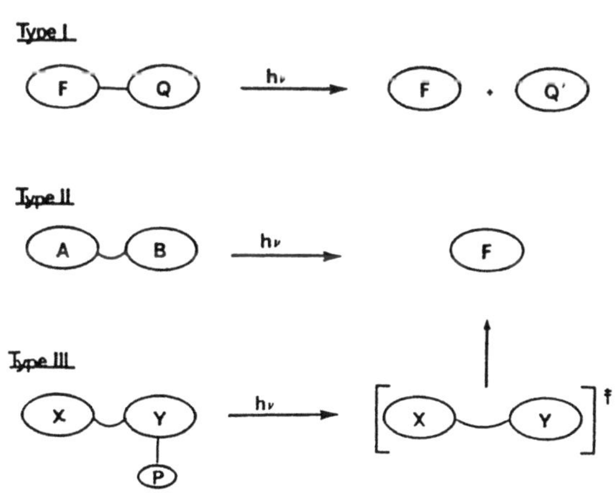

Figure 2. Schematic representation of three types of PAFs.

benzyl group, resulting in decarboxylative cleavage (Patchornick et al, 1970) (vide infra) to generate the free fluorescent amine.

The Type II PAF is an aroyl furyl chromone which is non-fluorescent due to its cross-conjugated π system. Irradiation induces a rearrangement to a furochromone which contains a fluorescent chromophore. The Type III PAF is a difunctionalized fluorescein derivative which exists in the closed lactone form, and thus is non-fluorescent. Photochemical restoration of one phenol permits opening of the lactone to generate the fluorescent xanthene chromophore responsible for fluorescein fluorescence.

We have prepared PAFs of each type illustrated in Figure 3. The N-fluoranthenyl nitrobenzyl carbamates (e. g., PAF I-2) were prepared by coupling nitrobenzyl chloroformates with 3-aminofluoranthene (Figure 4). Similarly, N-anthryl carbamates and N-naphthyl carbamates were prepared from 2-aminoanthracene and 2-aminonaphthalene, respectively. In order to attach the carbamate PAFs to larger molecular species, an aminopropyl linking arm attached to the carbamate nitrogen via alkylation with the disilyl protected aminopropyl bromide. After deprotection, the free amine could be

Type I

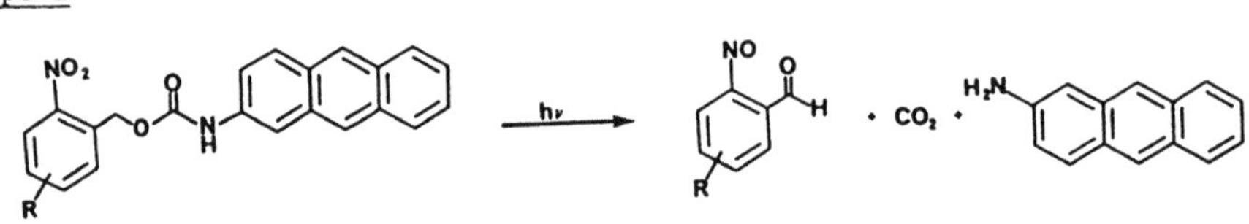

Type II

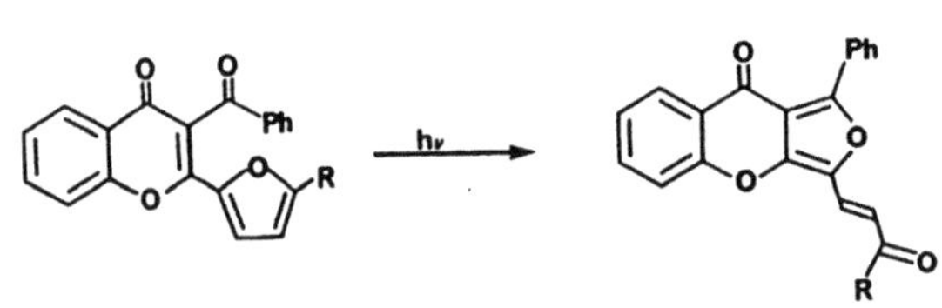

Type III

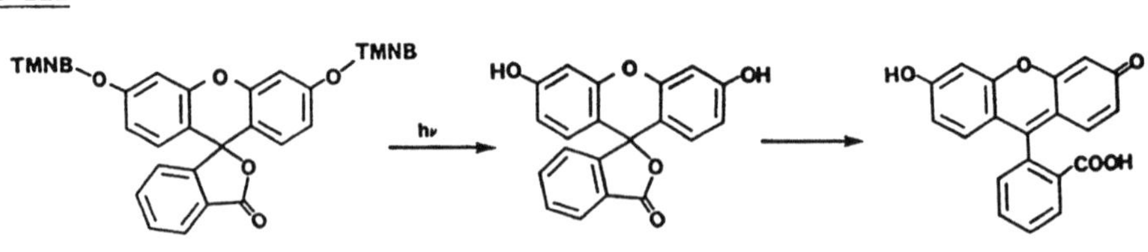

Figure 3. Examples of three types of PAFs and their photoactivation pathways to ultimate fluorophores.

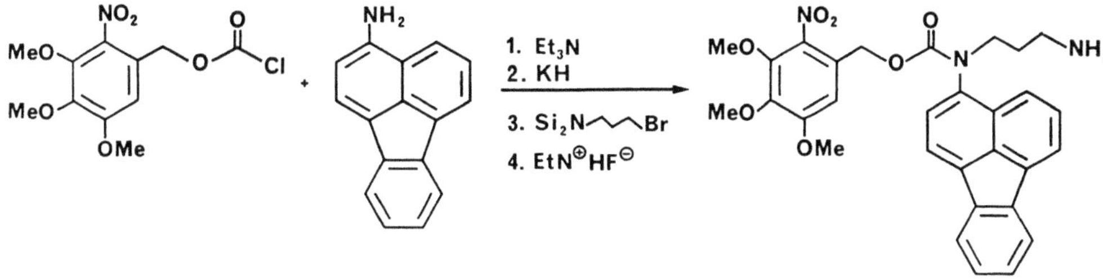

Figure 4. Synthesis of Type I carbamate PAFs

coupled to carboxyl functionality or converted to an iodoacetamide which permits attachment to amino or mercaptan functional groups.

The synthesis of the 3-aroyl-2-(2-furyl) chromones is outlined in Figure 5. Aldol condensation of the bis-aryl diketone with furaldehyde derivatives (catalytic piperidine in refluxing ethanol) provided the dihydrochromone derivatives which were oxidized to the chromones with selenium dioxide in refluxing dioxane. Aldol condensation with 5-hydroxymethyl furaldehyde followed by selenium dioxide oxidation generated the hydroxymethyl derivative PAF II-3 which has been converted to a variety of other functionalized derivatives.

The 3,4,5-trimethoxy-2-nitrobenzyl group is an effective quencher of the aminofluoranthene fluorescence, eliminating all emission at 515 nm and greater than 98 percent of the fluorescence emission at 410 nm as shown in Figure 6. This quenching presumably occurs by resonant energy transfer from the excited state of the

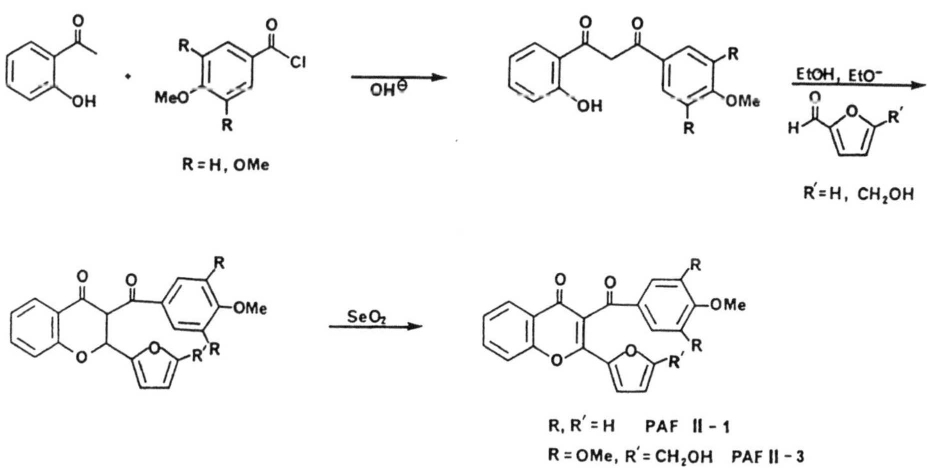

Figure 5. Synthesis of furyl aroyl chromones.

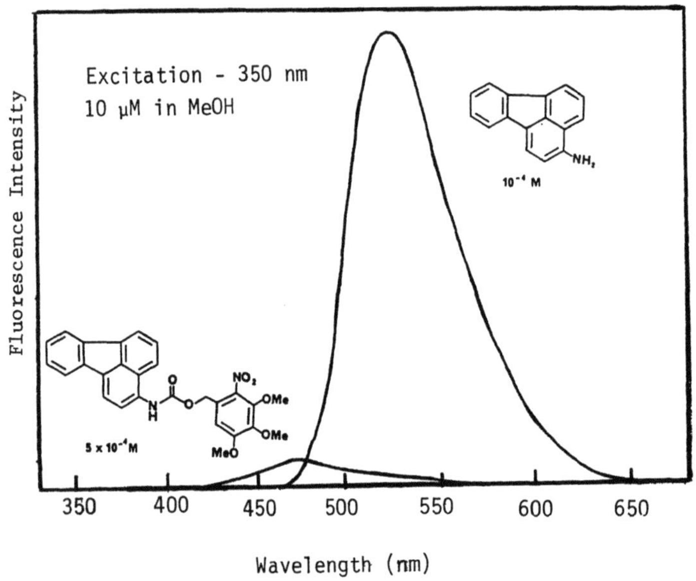

Figure 6. Fluorescence quenching in nitrobenzyl carbamates.

aminofluoranthene to the trimethoxynitrobenzyl chromophore. Quenching by less substituted nitrobenzyl chromophores was significantly less effective, due to poorer overlap of the energy-accepting transition of the quencher with the energy-donating excited state of the fluorophore.

Irradiation of PAF I-2 in methanol or water at 350 nm generates 3-aminofluoranthene, carbon dioxide and 3,4,5-trimethoxynitrosobenzaldehyde. The quantum efficiency for this activation was 0.19 ± 0.05. The mechanism that has been proposed for this reaction (Berson and Brown, 1955) is outlined in Figure 7. The excited aromatic nitro group abstracts a hydrogen atom to form a biradical, which rearranges to form the nitrosobenzaldehyde hemiacetal. Subsequent decarboxylation liberates aminofluoranthene and trimethoxynitrosobenzaldehyde. This decarboxylation step is a thermal reaction with kinetics that are influenced by solvent, pH, and temperature. FPD experiments of rapidly diffusing molecules require that the kinetics of this step be fast, relative to the photoactivation time and diffusion coefficient of the species being studied.

Irradiation of PAF II-1 in dichloromethane at 350 nm produces 1-p-anisylfurol[3,4b]chromone-3-acrolein, which fluoresces maximally at 527 nm. The quantum efficiency for this process is quite low (.03, Huffman et al, 1970). The proposed mechanism by which this

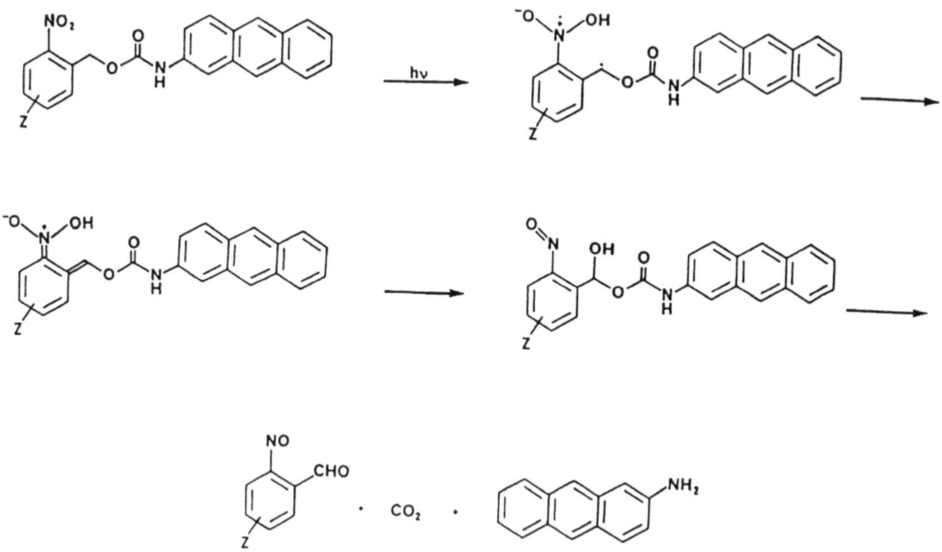

Figure 7. Mechanism of photoactivation of Type I nitrobenzyl carbamate PAFs.

conversion takes place (Huffman et al, 1970) is outlined in Figure 8. Fragmentation of the furan ring leads to a carbene intermediate, which can cyclize to the aroyl carbonyl to form the fluorescent product, or cyclize to the acrolein carbonyl to regenerate the original chromone. This pathway and other possible reaction pathways of the reactive carbene are undoubtedly responsible for the low reaction quantum efficiency.

We have prepared a number of furyl aroyl chromone derivatives in order to enhance the efficiency of photoactivation to the

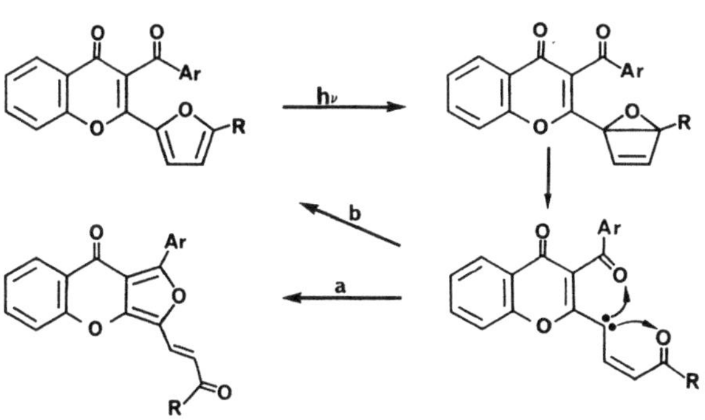

Figure 8. Mechanism for photoactivation of aroyl furyl chromones.

fluorescent furochromones. In accordance with the proposed mechanism (Figure 8) the desired pathway a should be favored over pathway b when the aroyl carbonyl oxygen is significantly more nucleophilic than the furan derived carbonyl oxygen. Thus, incorporation of three electron-releasing methoxyl groups on the aroyl phenyl ring results in significant enhancement of photoactivation efficiency relative to unsubstituted or monomethoxy substituted derivatives. Moreover, incorporation of an electron withdrawing carboxyl group at the 5-position of the furan ring further increases the photoactivation efficiency.

The synthesis of non-fluorescent fluoresceins is outlined in Figure 9. Photoactivation of the diprotected fluorescein PAF III-12 involves cleavage of both trimethoxynitrobenzyl groups to generate fluorescein. The intermediate mono protected fluorescein is fluorescent, but not to the extent of fluorescein itself. This is presumably due to some quenching of the fluorescence by the remaining trimethoxy nitrobenzyl group. Photoactivation of 3-O-methyl-6-O-trimethoxynitrobenzyl fluorescein (PAF III-2) requires only the single deprotection to generate 3-O-methylfluorescein which has nearly the same fluorescence intensity of fluorescein itself.

III. FPD APPARATUS

A schematic diagram of our FPD apparatus is shown in Figure 10. The photoactivating laser light is provided by the UV lines of a

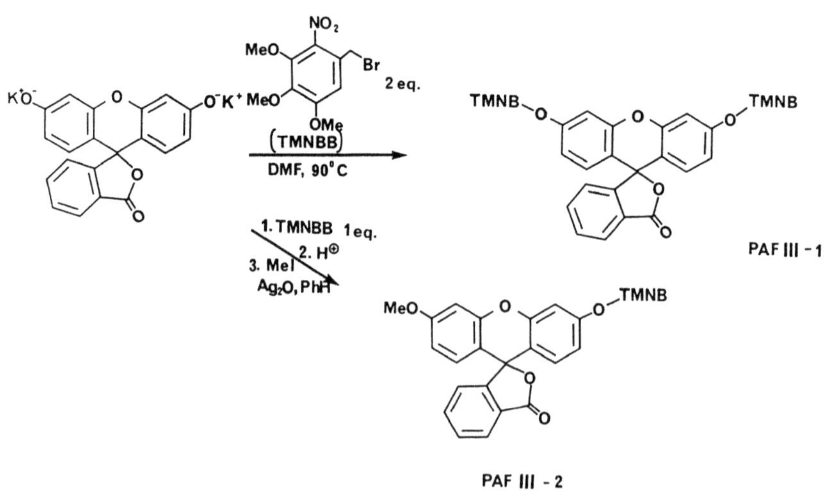

Figure 9. Synthesis of Type III fluorescein-based PAFs

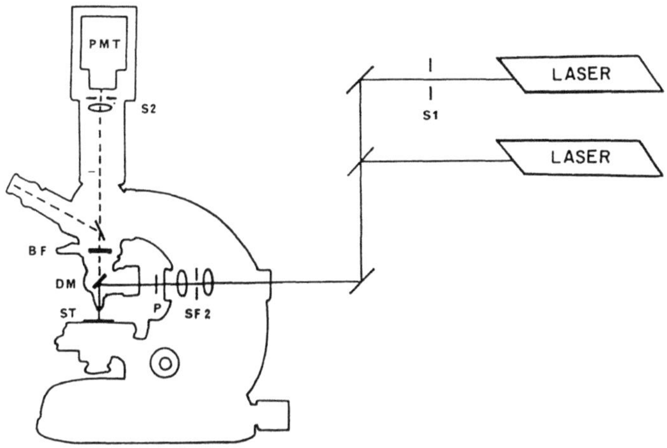

Figure 10. Schematic diagram of a fluorescence photoactivation and dissipation (FPD) instrument. The specimen is on the stage (ST) of a research microscope with epi-fluorescence optics. The activating laser is stopped by a shutter (S1) which is opened for the short period of activation. The monitor and activating beams are rendered coincident and directed to the microscope by a series of mirrors. After passing through a spatial filter (SF2) and a Ronchi ruling pattern (P) the beams are directed by a dichroic mirror (DM) to the stage. Fluorescence passes through the barrier filter (BF) to be detected by the photomultiplier tube (PMT).

Spectra Physics Model 2025 Ar^+ laser. The monitoring beam is usually one of the blue or green lines from a Spectra Physics Model 164 Ar^+ laser. The incident beams enter a Zeiss Universal Microscope and pass through a pattern (P), which is generally a Ronchi ruling (grating) of known period. A dichroic mirror reflects both the incident beams onto the microscope stage, on which is mounted the specimen of interest. During the collection phase fluorescence passes through the dichroic mirror and through a barrier filter onto a photomultiplier tube.

The sequence of an FPD measurement is as follows: With the grating translated at constant velocity, the monitoring beam illuminates the sample to detect any residual fluorescence due to the inherent fluorescence of the PAF or the fluorescence of any fluorophore that may have formed spontaneously. The grating is then stopped and held stationary while the shutter controlling the UV photoactivation beam is opened for a brief, measured time. The shutter is then closed, and again the sample is illuminated only with the visible monitoring beam while the grating is translated at constant velocity. The photoactivation in the sample then consists of a set of stripes spaced by the period of the image of the grating on the microscope stage. As the monitoring beam passes through the translated grating the photoactivated pattern and the stripes of the monitoring beam fall into and out of phase with a frequency that is proportional to the velocity of the translation of the grating. The output of the photomultiplier tube is hence a rising photocurrent that is modulated at the frequency determined by the translation of the pattern. Both the DC level and the modulated AC signal are processed separately, as has been described previously in our description of the modulation detection method of FPR (Lanni and Ware, 1982). Briefly, the AC signal is passed through a tuned amplifier and then is sent to an envelope detector that consists of a peak voltmeter and zero crossing detector in parallel. This device produces discrete values of the envelope function at the modulation frequency. The output of the envelope detector, the DC signal, and other timing sequence information are directed to parallel analog-to-digital inputs of a Nova 4C minicomputer for analysis.

The modulation AC envelope E(t) is a measure of the spatial Fourier component of the fluorescence intensity with wave vector K' given by

$$K' = 2\pi/L \qquad (1)$$

where L is the spacing of the projection of the grating lines in the specimen. For each distinct species with diffusion coefficient D, the relaxation of the envelope function will follow the simple diffusion equation:

$$E(t) = E(0)e^{-DK'^2t} \qquad (2)$$

In FPD, as in FPR, this modulation detection scheme has the advantages of (1) ease of determination of pattern size, (2) convenient functional form of data, (3) insensitivity to specimen motion, and (4) relative insensitivity to precision of superposition of photoactivation and monitor beams. The separate utilization of the DC signal may also be useful in the FPD measurement, though the form of the change in the DC signal is of course quite different. In the ideal FPD measurement, the DC signal will be raised after photoactivation from an initial value of nearly zero to some new finite value of photocurrent that remains constant over the time scale of dissipation of the spatial pattern. This condition, however, can only be met if there are many periods of the spatial pattern within the spatial domain of the size of the photoactivating beam. In practice, the DC level will rise over the time scale of the creation of the photoactivable fluorophore and then will fall due to diffusion of PAF out of the laser beam and possibly due to undesired photobleaching of the PAF by the monitoring beam. Hence, the DC signal is used primarily as a diagnostic for certain technical aspects of the measurement.

IV. FPD DATA

Although the design and synthesis of PAF molecules with optimized properties will require additional refinement, the PAFs already in hand are adequate to demonstrate the feasibility of the FPD technique. To date the greatest photoactivation fluorescence yield in an FPD experiment has been achieved with the furyl chromones (PAF II-1). Representative data for this molecule dissolved in CH_2Cl_2 are shown in Figure 11. The upper trace shows

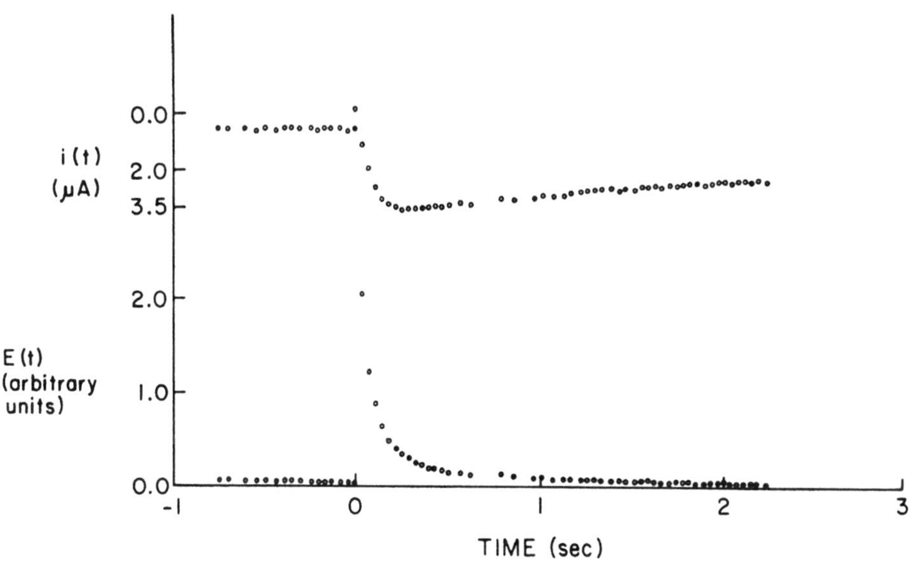

Figure 11. FPD data for the diffusion of PAF II-1 in CH_2Cl_2. The upper trace shows the photocurrent i(t) and the lower trace shows the modulation envelope E(t) as a function of time. The dye concentration was 1% by weight, the temperature was 5.5° C, and the experimental wave vector K' was 853 cm^{-1}

the DC photocurrent i(t), which is a measure of the total fluorescence intensity. The lower trace shows the modulation envelope E(t), which is the measured contrast of the photoactivated pattern, as described previously. Time 0 corresponds to the end of the photoactivation period, which in this measurement had a duration of 100 ms. For all measurements to be described here, the photoactivation intensity was about 10^3 W/cm^2 at 457 nm. The DC level in Figure 10 is seen to rise after the photoactivation over a time scale of about 200 msec, which is a result of the settling time of the current meter employed. This settling time does not influence the response time of the modulation signal. The modulation signal decays exponentially as predicted and with a very high signal-to-noise-ratio. The diffusion coefficient determined by computer fit to E(t) is 9.8 x 10-6 cm^2/s, which is a reasonable value for this temperature (5.5°). The long-term reduction of the DC photocurrent is caused by the diffusion of PAF out of the monitoring beam. The time constant of this change is much longer because the characteristic dimension of the beams is much greater than that of the grating pattern. Variation of the temperature of the PAF - CH_2Cl_2 system produced changes in the measured diffusion coefficient of the anticipated magnitude. Addition of a high-molecular-weight polymer to the system would be expected to reduce the measured diffusion coefficient of the tracer PAF molecule. This expectation was

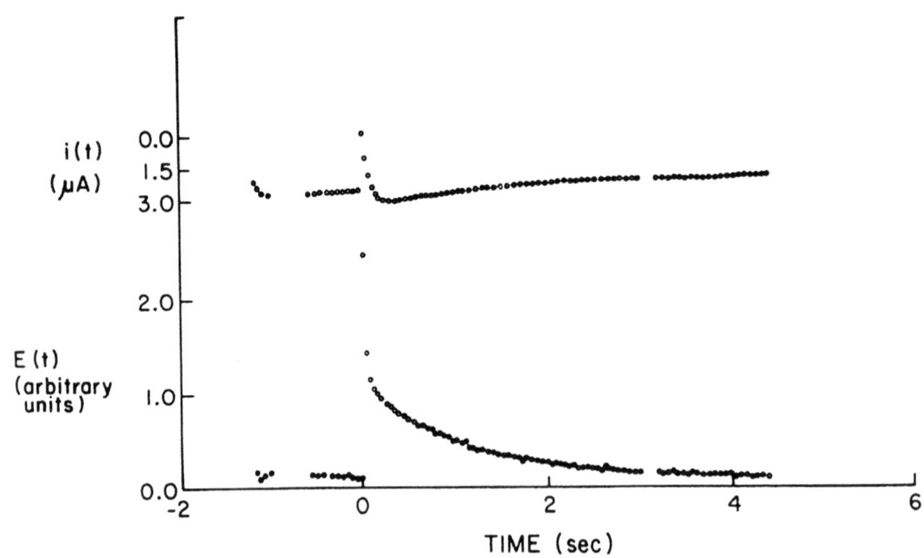

Figure 12. FPD data for the diffusion of PAF II-2 in CH_2Cl_2 containing 5% polystyrene by weight. The dye concentration was 2.6% by weight, the temperature was 15% C, and K' was 853 cm^{-1}.

verified by the addition of polystyrene (Polysciences, mol. wt. 6.7 x 10^5). The measured diffusion coefficients were reduced as anticipated. Sample FPD data for this experiment are shown in Figure 12. The measured diffusion coefficient from these data was 7.89 x 10^{-6} cm^2/s at 15°C. It was observed in this and other experiments that the addition of polystyrene to the PAF III-1 solution increased the photoactivated fluorescence yield by as much as a factor of four. It was also observed that degassing of samples produced up to twice as much photoactivated fluorescence and that many samples produced more photoactivated fluorescence at lower temperature. We have not yet determined whether these increased fluorescence yields are attributable to increased PAF production or to increased fluorescence yield through reduced quenching, but we speculate that the latter is more likely.

We have also obtained FPD data for the type I PAFs. The trimethoxynitrobenzyl fluoranthenyl carbamate (PAF I-3) was irradiated for 100 ms in dimethylformamide at 25° C. The diffusion coefficient determined by computer fit to E(t) is 3.70 x 10^{-6} ± .12 x 10^{-6} cm^2s^{-1}, which is reasonable for this temperature and solvent viscosity. The longer activation time for the PAF resulted in some diffusion while fluorophore was still being generated, as can be seen by the initially slow fluorescence dissipation. Further optimization of the quencher portion of these molecules should result in significantly faster photoactivation times.

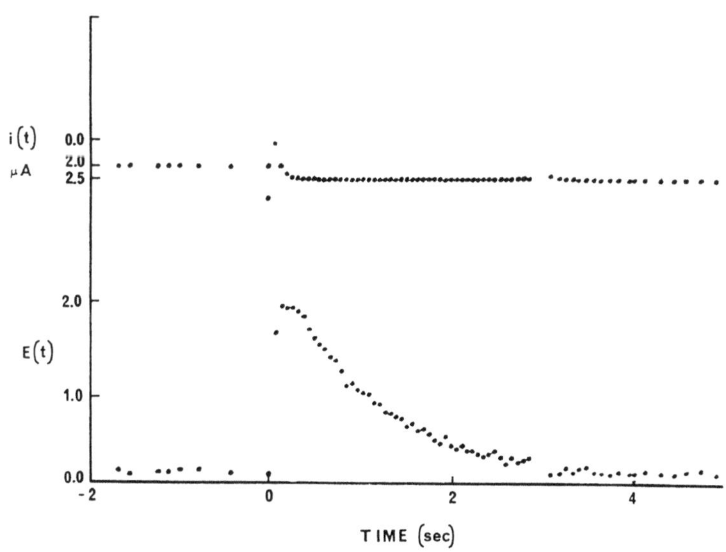

Figure 13. FPD data for the diffusion of PAF I-II in DMF at 25° C.

V. CONCLUSIONS

We have presented a description of a new technique for the measurement of the motion of specifically labelled particles. Fluorescence photoactivation and dissipation (FPD) is essentially the positive analog of fluorescence photobleaching recovery (FPD). The basis of this technique has been the successful design and synthesis of new fluorescent probes, photoactivable fluorophores (PAFs). All of the applications of FPR should be amenable to experimentation by FPD and additional applications involving the long-term monitoring of labelled species are made possible by the FPD method. Although the data presented here are clearly preliminary, it is evident that the FPD technique successfully measures translational diffusion of specific molecules with a high signal-to-noise ratio. The key to development of the technique will be the continued design and synthesis of photoactivable fluorophores (PAFs) with optimized photochemical and photophysical properties, and the successful attachment to macromolecular species of interest.

ACKNOWLEDGEMENTS

We are grateful to Dr. Frederick Lanni, formerly a member of our group and now at the Center for Fluorescence Research in Biomedical Sciences at Carnegie-Mellon University, for observations and comments that led us to this line of investigation. Portions of this work were first presented at the International Conference on the Application of Fluoresence in the Biomedical Sciences in Pittsburgh, Pennsylvania, in April, 1985, and will appear in the proceedings of that meeting. This work was supported by grant no. CA 35954 to G.A. Krafft from the National Institutes of Health and grant no. PCM-8306006 to B.R. Ware from the National Science Foundation. Acknowledgment is also made to the Donors of the Petroleum Research Fund, administered by the American Chemical Society, for partial support of this research. G.A. Krafft is a faculty fellow of the American Cancer Society.

References

1. Angelides K (1981): Fluorescent and photoactivable fluorescent derivatives of tetrodotoxin to probe the sodium channel of excitable membranes. Biochemistry 20:4107-4118.

2. Axelrod D, Koppel D, Schlesinger E, Elson E, Webb W (1976): Mobility measurement by analysis of fluorescence photobleaching recovery kinetics. Biophys J 16:1005-1069.

3. Bellers D (1971). Photo-Fries rearrangement and related photochemical [1,8]-shifts (J=3,5,7) of carbonyl and sulfonyl groups. Advan Photochem 8:109-159.

4. Cherry R (1979). Rotational and lateral diffusion of membrane proteins. Biochim Biophys Acta 559:289-327.

5. Edidin M, Zagyanski Y, Lardner T (1976). Measurement of membrane protein lateral diffusion in single cells. Science (Washington, D. C.) 191:466-468.

6. Feigl F, Feigl H, Goldstein O, (1955). A sensitive and specific test for coumarin through photocatalysis. J Amer Chem Soc 77:4162-4163.

7. Huffman D, Kuhn C, Zweig A (1970): Photoisomerization of 3-aroyl-2-(2-furyl) chromones. An example of quenching of a photochemical reaction by a product. J Amer Chem Soc 92:599-605.

8. Jacobson K, Wu E, Poste G, (1976): Measurement of the translational mobility of concanavalin A in glycerol-saline solutions, and on the cell surface by fluorescence recovery after photobleaching. Biochim Biophys Acta 433:215-222.

9. Lanni F, Ware B (1982): Modulation detection of fluorescence photobleaching recovery. Rev Sci Instrum 53:905-908.

10. Lewis G, Magel T, Lipkin D (1940): The absorption and emission light by "cis-" and "trans-" stilbenes and the efficiency of photochemical isomerization, J Amer Chem Soc 62:2973-2980.

11. Lippert E, Luder W, (1962): Photochemical "cis-trans" isomerization of "p"-dimethylaminocinamic acid nitrile. J Phys Chem 66:2430-433.

12. Mallory F, Mallory C, (1984): Photocylization of stilbenes and related molecules. Organic Reactions 30:1-456.

13. Mendenhall D, Kobayashi H, Shih F, Sternson L, Higuchi T, Fabian C, (1978): Clinical analysis of tamoxifen, an antineoplastic agent, in plasma. Clin Chem 24:1518-1524.

14. Peters R. (1981): Translational diffusion in the plasma membrane of single cells as studied by fluorescence microphotolyis. Cell Biol Intl Rep 5:733-760.

15. Peters R, Peters J, Tews KH, Bahr W, (1974): A microfluorimetric study of diffusion in erythrocyte membranes Biochim Biophys Acta 367:282-294.

16. Rogers D, Margerum J, Wyman G, (1957): Spectroscopic studies on dyes. IV The fluorescence spectra of thioindigo dyes. J Amer Chem Soc 79:2464-2468.

17. Sharafy S, Muszkat K, (1971): Viscosity dependence of fluorescence quantum yields. J Amer Chem Soc 93:4119-4125.

18. Singh D, Ullman E,(1967): Photochemical transposition of ring atoms in 3,5-diarylisoxazoles. An unusual example of wavelength control in a photochemical reaction of azirines. J Amer Chem Soc 89:6911-6916.

19. Uchida K, (1963): The photosensitized oxidation of leucouranine. III The kinetics of an acridine-sensitized aerated solution. Bull Chem Soc Jpn 36:1097-1106.

20. Ullman E, Singh B, (1966): Photochemical transposition of ring atoms in five-membered heterocycles. The photoarrangement of 3,5 diphenyl-isoxazole. J Amer Chem Soc 88:1844-1845.

21. Ware B, (1984): Fluorescence photobleaching recovery. Amer Lab 16(4):16-28.

22. Webb W, (1981): Luminescence measurements of macromolecular mobility. Ann NY Acad Sci 366:300-314.

23. Wheelock C, (1959): The fluorescence of some coumarins. J Amer Chem Soc 81:1348-1352.

24. White E, Friend C, Stein R, Maskill H, (1969): A probable "trans"-cyclooctatetraene. J Amer Chem Soc 91:523-525.

25. Zweig A, (1973): Photochemical generation of stable fluorescent compounds (photofluorescence) Pure Appl Chem 33: 389-410.

The Use of Electron-Opaque Tracers in Nuclear Transport Studies

C. M. Feldherr
Department of Anatomy, College of Medicine, University of Florida, Gainesville, FL 32610, USA

Introduction

Within the past several years it has become clear that there are selective mechanisms which facilitate the passage of certain endogenous macromolecules across the nuclear envelope. It is likely that a number of questions relating to this process, particularly those that involve the identification and characterization of specific exchange sites, can be answered by utilizing electron-opaque tracers.

These tracers, especially colloidal gold preparations, have been used to investigate a variety of problems in cell biology. The wide applicability of colloidal gold is based on several factors: First, particles with different diameters can be readily prepared. Second, the surface properties of the gold particles can be modified by the adsorption of different stabilizing agents. Third, the association of the particles with specific cellular structures can be precisely determined with the electron microscope.

In this report, we will review the procedures for preparing and stabilizing colloidal gold, and then consider specific applications of this methodology to problems concerning macromolecular transport between the nucleus and cytoplasm.

The preparation of gold sols

The preparation of colloidal gold involves the reduction of aqueous chloroauric acid to insoluble metal atoms, which then aggregate. This can be accomplished using a number of different reducing agents; however, the size of the particles produced by reducing with white phosphorus or trisodium citrate are especially suitable for biological studies.

The particles most commonly used in our laboratory range in diameter from 30-170 A and are prepared as follows (1, 2; also see re-

Nucleocytoplasmic Transport
Edited by R. Peters and M. Trendelenburg

ference 3): Six to 7 ml of 0.6% chloroauric acid and 0.75 ml of 0.72 N K_2CO_3 are added to 120 ml of ion-free water. One ml of reducing agent (one part of a saturated ether solution of phosphorus and 4 parts of ether) is then added to the solution. The mixture is allowed to stand at room temperature for 15 minutes, at which time it is boiled for 2-3 minutes. At this stage the sols are purple and have a pH ranging from 7.0 to 7.2. Particles smaller than 30-170 A can be obtained by simply decreasing the amount of chloroauric acid; for example, 25 to 55 A particles are produced when 2 ml rather than 6-7 ml of chloroauric acid are used. Even smaller particles can be prepared by further reduction of the chlorauric acid concentration.

By using trisodium citrate as a reducing agent it is possible to prepare monodisperse colloids with particles in the 120-1500 A size range (4, 5). The particle size in a given preparation is dependent on the amount of citrate that is added to the chloroauric acid solution. Large particles such as these, although useful in many types of investigations, are of marginal value in studying nucleocytoplasmic exchanges.

Properties of colloidal gold

Colloidal gold particles are lyophobic and have a net negative surface charge. The stability of the sols in water is due to repulsion of the ionic double layers that exist at the interfacial regions, i.e., between the surfaces of the particles and the bulk dispersion medium. The degree of stabilization is directly related to the electrical potential that exists across the double layer. The addition of electrolytes reduces the potential, and correspongingly reduces electrostatic repulsion among the particles. Therefore, at sufficiently high concentrations, electrolytes will cause flocculation of the colloidal particles.

Stabilization of gold particles

The flocculating action of electrolytes can be prevented by adding hydrophilic adsorbants to the gold sols. These hydrophilic agents, such as proteins or synthetic polymers, act by binding to the surface of the gold particles, which consequently acquire the characteristics of the stabilizing substances. Thus, it is possible to regulate not only the size but also the surface properties of the colloidal particles.

The specific mechanism(s) by which macromolecules bind to gold is not fully understood, although there is evidence that binding is affected by interfacial tension and the charge of the protective agent (6). In general, neutral and negatively charged substances are better stabilizing agents than basic molecules. The addition of basic substances usually results in flocculation of the gold, presumably by reducing the potential across the ionic double layer.

Gold sols can be stablized by simply adding the appropriate macromolecules and allowing several minutes for adsorption to occur. The concentration of the stabilizing agent required for protection is determined by adding increasing amounts of the material to small aliquots of gold sol. NaCl is then added to each sample to a final concentration of 1%. The samples which are not stabilized will turn blue. This color shift, which is a result of flocculation, can be evaluated spectrophotometrically or visually. The pH range for binding should also be determined for each protein (6). Following stabilization, large gold aggregates, which may have formed fortuitously, are removed by low speed centrifugation. The stabilized particles in the supernatant can then be collected at a higher speeds (the specific conditions for centrifugation will depend on the size of the particles).

Recently, DeRoe <u>et al</u>.(7), systematically investigated interactions between proteins and colloidal gold. They prepared 150 A particles using the citrate reduction method and stablilzed the colloid with a variety of radiolabeled proteins of increasing molecular weight from 6kd (insulin) to 500kd (trimeric IgA). By centrifuging the preparations, the amount of free (supernatant) and bound (pelleted) protein could be determined. It was found that the number of protein molecules bound to each gold particle decreased as the molecular weight increased, the values ranged from 220 (insulin) to 4 (IgA). Analysis of the data indicated that a linear relationship exists between binding and the inverse of $(MW)^{2/3}$. The dissociation constants were also found to decrease with increasing molecular weight.

In general, proteins that are bound to colloidal gold particles seem to maintain their normal functions and binding characteristics.

Thus, gold coated with enzymes (e.g., 6), antibodies (e.g., 5) lectins (e.g., 8), and protein A (e.g., 3) have been used successfully in cytochemical studies.

Nuclear Transport

This section will focus primarily on the use of colloidal gold in studies relating to the facilitated transport of endogenous proteins into the nucleus. However, as background it is necessary to briefly consider earlier results obtained with exogenous macromolecules. The nuclear uptake of various foreign proteins and synthetic polymers has been investigated in oocytes, protozoans and somatic cells (for reviews see references 9 and 10). It is clear that such substances are able to enter the nucleus, and it is generally believed that the mechanism of entry is passive diffusion. Using colloidal tracers (2) it was shown that uptake occurs through central channels located in the nuclear pores. The functional size of these "diffusion channels" varies somewhat in different cell types. Diameters of 90,100-110 and 120 A have been reported for amphibian oocytes (11), hepatocytes (12) and amoebae (2), respectively.

Theoretically, many endogenous nuclear proteins should also be capable of diffusing through channels which have these dimensions; however, there are a number of large karyophilic polypeptides with molecular weights of 150 kd and above that enter the nucleus more rapidly than can be accounted for by simple diffusion. There is now evidence that these molecules (and smaller proteins as well) are transported across the nuclear envelope (for reviews see references 13 and 14).

As an initial step toward understanding the transport mechanism it is important to identify the exchange sites. Specifically, one would like to know whether transport occurs through the pores or across the membranes of the envelope. Although the pores would seem to be the most likely pathways, the direct passage of proteins across membranes is well documented (15), and should also be considered as a possible route for nucleocytoplasmic exchanges. Furthermore, since transport is apparently required for the exchange of large polypeptides, one would also like to know whether molecular size is a limiting factor, as it is for diffusion.

Feldherr et al. (16) utilized colloidal gold as a means of resolving these questions. Thirty to 170 A particles were prepared by reducing gold chloride with phosphorus (see above). The gold was then stabilized with nucleoplasmin (approximately 10 ug/ml of gold sol), concentrated by centrifugration (26,000 g for 15 minutes) and dialyzed against oocyte intracellular medium. The thickness of the protein coat, as determined by negative staining, was about 15A. Nucleoplasmin is a major karyophilic protein found in Xenopus oocytes. It has a molecular weight of 165 kd and is composed of five 33 kd subunits. Dingwall et al. (17) have clearly demonstrated that this protein is selectively transported across the nuclear envelope. In addition, the polypeptide domains that are required for transport can be removed by digestion with trypsin.

Nucleoplasmin-coated gold particles were microinjected into the cytoplasm of stage 6 Xenopus oocytes at a point just adjacent to the nucleus. The cells were fixed either 15 minutes or 1 hour later and subsequently examined with the electron microscope. It was found that the gold particles had accumulated along the nuclear envelope, especially in the areas adjacent to the pores and within the central regions of these structures (Figure 1). Furthermore, the size distribution of the gold particles present in the nucleoplasm was not

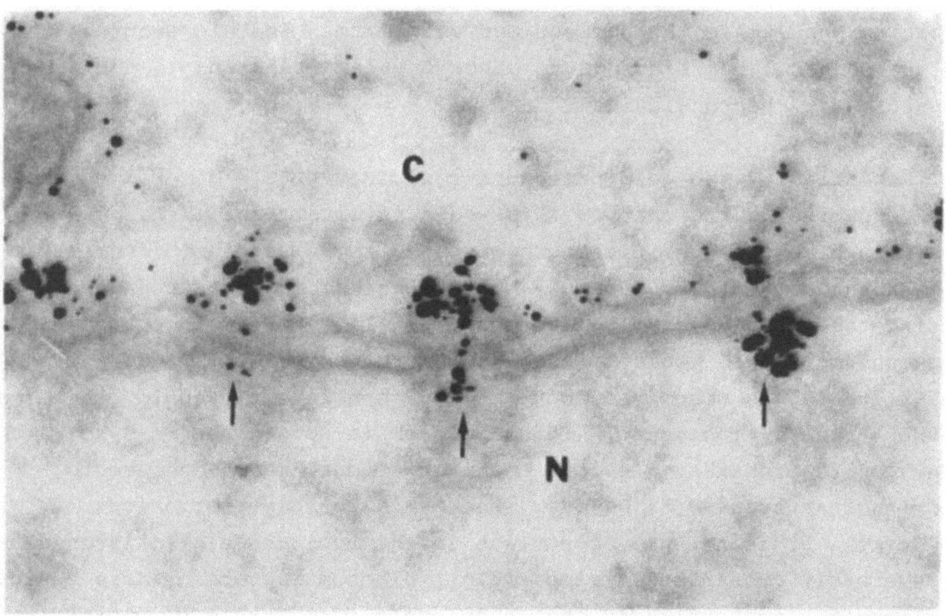

Figure 1: An electron microgrph of an oocyte fixed 15 minutes after being injected with nucleoplasmin-coated gold particles. Particles frequently accumulated just adjacent to, and within the centers of the nuclear pores (arrows). N, nucleus; C, cytoplasm.

significantly different than that in the cytoplasm. This demonstrates that particles up to 200 A in diameter (including the protein coat) can readily cross the envelope. To show that these results were due to the specific properties of the nucleoplasmin, controls were performed in which gold sols were coated with polyvinylpyrrolidone (PVP), trypsin-digested nucleoplasmin or cell homogenates. In all instances, the control particles failed to accumulate in the pore areas and were essentially excluded from the nucleus.

It was concluded that nuclear transport occurs through the centers of the pores, as does simple diffusion. However, the "transport channels" are considerably larger, at least 200 A versus 90 A for the "diffusion channels". It is not known how these two processes are able to occur within the same region of the pores.

Prospective uses for colloidal gold

The experiments just described show that it is possible, using colloidal tracers, to establish morphological criteria for nuclear transport. The criteria are based on 1) the distribution of the gold particles within individual pores and 2) the size distribution of the particles present in the nucleus and cytoplasm. In this section, the application of these findings to other problems relating to nuclear transport will be considered.

Potentially, the gold procedure could serve as a general method for demonstrating transport. This would involve stabilizing gold particles with the specific macromolecules to be tested (presumably endogenous proteins or, perhaps, nucleic acids) and subsequently injecting the colloids into an appropriate cell type. Gold coated with exogenous molecules would be used in control experiments. Morphological criteria, as described above, could then be employed to determine whether transport occurred. This approach is more direct and has a wider application than previous techniques that have been used to study nuclear transport; however, there are certain limitations. Some of the proteins that accumulate in the nucleus (e.g., histones) or reversibly cross the envelope (e.g., ribosomal proteins) are basic. As a result, it is unlikely that they would form stable complexes with gold. It might be possible, however, to couple these proteins to molecules that are effective stabilizing agents. It should also be kept in mind that the proposed experiments are focused specifically on transport through the pores. Although there is considerable evidence that these structures are the major, if not the exclusive sites for nucleocytoplasmic exchange, the possibility remains that some proteins might be transported directly across the membranes of the envelope. This process would be more difficult to detect morphologiclly and, in fact, might be inhibited by the presence of the gold particles.

There is evidence, obtained with exogenous macromolecules, that the properties of the pores vary among different cell types (see previous section). Variations also occur during different stages of the cell cycle. It was found that the size of PVP-coated gold particles capable of entering the nucleus decreased with increasing time after cytokinesis in amoebae (18). Presumably, the above results reflect the variability of the "diffusion channel." In determining whether the pores have a direct role in regulating cell activity, it is important to know if the characteristics of the "transport channel" also differ. In this regard, gold particles coated with transportable macromolecules should be appropriate tracers for comparative studies on diverse cell types. Furthermore, since experiments utilizing colloidal gold can be performed in 15 minutes or less, this method should also be useful for investigating transport during specific periods of activity in a given cell type.

Finally, there are a number of questions concerning the transport process that relate to the functional capabilities of individual pores. For example: Are all pores actively involved in transport? Can all transportable proteins enter the nucleus through the same pores, or are there functionally different classes of "transport channels?" Similarly, can RNA efflux and protein uptake occur through the same pores? Colloidial gold should be an ideal tracer for studies of this type. Not only could individual pores be analyzed by this method, but two substances could be studied simultaneously by complexing the molecules to different size gold particles.

Summary

Colloidal gold has proven to be an especially useful tracer for studying a number of different problems in cell biology. In a recent investigation, gold particles coated with nucleoplasmin (a major karyophilic protein) were used to investigate facilitated nuclear transport in Xenopus oocytes. The particles were injected into the cytoplasm and subsequently localized with the electron microscope. It was found that particles up to 200 A in diameter were transported across the nuclear envelope through central channels in the pores. These results suggest that morphological criteria, based on the distribution of colloidal gold, could be used as a means of demonstrating transport. Possible applications of these findings to other problems relating to nuclear transport are considered.

References

1. Weiser, H.B. (1933). Inorganic Colloid Chemistry, Vol.1, John Wiley and Sons, New York.
2. Feldherr, C.M. (1965). The effect of the electron-opaque pore material on exchanges through the nuclear annuli. J. Cell Biol. 25, 43-53.
3. Roth, J. (1982). The protein A-gold (pAg) technique -- A qualitative and quantitative approach for antigen localization on thin sections. In "Techniques in Immunocytochemistry" (G.R. Bullock and P. Petrusz, eds.), Vol. 1, pp 107-133, Academic Press, New York.
4. Frens, G. (1973). Controlled nucleation for the regulation of the particle size in monodisperse gold suspensions. Nature Phys. Sci. 241, 20-22.
5. Horisberger, M., and Rosset, J. (1977). Colloidal gold, a useful marker for transmission and scanning electron microscopy. J. Histochem. Cytochem. 25, 295-305.
6. Geoghegan, W.D., and Ackerman, G.A. (1977). Adsorption of horse - radish peroxidase, ovomucoid and anti-immunoglobulin to colloidal gold for the indirect detection of concanavalin A, wheat germ agglutinin and goat anti-human immunoglobulin G on cell surfaces at the electron microscope level: A new method, itheory and application. J. Histochem. Cytochem. 25, 187-1200.
7. DeRoe, C., Courtoy, P.J., Quintart, J., and Baudhuin, P. (1984). Molecular aspects of the interactions between proteins and colloidal gold. J.Cell biol. 99, 57a.
8. Roth, J., and Wagner, M. (1977). Peroxidase and gold complexes of lectins for double labelling of surface-binding sites by electron microscopy. J. Histochem. Cytochem. 25, 1181-1184.
9. Bonner, W.M. (1978). Protein migration and accumulation in nuclei. In "The Cell Nucleus. Chromatin, Part C" (H. Bush, ed.), Vol. 6, pp 97-148, Academic Press, New York.
10. Paine, P.L., and Horowitz, S.B.(1980). The movement of material between nucleus and cytoplasm. In "Cell Biology: A Comprehensive Treatise" (D.M. Prescott and L. Goldstein, eds.), Vol. 4, pp 299-338, Academic Press, New York.
11. Paine, P.L. Moore, L.C., and Horowitz, S.B. (1975). Nuclear envelope permeability. Nature 254, 109-114.
12. Peters, R. (1984). Nucleo-cytoplasmic flux and intracellular mobility in single hepatocytes measured by fluorescence microphotolysis. EMBO Journal 3, 1831-1836.
13. Dingwall, C. (1985). The accumulation of proteins in the nucleus. Trends Biochem. Sci. 10, 64-66.
14. Feldherr, C.M. (1985). The uptake and accumulation of proteins by the cell nucleus. BioEssays 3, 52-55.

15. Sabatini, D.D., Kreibich, G. Morimoto, T., and Adesnik, M. (1982). Mechanisms for the incorporation of proteins in membranes and organelles. J. Cell Biol. 92, 1-22.
16. Feldherr, C.M., Kallenbach, E., and Schultz, N. (1984). Movement of a karyophilic protein through the nuclear pores of oocytes. J. Cell Biol. 99, 2216-2222.
17. Dingwall, C., Sharnick, S.V., and Laskey, R.A. (1982). A polypeptide domain that specifies migration of nucleoplasmin into the nucleus. Cell 30, 449-458.
18. Feldherr, C.M. (1966). Nucleocytoplasmic exchanges during cell division. J. Cell Biol. 31, 199-203.

Two-Way Traffic Between Nucleus and Cytoplasm: Cell Surgery Studies on *Acetabularia*

G. Neuhaus and H.-G. Schweiger
Max-Planck-Institut für Zellbiologie, D-6802 Ladenburg bei Heidelberg, FRG

In contrast to prokaryotes the eukaryotic cell is distinguished by a nuclear membrane separating the cell nucleus where replication and transcription occurs and the cytoplasm where the mRNAs originating from the nucleus are translated. Why eukaryotic cells developed this compartmentation cannot as yet be definitely answered. One possible function for the nuclear membrane may be a controlled transport from the nucleus into the cytoplasm or vice versa and therefore may play a role in gene expression.

Experiments on this two-way traffic between cell nucleus and cytoplasm are impaired by the fact that there are only a few types of cells which allow clean experimental dissection of nucleus and cytoplasm. One of the very few types of cells which fulfills these requirements in an almost ideal way is the unicellular and uninuclear marine green alga Acetabularia (Schweiger and Berger 1979). Due to its size it can easily be handled and grown under controlled conditions in an artificial medium (Schweiger et al. 1974, 1977). Furthermore Acetabularia is quite insensitive to a number of cell surgical techniques. One of these techniques which has been used in classic experiments of cell biology is the amputation of the rhizoid leading to an anucleate cell (Hämmerling 1932, 1935, 1943). Acetabularia cells subjected to this type of surgery are not only capable of surviving but also of performing species specific morphogenesis which is continued several weeks (Hämmerling 1932). Other surgical techniques which may be performed on Acetabularia, are the isolation of the nucleus and the implantation of an isolated nucleus into an anucleate fragment (Hämmerling 1955, Sandakhchiev et al. 1973, Schweiger and Berger 1979).

A number of results illustrating nuclear effects on the cytoplasm were obtained by using different combinations of cell surgical methods. In some experiments the release of mRNA and rRNA from the nucleus into the cytoplasm as well as their transport through the cell were studied (Kloppstech and Schweiger 1975a, 1975b, Schweiger 1977). Other experiments performed more recently revealed a function of

Nucleocytoplasmic Transport
Edited by R. Peters and M. Trendelenburg

the cell nucleus which was as yet unknown. They demonstrated that the nucleus is capable of suppressing the regulation of enzymes involved in the deoxyribonucleotide metabolism and that this suppressing effect is species specific (Bannwarth and Schweiger 1983, unpublished data). Another set of experiments showed that the nucleus is also capable of shifting the phase of the circadian rhythm (Schweiger et al. 1964).

The effect of the cytoplasm on the cell nucleus was illustrated by experiments showing that the cytoplasm is capable of affecting the onset of nuclear divisions in Acetabularia (Hämmerling 1939). The ultrastructure of the cell nucleus is influenced by the cytoplasm as well (Franke et al. 1974, Berger and Schweiger 1975).

A more recent approach took advantage of the fact that isolated nuclei are insensitive to a number of different types of manipulation (Schweiger and Berger 1979). In these experiments isolated cell nuclei were microinjected with nucleic acids and eventually recombined with cytoplasm.

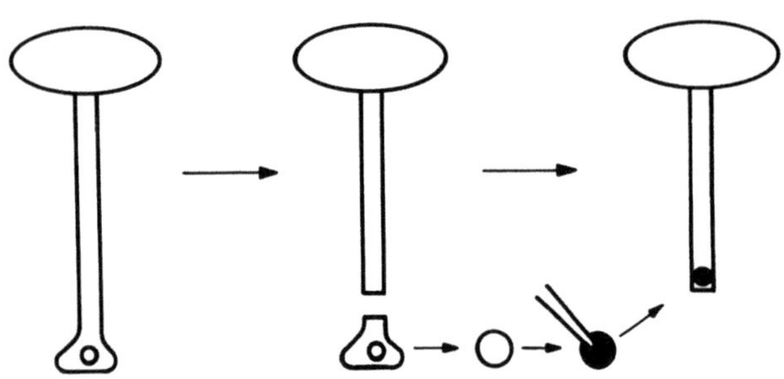

Figure 1. Introduction of foreign nucleic acids into an Acetabularia cell by microinjection.

Two different techniques were used for combining a microinjected nucleus and cytoplasm: Implantation of the microinjected nucleus into an anucleate fragment or fusion of the microinjected nucleus with a cytoplast (Gibor 1965, Primke et al. 1978). Handling an isolated Acetabularia nucleus is facilitated by its size. It may reach a diameter of 50 - 100 µm. Since an isolated Acetabularia nucleus is capable of surviving microinjection and implantation one may ask whether Acetabularia can express microinjected foreign genes.

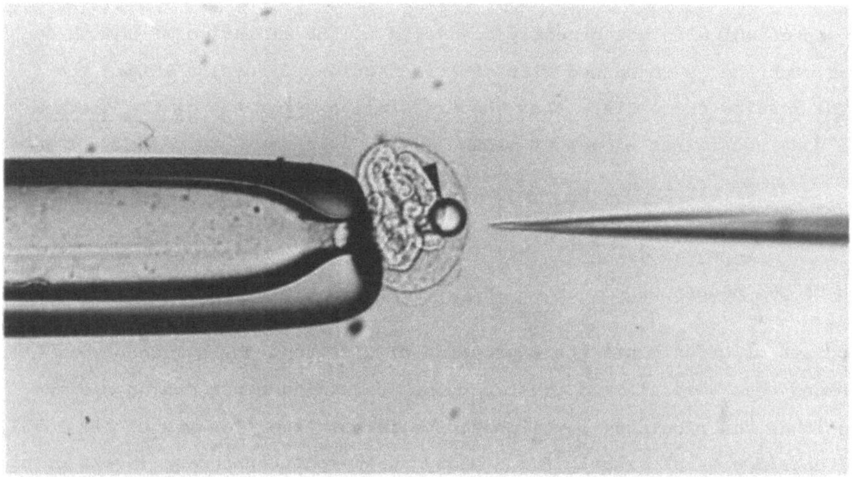

Figure 2. Micrograph of a microinjected nucleus of Acetabularia. The isolated nucleus is held by smoothly lowered pressure in a holding capillary (left). The tip of the injection capillary (right) has a diameter < 1 μm. In this case approximately 400 pl of oil (arrow) were injected into the nucleus.

Expression of RNA sequences

Among the RNAs tested for expression in Acetabularia cells are tobacco-mosaic-virus RNA (Cairns et al. 1978a), mengo-virus RNA (Cairns et al. 1978b) as well as zein mRNA (Langridge et al. 1986). In all three cases expression was obtained as evidenced by indirect immunofluorescence using antibodies against the corresponding translation product. Fluorescence was observed within 3 - 5 days after reimplantation of the microinjected nucleus. Since Acetabularia has a very high RNAase activity (Schweiger 1966), it can be concluded from these experiments that the microinjected nucleus is capable of protecting as well as releasing the introduced RNA intact. The evidence obtained by indirect immunofluorescence only proves the existence of antigenic epitopes on the translation products. This does not necessarily mean that the expression products are functional proteins. In the case of experiments with injected zein mRNA the immunofluorescence of the expression product revealed a particular structure which resembled that obtained in maize endosperm visualized by a similar method (Dierks-Ventling and Ventling 1982, Langridge et al. 1986).

A further aspect which is to be discussed in this context is that the corresponding RNA had to pass the nuclear membrane in order to be translated. The results obtained indicate that the transport of the RNA through the nuclear membrane does not-or only to a limited extent-underlie species specific control mechanisms.

A further aspect which is not directly connected to the migration of RNA through the nuclear membrane is concerned with the translation of foreign RNA in Acetabularia. The results demonstrate that the translation apparatus of the Acetabularia cell does not reject exogenous mRNA.

Expression of DNA sequences

In a second set of experiments the expression of different DNA sequences was studied. Among them were plasmid constructions containing genes coding for the small subunit of the ribulose-1,5-bisphosphate carboxylase (Neuhaus et al. 1983), for zein (Langridge et al. 1986), for a neomycin phosphotransferase (Weeks et al. 1985), as well as adenovirus DNA (Cairns et al. 1978c) and SV40 DNA (Neuhaus et al. 1984). In all cases the microinjected DNA was expressed in Acetabularia as shown by indirect immunofluorescence or through resistance to the antibiotic. This result suggests that these genes are expressed at the level of both transcription and translation. Therefore similar to its translation machinery the transcription apparatus of the Acetabularia cell presumably is unspecific since it can not discriminate between endogenous and exogenous DNA. Another important finding demonstrating that the Acetabularia cell is capable of recognizing enhancer sequences of the SV40 T-antigen should be mentioned in this context. This was concluded from experiments using gene constructions coding for the T-antigen with and without enhancer sequences (Neuhaus et al. 1984). These experiments suggest that although the Acetabularia cell is incapable of distinguishing between endogenous and exogenous DNA it is capable of recognizing regulatory sequences.

The expression of DNA sequences has primarily been demonstrated on the grounds of indirect immunofluorescence and in the case of the neomycin phosphotransferase also by induction of enzyme activity. So far the existence of the corresponding transcripts could not be shown, because the isolation of sufficient amounts of intact RNA from injected Acetabularia cells proved impossible. This may be due to the fact that the transcription rate of these sequences was rather low and the ribonuclease activity in the cell is rather high (Schweiger 1966).

As already discussed in detail for the expression of different RNAs the experiments with DNA lead to the same conclusion: the release of transcripts from the cell nucleus into the cytoplasm represents a rather unspecific process or in other words the nuclear membrane does not possess species-specific selecting capabilities. However, species-specific selection does occur in Acetabularia at a different level as has been shown by microinjection experiments with genomic genes coding for the small subunit of carboxylase. Although transcripts of this

DNA were released into the cytoplasm and translated into protein, no evidence could be found for the transport of heterologous translation products into the chloroplasts of Acetabularia, as is the case with homologous small subunit (unpublished data).

All experiments mentioned above considered the release of macromolecules from the nucleus into the cytoplasm. By injecting SV40 DNA further events could be studied. As has been shown for animal cells (Tegtmeyer et al. 1975, Das et al. 1985) the early gene product, the large T-antigen, also accumulates in the Acetabularia cell nucleus (Neuhaus et al. 1984, Weeks et al. 1985).

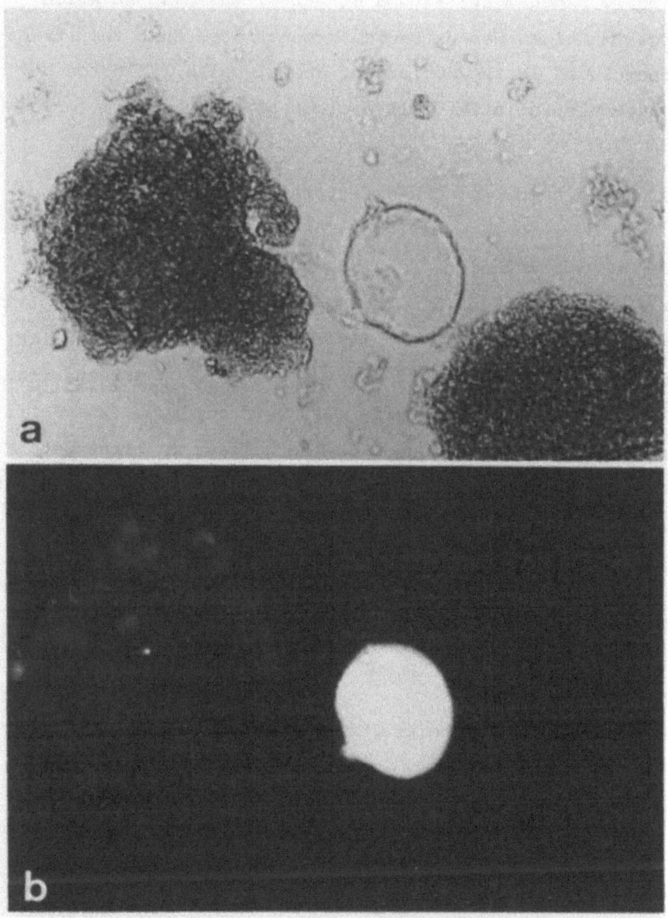

Figure 3. T-antigen specific immunofluorescence in the nucleus of a SV40 DNA injected Acetabularia cell 24 h after microinjection. (a): white light image, the nucleus and the surrounding cytoplasm can be distinguished. (b): immunofluorescence.

The sequence of events after injection of the SV40 DNA and implantation of the nucleus is the transcription of the gene in the nucleus and the release of the transcription product into the cytoplasm, where it is translated. The translation product, i.e. the large T-antigen, then migrates from the cytoplasm into the nucleus where it is accumulated.

The experiments on the large T-antigen of the SV40 deserve interest also for another reason. There seems to be little doubt that the accumulation of the large T-antigen in the cell nucleus, as it has been demonstrated in animal cells, is accompanied by the binding of specific sites of the protein to the nuclear DNA (Tooze 1980, Das et al. 1985). Experiments with a mutant of the SV40 in which one amino acid of the large T-antigen was exchanged demonstrated that this translation product was not accumulated in the nucleus of animal cells (Kalderon et al. 1984) neither was it accumulated in the cell nucleus of Acetabularia (unpublished data).

In experiments on Acetabularia a number of basic features of the relationships between the cell nucleus and the cytoplasm have been demonstrated. The recent experiments using the microinjection technique focused on the specificity of the release of macromolecules from the nucleus and their migration into the cytoplasm by passing the nuclear membrane. Summarizing one may conclude that this barrier which separates the genome with its transcription and replication apparatus from the cytoplasm and the site of translation exhibits little if any species specificity for nucleic acids or T-antigen. The function of the nuclear membrane with respect to the migration of the macromolecules into and out of the nucleus is not yet understood. Further experiments should also consider other aspects of material transport through the nuclear membrane.

Acknowledgements

We are grateful to Dr. G. Neuhaus-Url for helpful discussions, to Mrs. B. Gernert for critically reading the manuscript, to Mrs. R. Fischer for the art work and Ms. C. Schardt for typing the manuscript.

References

Bannwarth, H. and Schweiger, H.G. (1983) The influence of the nucleus on the regulation of dCMP deaminase in Acetabularia. Cell Biol. Intern. Rep. 7, 859 - 868.

Berger, S. and H.G. Schweiger (1975) Cytoplasmic induction of changes in the ultrastructure of the Acetabularia nucleus and perinuclear cytoplasm. J. Cell Sci. 17, 517 - 529.

Cairns, E., Sarkar, S. and Schweiger, H.G. (1978a) Translation of tobacco-mosaic-virus RNA in Acetabularia cell cytoplasm. Cell Biol. Int. Rep. 2, 573 - 578.

Cairns, E., Gschwender, H.H., Primke, M., Yamakawa, M., Traub, P. and Schweiger, H.G. (1978b) Translation of animal virus RNA in the cytoplasm of a plant cell. Proc. Natl. Acad. Sci. USA, 75, 5557 - 5559.

Cairns, E., Doerfler, W., Schweiger, H.G. (1978c) Expression of an DNA animal virus genome in a plant cell. FEBS Lett. 96, 295 - 297.

Das, G.C., Niyogi, S.K. and Salzman, N.P. (1985) SV40 promoters and their regulation. Progress Nucl. Acid Res. 32, 217 - 236.

Dierks-Ventling, Ch. and Ventling D. (1982) Tissue-specific immunofluorescent localization of zein and globulin in Zea mays (L.) seeds. FEBS Lett. 144, 167 - 172.

Franke, W.W., Berger, S., Falk, H., Spring, H., Scheer, U., Herth, W., Trendelenburg, M.F. and Schweiger, H.G. (1974) Morphology of the nucleo-cytoplasmic interactions during the development of Acetabularia cells. I The vegetative phase. Protoplasma 82, 249 - 282.

Gibor, A. (1965) Surviving cytoplasts in vitro. Proc. Natl. Acad. Sci. USA 54, 1527 - 1531.

Hämmerling, J. (1932) Entwicklung und Formbildungsvermögen von Acetabularia mediterranea. II. Das Formbildungsvermögen kernhaltiger und kernloser Teilstücke. Biol. Zentralblatt 52, 42 - 61.

Hämmerling, J. (1935) Über Genomwirkungen und Formbildungsfähigkeit bei Acetabularia. Roux' Arch. Entwicklungsmech. 132, 424 - 462.

Hämmerling, J. (1939) Über die Bedingungen der Kernteilung und der Zystenbildung bei Acetabularia mediterranea. Biol. Zentralblatt 59, 158 - 193.

Hämmerling, J. (1943) Ein- und zweikernige Transplantate zwischen Acetabularia mediterranea und Acetabularia crenulata. Z. Indukt. Abstammungs-Vererbungsl. 81, 114 - 180.

Hämmerling, J. (1955) Neuere Versuche über Polarität und Differenzierung bei Acetabularia. Biol. Zentralblatt 74, 545 - 554.

Kalderon, D., Richardson, W.D., Markham, A., Smith, A.E. (1984) Sequence requirements for nuclear location of simian virus 40 large T-antigen. Nature 311, 33 - 35.

Kloppstech, K. and Schweiger, H.G. (1975a) 80S ribosomes in Acetabularia major. Distribution and transportation within the cell. Protoplasma 83, 27 - 40.

Kloppstech, K. and Schweiger, H.G. (1975b) Polyadenylated RNA from Acetabularia. Differentiation 4, 115 - 123.

Langridge, P., Brown, J.W.S., Pintor-Toro, J.S., Feix, G., Neuhaus, G., Neuhaus-Url, G. and Schweiger, H.G. (1986) Expression of zein genes in Acetabularia mediterranea. Europ. J. Cell Biol., in press.

Neuhaus, G., Neuhaus-Url, G., Schweiger, H.G., Chua, N.H., Broglie, R., Coruzzi, G., Chu, N. (1983) Expression of higher plant genes in the green alga Acetabularia mediterranea. Europ. J. Cell Biol. Suppl. 2, 30.

Neuhaus, G., Neuhaus-Url, G., Gruss, P., Schweiger, H.G. (1984) Enhancer-controlled expression of the simian virus 40 T-antigen in the green alga Acetabularia. EMBO J. 3, 2169 - 2172.

Primke, M., Berger, S. and Schweiger, H.G. (1978) Protoplasts from Acetabularia: Isolation and fusion. Cytobiologie 16, 375 - 380.

Sandakhchiev, L., Niemann, R. and Schweiger, H.G. (1973) Kinetics of changes of malic dehydrogenase isozyme pattern in different regions of Acetabularia hybrids. Protoplasma 76, 403 - 415.

Schweiger, H.G. (1966) Ribonuklease-Aktivität in Acetabularia. Planta 68, 247 - 255.

Schweiger, H.G. (1977) Transcription of the nuclear genome of Acetabularia. In: L. Bogorad and J.H. Weil (eds.) Nucleic acids and protein synthesis in plants. Plenum Publishing Corporation, New York, 65 - 83.

Schweiger, E., Wallraff, H.G. and Schweiger, H.G. (1964) Endogenous circadian rhythm in cytoplast of Acetabularia: influence of the nucleus. Science 146, 658 - 659.

Schweiger, H.G., Berger, S., Kloppstech, K., Apel, K. and Schweiger, M. (1974) Some fine structural and biochemical features of Acetabularia major (Chlorophyta, Dasycladaceae) grown in the laboratory. Phycologia 13, 11 - 20.

Schweiger, H.G., Dehm, P., Berger, S. (1977). Culture conditions for Acetabularia. In: C.L.F. Woodcock (ed.) Progress in Acetabularia Research. Academic Press New York 319 - 330.

Schweiger, H.G. and Berger, S. (1979) Nucleocytoplasmic interrelationships Acetabularia and some other Dasycladaceae. Int. Rev. Cytol., Suppl. 9, 11 - 44.

Tegtmeyer, P., Schwartz, M., Collins, J.K. and Rundell, K. (1975) Regulation of tumor antigen synthesis by simian virus 40 gene A. J. Virol. 16, 168 - 178.

Tooze, J. (ed.) (1980) Molecular biology of tumor viruses, DNA tumor viruses Pt. 2: 2nd ed., Cold Spring Harbor, N.Y.

Weeks, D., Brunke, K., Beerman, N., Anthony, J., Neuhaus, G., Neuhaus-Url, G. and Schweiger, H.G. (1985) Promoter regions of the four coordinately regulation tubulin genes of Chlamydomonas and their use in the construction of fused genes which are expressed in Acetabularia. In: M. Freeling (ed.) Plant Genetics. Alan R. Liss, New York 477 - 490.

Nuclear Envelope Vesicles as an In Vitro Model System to Study Nucleocytoplasmic Transport of Proteins and Ribonucleic Acids

N. Riedel, M. Bachmann, D. Prochnow, and H. Fasold
Institut für Biochemie der Johann-Wolfgang-Goethe-Universität,
Theodor-Stern-Kai 7, D-6000 Frankfurt am Main 70, FRG

We describe a procedure for the preparation of nuclear envelope vesicles from rat liver nuclei. These vesicles contain the components of the nuclear envelope, e.g. the pore-complex-lamina fraction, and a residual DNA content of 1.5%. Typical preparations consist of about 90% vesicles with the vesicular character of these envelopes shown by means of microscopic and biochemical studies. Uptake measurements using nuclear and non-nuclear proteins show that the vesicles have a high affinity only for proteins that are specific for the nuclear compartment and indicate that the nuclear envelope affects the uptake kinetics and increases the capacity for nuclear proteins. The nuclear envelope vesicles contain the translocation mechanism for mRNA and show an unidirectional transport of mRNA from the vesicular interior into the medium; this efflux can be stimulated by ATP. Because the vesicles are virtually free of the components of the nuclear interior, but retain properties of intact nuclei, we believe that they are a valuable model system to investigate the structures and mechanisms involved in the regulation of nucleocytoplasmic transport processes.

INTRODUCTION

The nuclear envelope is a characteristic structure of the eucaryotic cell. It consists of an inner and an outer membrane which are separated by the nuclear space (2). The outer membrane is connected to the endoplasmic reticulum, while the inner membrane faces the nucleoplasm and is tightly associated with the fibrous lamina (22). The lamina is spaced between the inner nuclear membrane and the peripheral chromatin and is found in most eucaryotes (54). The lamina is restricted to the periphery of the interphase nucleus and is not a component of the internal nuclear matrix (17,28). It is believed to be involved in the association of the peripheral chromatin (20,26,27,28,51,61) and in stabilizing the nuclear periphery. The peripheral chromatin is resistent to DNase I-treatment and cannot be removed by either detergents or high salt concentrations (25).

Another characteristic feature of the nuclear envelope are the pore complexes. The structure and composition of the pore complexes have not been completely analyzed. Under electron microscopy the pores appear as cylinders

Nucleocytoplasmic Transport
Edited by R. Peters and M. Trendelenburg

which span the nuclear envelope. They show an octagonal symmetry around a central axis perpendicular to the membrane. Eight globular subunits with diameters of approximately 22 nm form an outer and an inner ring around the pore. A central "plug" is located in the center of the pore and is connected with the periphery via "spokes". The whole pore complex, which has an outer diameter of 120 nm (31,58) can be described as a "triple-layered sandwich" consisting of the spokes and the central plug between the outer and inner ring. Some of the proteins of the pore complex have recently been purified. Preparations of oocytes, which have a very high density of pore complexes, are enriched in two proteins of molecular weights of 73,000 d and 150,000 d (39). A protein of 68,000 d could be shown to be pore complex specific in X. laevis oocytes (55), whereas in rat liver a glycoprotein of 190,000 d was found to be a component of the pore complex (29).

The nuclear envelope is of particular interest because of its likely role in nucleocytoplasmic transport processes. The pore complexes are the most prominent candidates as pathways mediating the exchange of molecules between the nucleus and the cytoplasm. Early microinjection experiments with amoeba using gold particles showed that these were taken up into the nucleus through the pore complexes (23). Measurements with oocytes using labeled proteins (9,10,42) or dextrans of different size (43) showed that the diameter of the pore is in the range of 9 nm (43).

Small molecules and ions can move across the nuclear envelope via diffusion (33,44). Measurements with larger molecules, like myoglobin, γ-globulin, ovalbumin and bovine serum albumin showed that the size of the protein seems to be a decisive parameter for passage through the nuclear envelope (9). Proteins with a size in the range of the pore diameter need hours to enter the nucleus (42). However, the uptake of at least some large nuclear proteins seems to be mediated and regulated by the pore complexes, which is shown by the fact that their uptake into the nucleus is about 20 times faster than could be explained by a diffusional process (24). For nucleoplasmin, a pentamer of molecular weight of 165,000 d, it could be demonstrated that a region of 12,000 d of the monomer is necessary for selective transport into the nucleus (18). For the nuclear proteins N1 - N4 it was shown that the cytoplasmic precursors and the accumulated proteins in the nucleus are identical, indicating that the signal sequence that directs these proteins to the nucleus is an internal part of the translational product (16). This is also supported by experiments using SV 40 large T antigen, which show that a small region of the protein serves as a signal sequence for directed transport into the nucleus. Fusion proteins containing the signal sequence and typical cytoplasmic or bacterial proteins were accumulated in the nucleus (36).

The export of ribosomal and messenger RNA from the nucleus is poorly understood. Again, a simple diffusion process can be excluded, because the diameter of RNP particles exceeds the diameter of the pores (13). The efflux is energy-dependent, and a nuclear envelope-associated Mg^{2+}-dependent ATPase activity has been described by many laboratories (4,12,38). The exact localization and characteristics of the enzyme are not clear, mostly because subfractions of the nuclear envelope always contain components of the lamina and the pore complexes. The nuclear envelope-associated Mg^{2+}-ATPase can be stimulated by RNA and synthetic polynucleotides, where poly(A) has the strongest effect (3,7). Antibodies directed against lamin B inhibit the ATPase activity and the transport of RNA in isolated nuclei (6), and a number of laboratories have shown that the release of mRNA from isolated nuclei is ATP-dependent (37,45,52), although this is not the case for nuclei derived from certain tumor cells (52,56). For the transport of rRNP particles no energy dependency could be established, which led to the idea that the poly(A) tail of mRNA might be the signal sequence for the enzyme (5,7).

Here we describe the preparation and characterization of vesicles of the nuclear envelope. These vesicles contain all the components of the nuclear envelope, e.g. the lamina and the pore complexes, in the same orientation as in whole nuclei. In contrast, they contain only 1.5% of the DNA of the nucleus and are mostly free of the components of the nuclear interior. By means of microscopic and biochemical procedures we show that the preparations contain about 90% tight vesicles and are enriched in the Mg^{2+}-dependent ATPase that mediates nucleocytoplasmic transport processes. We performed transport measurements with these vesicles using radiolabeled nuclear and non-nuclear proteins and could show that the characteristics for their uptake are very similar to those of whole nuclei. Measurements with poly(A) and mRNA show that the vesicles also contain the translocation mechanism for mRNA. We believe that these vesicles are a very valuable model system for further elucidating the mechanism and regulation of transport processes across the nuclear envelope.

RESULTS AND DISCUSSION

Evidence for the preparation of closed nuclear envelope vesicles

We have previously shown (38) that the ATP analog S-dinitrophenyl-6-mercaptopurine riboside triphosphate, here referred to as TPA, can be covalently linked to and inhibit a nuclear envelope-associated Mg^{2+}-dependent ATPase and that it cannot penetrate the nuclear envelope. The synthesis and characteristics of this ATP analog were described earlier (21,38).

Nuclei were prepared according to Blobel and Potter (8) and used as the starting material to prepare nuclear envelopes. Incubation of nuclear envelopes derived by DNase I-treatment (19) with TPA at a concentration of 2×10^{-5} M for 30 min. led to an inhibition of the enzyme of up to 50%. However, a distinct inhibition of the enzyme was already measurable at a TPA concentration of 1 x 10^{-6} M (38). Surprisingly, an incubation of whole nuclei with TPA resulted in no inhibition of the ATPase activity, even at concentrations of 1 x 10^{-4} M (38). Knowing that the diameter of the pore complexes is in the range of 9 nm (43), one would believe that an ATP analog can easily permeate the nuclear envelope by diffusion. That this is not the case can be shown by incubation of whole nuclei with α-labeled ^{32}P-TPA. After an incubation for 30 min. at a concentration of 2 x 10^{-5} M the nuclei were washed in 0.25 M STKMC-buffer (38) until the supernatant was free of radioactivity. Only 6% of the added labeled TPA was found to be associated with the nuclear fraction. All this radioactivity could then, however, be removed by incubating the nuclei with 1% Triton X-100, which is known to remove the outer and most of the inner nuclear membrane (1,41,56). A preparation of nuclear envelopes using heparin to extract the nuclear contents after opening the nuclei in hypotonic media was originally described by Bornens and Courvalin (11). The composition of these nuclear envelopes is very similar to that of nuclear envelopes derived using DNase I-treatment (19). They contain the components of the pore-complex-lamina fraction and have a residual DNA content of 1.5% (11; data not shown). The procedure of Bornens and Courvalin (11) was slightly modified and the nuclear envelopes were washed and resuspended in 0.25 M STKMC-buffer. After incubation of these nuclear envelopes with TPA at concentrations of up to 1 x 10^{-4} M, we were not able to detect an inhibition of the ATPase of more than 10%. Thus, these results indicated that the ATPase was not accessible to the ATP analog, a behavior identical to that of whole nuclei. We then added the TPA at a concentration of 2 x 10^{-5} M to the lysis buffer during the preparation of the nuclear envelopes with heparin and measured the ATPase activity in comparison to a heparin nuclear envelope fraction that was subsequently or not at all incubated with TPA. We found that after an inclusion of TPA during the preparation the ATPase was covalently inhibited by TPA by 40%.

Phase contrast microscopy was used to elucidate the morphology of the nuclear envelope preparation. As is shown in Fig.1, our nuclear envelopes prepared using heparin consist almost entirely of nuclear envelope vesicles. The preparation is uniform and from analyzing about 100 vesicles, we found that the size of the vesicles is in the range of 4.0 µm to 9.2 µm, 72% of which are in the range of 6.5 µm to 7.5 µm. Thus, the size is comparable to that of rat liver nuclei which are in the range of 7.5 µm. The insert in Fig.1 is a transmission electron micrograph of the vesicular periphery showing the presence and preservation of pore complexes and inner and outer nuclear membranes.

The microscopy and the measurements with the ATP analog prove the vesicular character of the nuclear envelopes, show that the nuclear envelope itself is in the same orientation as in intact nuclei and indicate that an inclusion of molecules can be achieved during the preparation of these vesicles. It was also

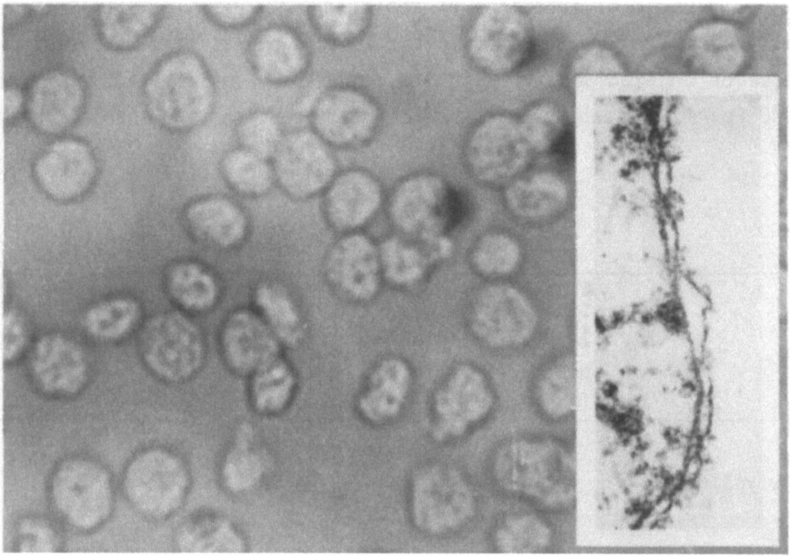

Fig. 1: Phase contrast light micrograph of rat liver nuclear envelope vesicles prepared by the modified heparin extraction method. The insert shows an electronmicroscopic view of the pore complexes and inner and outer nuclear membranes (magnification 220,000 X).

shown that a Mg^{2+}-dependent ATPase is located on the inner side of the nuclear envelope, as would be expected for an enzyme that is involved in nucleocytoplasmic transport processes. However, these experiments tell nothing about the stability and tightness of the vesicles. We therefore included ^{14}C-labeled ferritin into the vesicles during their preparation. Previous studies showed that this protein (MW 465,000 d) cannot penetrate the nuclear envelope (32,42,43). Incubation of nuclear envelope vesicles with radiolabeled ferritin confirmed this result and showed that no uptake could be measured over a time period of 4 hours. For the inclusion of ferritin into the vesicles we then added 1 mg/ml of ^{14}C-labeled ferritin to a suspension of whole nuclei. After dividing the nuclear suspension into 2 fractions, one was used in the preparation of nuclear envelope vesicles using heparin, the other in the preparation of nuclear envelopes using DNase I. At the end of the preparations the nuclear envelope fractions were resuspended in 0.25 M STKMC-buffer and 0.25 M STKM-buffer, respectively, and centrifuged through a cushion of 30% sucrose in the appropriate buffer until the supernatants were completely free of radioactivity. The radioactivity of each fraction was measured and normalized relative to the number of nuclear envelopes. The result of three different experiments was a five-fold higher ferritin content in the nuclear envelope vesicle fraction which we explain by an inclusion of ferritin in the vesicles during the preparation.

To determine the tightness of the vesicles, a vesicle fraction that was loaded with ^{14}C-labeled ferritin was resuspended in STKMC-buffer at 5,000 cpm/ml.

At different time points 1 ml fractions of the suspension were centrifuged and the radioactivity in the supernatant and in the vesicle pellet were measured. We found an increase of the radioactivity in the supernatant of only 5% during a time period of 48 hours, indicating that the vesicles are tight and stable for at least two days.

fraction	poly(A) in µg/ml addition	poly(A) in µg/ml inclusion	ATPase-activity in %
nuclear envelope vesicles	-	-	100
	1	-	105
	3	-	101
	6	-	101
	10	-	111
	50	-	108
	200	-	111
	-	50	129
	-	600	153
DNase I-treated nuclear envelopes	-	-	100
	60	-	140

Table 1: Effect of poly(A) on the activity of the nuclear envelope-associated magnesium-dependent ATPase of nuclear envelope vesicles and nuclear envelopes prepared using DNase I. The enzyme activities of the controls were set as 100%. Each value represents the average of three separate measurements.

We further analyzed the tightness and the correct orientation of the vesicles by measuring the nuclear envelope-associated Mg^{2+}-dependent ATPase activity in the presence of poly(A). As previously mentioned, this enzyme is thought to be involved in nucleocytoplasmic transport processes, especially for mRNA (5,37,45,47,53). The enzyme can be stimulated by RNA and synthetic polynucleotides (3,7) and a comparison of different polynucleotides showed that poly(A) is the most potent activator of the enzyme. An important prerequisite is that the polynucleotide must have a chain length of at least 15 AMP-residues (3,7). For the measurements described below we used poly(A) of a molecular weight of 100,000 d (Sigma Co; approximately 285 AMP-residues). Nuclear envelopes were preincubated with increasing amounts of poly(A) for 30 minutes, followed by measuring the ATPase activity. The result is shown in Table 1.

Increasing amounts of poly(A) stimulate the ATPase of nuclear envelope vesicles by 11% maximally, which we believe is due to the fact that the vesicle suspension consists of approximately 90% vesicles, leaving the remaining 10% of the nuclear envelopes accessible for poly(A). This result is similar to that obtained with whole nuclei, where no increase in the ATPase activity of more than 10% could be measured (data not shown). However, addition of poly(A) to the lysis buffer during the preparation of the vesicles causes a clear increase of the ATPase activity in the subsequent enzyme assay. Poly(A) in a concentration of 50 µg/ml led to an increase in the enzyme activity of 29%, and 600 µg/ml to an increase of 53% compared to a control preparation without poly(A). These numbers are comparable to those using nuclear envelopes prepared with DNase I, where externally added poly(A) in a concentration of 60 µg/ml increased the ATPase activity by 40%. Thus, these measurements with poly(A) clearly indicate that the nuclear envelopes prepared using heparin consist almost entirely of vesicles with the nuclear envelope in the same orientation as in whole nuclei. These vesiclesare tight and stable for at least two days.

Flux measurements across the nuclear envelope of the vesicles

Whole nuclei contain chromatin and the nuclear matrix. Microinjection experiments and experiments with isolated whole nuclei therefore make it difficult to determine whether a nuclear accumulation of proteins is caused by specific binding to components of the nuclear interior or by transport processes regulated by components of the nuclear envelope itself.

Nuclear envelope vesicles have the important advantage that they are free of most of the nuclear contents, but retain the components of the nuclear envelope. Therefore, we wanted to investigate the uptake kinetics of some nuclear and non-nuclearproteins.

Kinetics of histone uptake into nuclear envelope vesicles

Histones were prepared according to Johns (34) and their purity examined by polyacrylamide gel electrophoresis. The mixture used consisted of the histones H1, H2A, H2B, H3 and H4, which were labeled with ^{14}C-acetic anhydride.

The kinetics of histone uptake into nuclear envelope vesicles is shown in Fig.2 in comparison to nuclear envelopes prepared using DNase I. As can be seen, the uptake of histones is a very fast process, having its maximum at 2 minutes, after which a plateau is reached. Based on the DNA content of the vesicles we can estimate that the amount of histones taken up is 2.5 µg per µg DNA under the conditions used. Nuclear envelopes prepared with DNase I do not show similar kinetics. Whereas the uptake of histones into the vesicles increased by 140% in the first 300 seconds, this was only 48% in the case of the DNase I-nuclear envelopes, thus probably reflecting an absorption to components of the nuclear envelope, which reached a plateau after 50 seconds. Based on the DNA content, the capacity for histones was 45% higher in the case of the vesicles. Accumulation by binding to the vesicular interior cannot alone account for the observed uptake kinetics, because the non-vesicular nuclear envelopes prepared by DNase I should then show very similar kinetics. Thus, we believe that the nuclear envelope itself contributes to the differences in uptake kinetics and the much higher capacity of the vesicles for histones (see below).

We tested a variety of nucleotides, e.g. GTP, GDP, GMP, cGMP, guanosine, ATP, cAMP and CTP for their effect on histone uptake. As is shown in Fig.2, GTP at a concentration of 10 mM stimulated the uptake of histones into the vesicles,

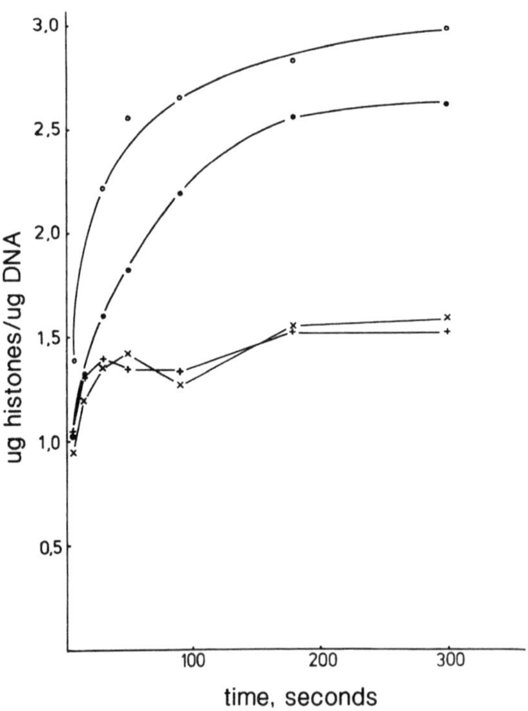

Fig.2: Kinetics of histone uptake into nuclear envelope vesicles in the presence (o) and absence (•) of 10 mM GTP. In contrast to the vesicles, the kinetics for the DNase I-treated nuclear envelopes is very different (+), shows no GTP effect (x) and seems to reflect a simple adsorption process.

while no effect on the DNase I-treated nuclear envelopes was measurable. None of the other nucleotides tested had a clear effect.

Nuclear envelope vesicles were incubated with radiolabeled histones for 20 minutes in the presence or absence of GTP to obtain maximum uptake (set as 100%). The exchange rate was then measured by adding a ten-fold excess of non-radioactive histones followed by separation of the vesicles by centrifugation at different time points between 5 and 300 seconds and measurement of the radioactivity in the vesicle pellets. In the absence of GTP the exchange-efflux of radioactivity from the vesicles reached a plateau after approximately 120 seconds, where only 65% of the maximum radioactivity remained associated with the vesicular pellet. This indicates that approximately 35% of the radioactivity taken up by the vesicles were not stably bound to components of the vesicular interior. The remaining radioactivity is in the range of the value observed for the DNase I-nuclear envelopes, indicating similar binding capacities for the two preparations. We estimate that the internal volume of the vesicles in the incubation medium is approximately 1.7%. Thus, these results suggest that the nuclear envelope of the vesicles is involved in the accumulation of histones. The exchange-efflux in the presence of GTP was only 5% of the maximum radioactivity over the time course of 300 seconds, showing that GTP causes a retention of the histones in the vesicular interior.

We do not know how GTP stimulates histone uptake and inhibits histone efflux. We are presently using radiolabeled GTP and GTP analogs followed by a subfractionation of the vesicles in order to find components of the nuclear envelope that have affinity for GTP and might be responsible for the increased uptake of histones.

To show that the observed uptake of histones is not due to an adsorption to the outside of the vesicles, nuclear envelope vesicles were first incubated with ^{14}C-labeled histones for 20 minutes in the presence of GTP to complete the uptake. All excess radioactivity was then removed by centrifugation through 30% sucrose cushions as described. After resuspension of the vesicles, the radioactivity was measured and set as 100%. The vesicles were subsequently incubated with 1% Triton X-100 at $4^{0}C$ for 10 minutes to remove the nuclear membranes, followed by a second centrifugation. Only 6.7% of the radioactivity was associated with the membrane-containing supernatant, while the remainder was still associated with the vesicular suspension.

Uptake measurements with histones using vesicles that were first treated with Triton X-100 to remove the membranes were also performed. The uptake kinetics obtained were very similar to those shown in Fig.2, indicating a minor role for the nuclear membranes.

Uptake of high-mobility-group proteins (HMG-proteins) and acidic chromosomal proteins

In mammalian cells, HMG proteins form a group of 4 different proteins, HMG1, 2, 14 and 17. HMG proteins are associated with approximately 10% to 20% of the nucleosomes, with this association preferably (59) but not exclusively (30,40,48) taking place with actively-transcribed chromatin. HMG 14 and HMG 17 were found to have specific binding sites for nucleosomes (40,57) and HMG 1 and HMG 2 seem to play a direct role in destabilizing double-stranded DNA. In addition, it was shown that HMG 14 and HMG 17 are associated with the nuclear matrix and the nuclear envelope (49).

In contrast to the well-characterized HMG proteins, the acidic chromosomal proteins comprise a broad spectrum of polypeptides in the molecular weight range of 20,000 d to 200,000 d. They bind specifically to DNA, but their function is poorly understood.

In comparison to the histones, the uptake of HMG proteins prepared according to Sanders (50) and acidic chromosomal proteins prepared according to Wilson and Spelsberg (60) continued for approximately 4 hours before a plateau was reached. Also, the binding capacity of the vesicles for these proteins was lower and no GTP or GDP effect could be observed.

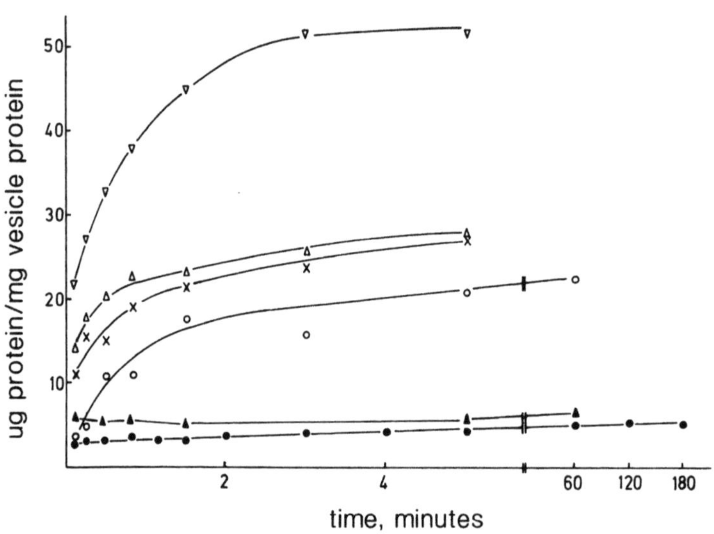

Fig.3: Uptake or binding kinetics of non-nuclear proteins in nuclear envelope vesicles. Histones (▽), polylysine (Δ), histone H1 (x), myoglobin (o), γ-globulin (▲) and cytochrome c (•) were used at concentrations of 100 μg per mg vesicle protein and each value represents the mean of 2 separate measurements.

Uptake measurements of non-nuclear proteins

To judge whether the described uptake kinetics are a specific process, it is necessary to determine whether the affinity of the vesicles is specific for nuclear proteins and if proteins which have no known function in the nucleus are not at all, or at least to a much lower extent, accumulated.

In the following experiments we used different non-nuclear proteins to measure their affinity for the nuclear envelope vesicles. The results are shown in Fig.3.

γ-globulin has a molecular weight of 150,000 d and is too large to traverse the nuclear envelope of oocyte nuclei (9). After microinjection, the nucleus:cytoplasm ratio has a value of 0.08 (9). As is shown in Fig.3, we were unable to measure an uptake of γ-globulin during a time period of 3 hours. The amount of radioactivity bound was only 6% and probably reflects an adsorption to the outside of the nuclear membrane. A value of approximately 6% was also found for the adsorption of histones and the ATP analog TPA.

Myoglobin (17,500 d) was shown to be accumulated in the nucleus after microinjection into oocytes (9,42). In the course of 2 hours a nucleus:cytoplasm ratio of 2 can be measured and the uptake is due to a binding to intranuclear components (42). Similar results were obtained using the nuclear envelope vesicles, with the amount of myoglobin taken up reaching approximately 40% of the value for histones.

The uptake kinetics for polylysine and histone H1, which has the highest lysine content of all the histones, was very similar. It is noteworthy that the uptake kinetics for the histone mixture and histone H1 alone are different.

Cytochrome c with a molecular weight of 12,500 d and an isoelectric point of 10 is a small and very basic protein which seems to be especially suitable for a comparison with histones. Early experiments using oocytes showed that this protein has no affinity for nuclei (42). After microinjection, a nucleus:cytoplasm ratio of 1 is reached very rapidly and is then stable for several hours, reflecting a diffusion into the nucleus to equilibrium. The measurements using nuclear envelope vesicles are consistent with this result. The increase of radioactivity from 2.7% to 4.7% probably represents a diffusion into the vesicles which ends after 5 minutes.

The uptake measurements of all of the above mentioned proteins were also performed in the presence of 10 mM GTP; however, no stimulating effect for any of the proteins was found.

In conclusion, these results show that the vesicles have a high affinity for nuclear proteins. The measurements using this in vitro system yield results that are similar to those described in in vivo experiments using oocytes (9,42) and in experiments using HeLa cell nuclei (15).

Unidirectional transport of RNAs

As mentioned above, poly(A) is a potent activator of the Mg^{2+}-dependent nuclear envelope-associated ATPase, and Table 1 shows that poly(A) only stimulates the ATPase of the nuclear envelope vesicles when introduced during their preparation. Similar experiments were performed with rRNA and mRNA, which were isolated from rat liver polysomes according to standard procedures (35,46) followed by iodination with ^{131}J using the procedure of Commerford (14). In analogy to the measurements with poly(A), we first incubated DNase I-treated nuclear envelopes and nuclear envelope vesicles with increasing concentrations of rRNA and mRNA. While a stimulation of the ATPase of up to 50% could be measured for the DNase I-nuclear envelopes, the effect on the vesicles was again 10% maximally, which is due to the amount of leakÿ vesicles in the preparation. Incubation of whole nuclei and nuclear envelope vesicles with labeled rRNA and mRNA in the presence and absence of ATP over a time course of 60 minutes confirmed our results obtained when measuring the ATPase activity and showed that no uptake into nuclei or vesicles occurred. No more than 0.4% of the radioactivity added was associated with the nuclei or vesicles and an increase during a 60 minutes period could not be measured. We then included rRNA and mRNA during the preparation of vesicles. The vesicles were washed in transport buffer

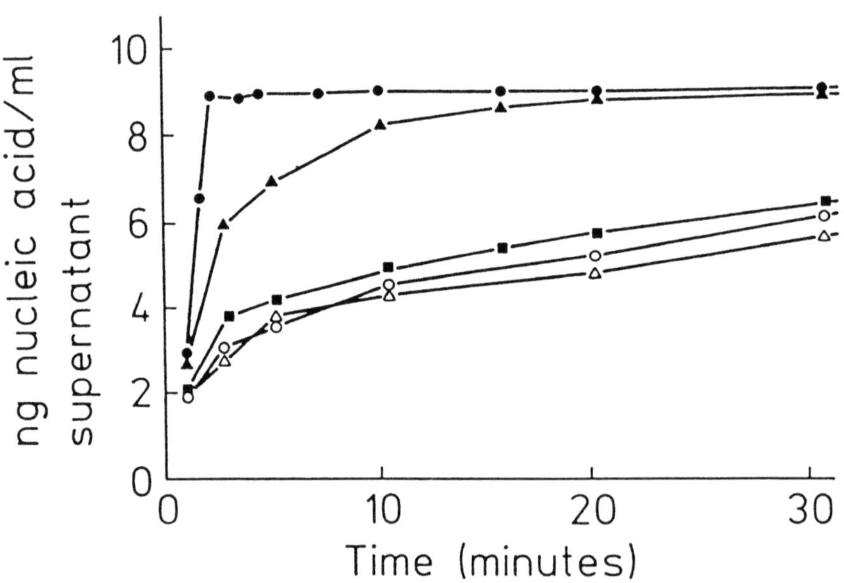

Fig. 4: Efflux of various ribonucleic acids from nuclear envelope vesicles: (Δ), rRNA in the absence and (o), in the presence of 2 mM ATP; (■), mRNA in the absence and (▲), in the presence of 2 mM ATP; (•), poly(A) in the absence of ATP.

and resuspended at a protein concentration of 2 mg/ml. Efflux measurements were then performed by taking aliquots of 0.15 ml at time points between 5 seconds and 30 minutes, followed by centrifugation and measurement of the radioactivity in the supernatant. The result, which will be described in detail elsewhere (Riedel et al., submitted) are shown in Fig. 4.

While the vesicles show only a slow efflux of rRNA which cannot be stimulated by ATP, the efflux of poly(A) and mRNA is clearly measurable and the complete efflux of mRNA can be stimulated in the presence of 2 mM ATP. Thus, these nuclear envelope vesicles contain the translocation mechanism for mRNA and show a unidirectional transport of poly(A) and mRNA.

In summary, the nuclear envelope vesicles described represent a valuable model system to study the mechanism and regulation of nucleocytoplasmic transport and to analyze the components of the nuclear envelope that are involved in these processes.

REFERENCES

1. Aaronson R.P. and Blobel G. (1974). J. Cell Biol. 62, 746-754.
2. Afzelius B.A. (1955). Exp. Cell Res. 8, 147-158.
3. Agutter P.S., Harris J.R. and Stevenson I. (1977). Biochem. J. 162, 671-679.
4. Agutter P.S., Cockrill J.B., Lavine J.E., McCaldin B. and Sim R.B. (1979). Biochem. J. 181, 647-658.
5. Agutter P.S. (1980). Biochem. J. 188, 91-97.
6. Baglia F.A. and Maul G.G. (1983). Proc. Natl. Acad. Sci. USA 80, 2285-2289.
7. Bernd A., Schroeder H.C., Zahn R.K. and Mueller W.E.G. (1982). Europ. J. Biochem. 129, 43-49.
8. Blobel G. and Potter V.R. (1966). Science 154, 1662-1665.
9. Bonner W.M. (1975). J. Cell Biol. 64, 421-430.
10. Bonner W.M. (1975). J. Cell Biol. 64, 431-437.
11. Bornens M. and Courvalin J.C. (1978). J. Cell Biol. 76, 191-206.
12. Clawson G.A., James J., Woo C.H., Friend D.S., Moody D. and Smuckler E.A. (1980). Biochemistry 19, 2748-2756.
13. Clawson G.A. and Smuckler E.A. (1982). J. Theor. Biol. 95, 607-613.
14. Commerford S.L. (1971). Biochemistry 10, 1993-1999.
15. Cox G.S. (1982). J. Cell Sci. 58, 363-384.
16. Dabauvalle M-C., Franke W.W. (1982). Proc. Natl. Acad. Sci. USA 79, 5302-5306.
17. Detke S. and Keller J.M. (1982). J. Biol. Chem. 257, 3905-3911.
18. Dingwall C.S., Sharnick V. and Laskey R.A. (1982). Cell 30, 449-458.
19. Dwyer N. and Blobel G. (1976). J. Cell Biol. 70, 581-591.
20. Fais D., Prusov A.N. and Polyakov V.Yu. (1982). Cell Biology Int. Rep. 6, 433-441.
21. Fasold, H., Hulla F.W., Ortanderl F., Rack M. (1977) in Methods in Enzymology XLVI, 289-295 (eds. Jakoby W.B. and Wilshek M., Acad.Press N.Y.).
22. Fawcett D.W. (1966). Am. J. Anat. 119, 129-146.
23. Feldherr C.M. (1965). J. Cell Biol. 25, 43-51.
24. Feldherr C.M., Cohen R.J. and Ogburn J.A. (1983). J. Cell Biol. 96, 1486-1490.
25. Franke W.W., Deumling B., Zentgraf H., Falk H. and Rae P.M.M. (1973). Exp. Cell Res. 81, 365-392.
26. Franke W.W. (1977) Biochem. Soc. Symp. 42, 125-135.
27. Franke W.W., Scheer U., Krohne G., Jarasch E.D. (1981). J. Cell Biol. 91, 39s-50s.
28. Gerace L., Blum A. and Blobel G. (1978). J. Cell Biol. 79, 546-566.

29. Gerace L., Ottaviano Y. and Kondor-Koch C. (1982). J. Cell Biol. 95, 826-837.
30. Goodwin G.H. and Johns E.W. (1978). Biochim. Biophys. Acta 519, 279-284.
31. Green N.M. (1982). Nature 297, 287-288.
32. Gurdon J.B. (1970). Proc. Roy. Soc. Lond.B 176, 303-314.
33. Horowitz S.B. (1972). J. Cell Biol. 54, 609-625.
34. Johns E.W. (1964). Biochem. J. 92, 55-59.
35. Kaempfer R. (1979) in Methods in Enzymology LX, 380-392 (eds. Colowick J.P. and Kaplan N.O., Acad. Press N.Y.)
36. Kalderon D., Roberts B.L., Richardson W.D. and Smith A.E. (1984). Cell 39, 499-509.
37. Kletzien R.F. (1980). Biochem. J. 192, 753-759.
38. Kondor-Koch C., Riedel N., Valentin R., Fasold H. and Fischer H. (1982). Europ. J. Biochem. 127, 285-289.
39. Krohne G., Franke W.W. and Scheer U. (1978). Exp. Cell Res. 116, 85-102.
40. Mardian J.K.W., Paton A.E., Bunick G.J. and Olins D.E. (1980). Science 209, 1534-1536.
41. Maul G.G. and Avdalovic N. (1980). Exp. Cell Res. 130, 229-240.
42. Paine P.L. and Feldherr C.M. (1972). Exp. Cell Res. 74, 81-98.
43. Paine P.L., Moore L.C. and Horowitz S.B. (1975). Nature 254, 109-114.
44. Paine P.L. (1975). J. Cell Biol. 66, 652-657.
45. Palayoor T., Schumm D.E. and Webb T.E. (1981). Biochim. Biophys. Acta 654, 201-210.
46. Palmiter R.D. (1974). Biochemistry 13, 3606-3615.
47. Purrello F., Vigneri R., Clawson G.A. and Goldfine I.D. (1982). Science 216, 1005-1006.
48. Rabbani A., Goodwin G.H. and Johns E.W. (1978). Biochem. Biophys. Res. Comm. 81, 351-358.
49. Reeves R. and Chang D. (1983). J. Biol. Chem. 258, 679-687.
50. Sanders C. (1977). Biochem. Biophys. Res. Comm. 78, 1034-1042.
51. Schatten G. and Thomas M. (1978). J. Cell Biol. 77, 517-535.
52. Schumm D.E. and Webb T.E. (1975). Biochem. Biophys. Res. Comm. 67, 706-713.
53. Schumm D.E. and Webb T.E. (1981). Arch. Biochem. Biophys. 210, 275-279.
54. Shelton K.R., Egle P.M. and Cochran D.L. (1981). Biochem. Biophys. Res. Comm. 103, 975-981.
55. Stick R. and Krohne G. (1982). Exp. Cell Res. 138, 319-330.
56. Stuart S.E., Clawson G.A., Rottmann F.M. and Patterson R.J. (1977). J. Cell Biol. 72, 57-66.
57. Uberbacher E.C., Mardian J.K.W., Rossi R.M., Olins D.E. and Bunick G.J. (1982). Proc. Natl. Acad. Sci. USA 79, 5258-5262.
58. Unwin P.N.T. and Milligan R.A. (1982). J. Cell Biol. 93, 63-75.
59. Weisbrod S.T. (1982). Nucleic Acids Res. 10, 2017-2042.
60. Wilson E.M. and Spelsberg T.C. (1975) in Methods in Enzymology 40, 171-176 (eds. Colowick S.P. and Kaplan N.O.), Acad.Press N.Y.
61. Zbarsky I.B. (1978). Int. Rev. Cytol. 54, 295-363.

Giant Polysome Formation in *Chironomus* Salivary Gland Cells as Studied by Microdissection Techniques

J.-E. Edström
Department of Genetics, University of Lund, 223 62 Lund, Sweden

Microdissection procedures can be used to study nucleocytoplasmic transport and to follow the fate of RNA molecules in the cytoplasm after their release from the nucleus (1, 2). After administration of an RNA precursor cells are fixed and microdissected. The distribution of RNA molecules in different parts of the cell is followed as a function of time after precursor administration. The advantage of the procedure over cell fractionation is that results reflect in vivo conditions. Redistributions and losses that can follow upon cell fractionation of fresh material are avoided and results can be related to a defined cytoplasmic region. The techniques have been applied to salivary gland cells of Chironomus larvae. Two different dissection schedules have been used. In one procedure (Fig. 1) the appearance in the cytoplasm of newly labelled RNAs is followed as a function of the distance from the nucleus (1). In another approach (2) cytoplasm with different amounts of endoplasmic reticulum (Fig. 2) is isolated (Fig. 3). The information obtained by these techniques can be used to draw conclusions regarding the mode of formation and function of the giant, 80 ribosome-polysomes dominating the polysome population of these cells.

Export of Balbianiring derived 75S mRNA

The dominant mRNA in Chironomus salivary gland cells is a 75S group of molecules transcribed by Balbiani ring (BR) genes (3) and coding for giant secretory proteins, the sp-I family (4). This mRNA goes to polysomes (5) containing about 80 ribosomes (6). The 75S RNA appears during transcription in large RNP granules which can be followed in the nuclear sap and be seen traversing the nuclear envelope. They can be observed in the cytoplasmic zone closest to the nucleus but are missing throughout the remainder of the cytoplasm (7, 8). Does this mean that 75S RNA enters polysomes in the perinuclear area or that it spreads through the whole cytoplasm in a form that is different from the BR-granule before joining the polysomes. The answer to this question could be obtained through comparative analysis of a thin peripheral cytoplasmic zone with a low content of rough endoplasmic reticulum (RER) and a more central zone (representing the bulk of the

Nucleocytoplasmic Transport
Edited by R. Peters and M. Trendelenburg

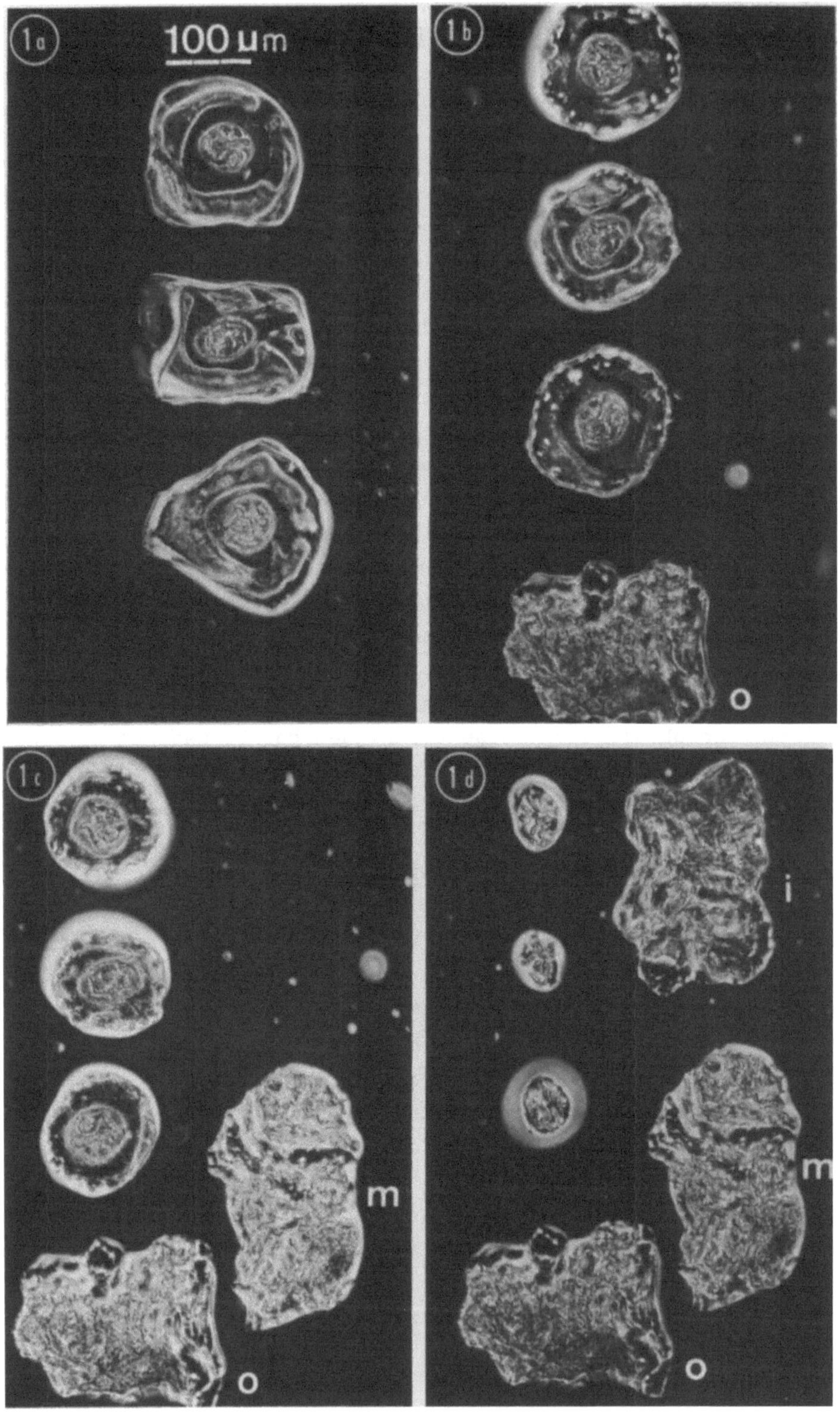

Figure 1 - Three salivary gland cells isolated by micromanipulation (a), after removal of the outer (b), middle (c) and inner zone (d). Letters i, m and o denote material from inner, middle and outer zone, respectively. From ref. (1).

cytoplasm) with a high content of RER (2) (Fig. 2, Fig. 3). Briefly the results were that 75S RNA labelled within the last 18 hrs was present in both zones with an about 3-fold enrichment in the inner RER rich area over the outer RER poor zone. After longer times, 6 days, the association to the RER was, however, much more pronounced, with a 5-fold enrichment. In other words, the 75S RNA seemed to be pulled from areas with little RER to RER rich areas, or, from cytosol to RER. In the present case the net direction of movement was towards the nucleus due to the peripheral localization of the cytosol rich region. Therefore 75S RNA probably spreads in a form dif-

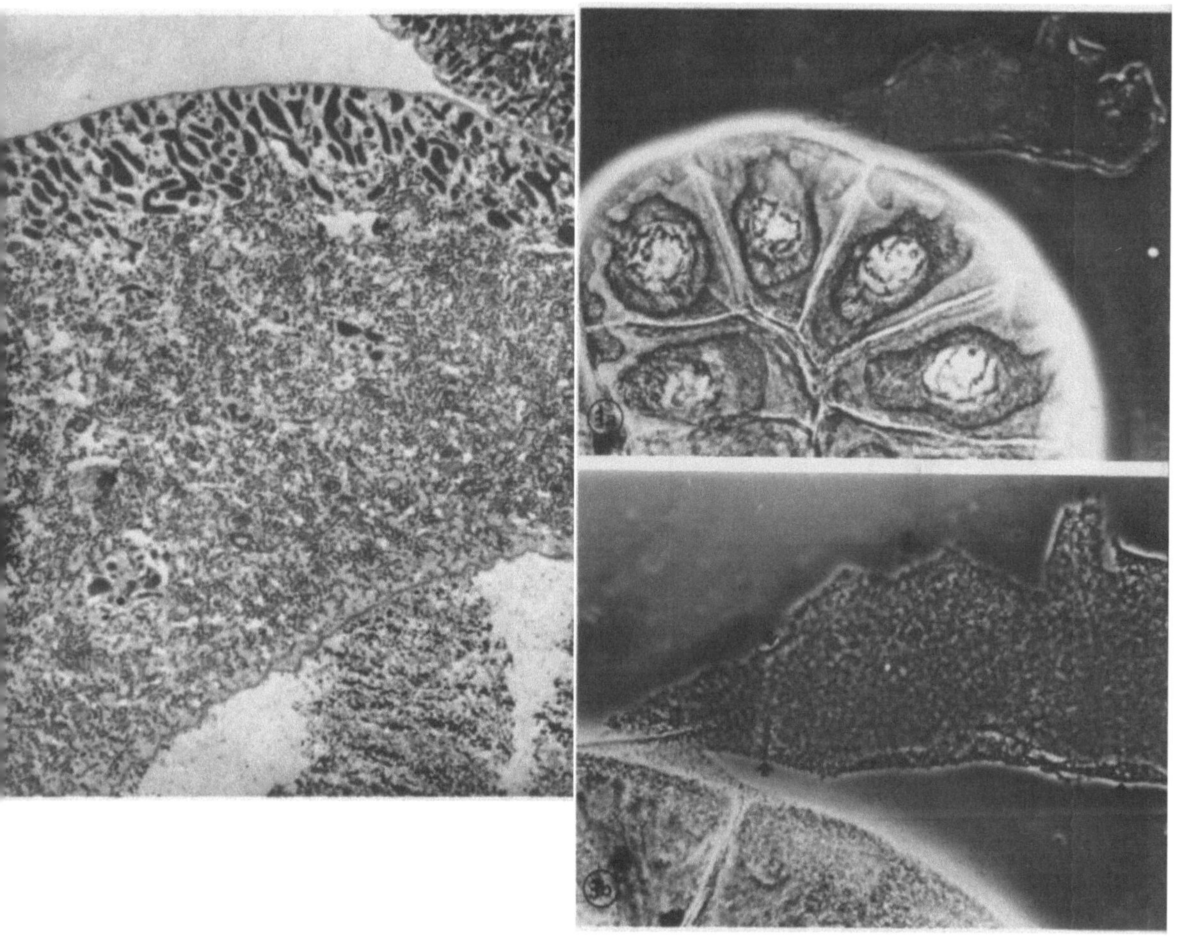

Figure 2 - Electron micrograph of salivary gland cell dominated by tubular rough endoplasmic reticulum. Basal mitochondria rich zone is seen upwards and nucleus at lower right.
Figure 3 - Basal zone released from a salivary gland as seen in the oil chamber under phase contrast at 133x (upper) and 257x (lower) magnification. From ref. (2).

ferent from the RNP-granule (BR-granule) through the cytoplasm before it forms polysomes.

The centripetal movement observed between 18 hrs and 6 days (that was due to differences in RER distribution between a thin peripheral zone and the bulk of the cytoplasm) does of course not contradict the fact that the RNA originally spreads from the nucleus to the cytoplasm. Proximodistal 75S RNA gradients can be observed after shorter times if cytoplasm (mainly RER rich regions) is dissected on the basis of its distance from the nucleus (1) (Fig. 4). These gradients flatten out with time and may be due to slow diffusion of the components containing 75S RNA.

Export of ribosomal subunits

Proximodistal gradients can be used for conclusions about the fate of ribosomal subunits (RSU) after their release from the nucleus. Both

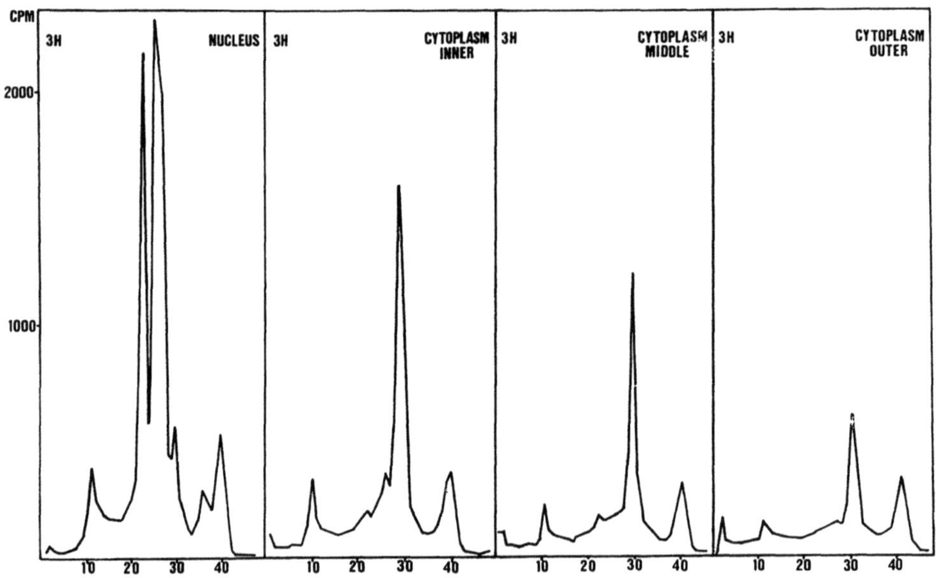

Figure 4 - Electrophoretic separations in 1% agarose gels of RNA extracted from 12 cells of an animal injected with tritiated uridine and cytidine 3 hrs before sacrifice. The cells were dissected into nuclei and three concentric cytoplasmic zones. The peak around slice 10 is 75S Balbiani ring derived RNA. The cytoplasmic peak around slice 30 represents 28S + 18S ribosomal RNA (comigrating after denaturation) and the peak around slice 40 4S RNA. The two largest peaks in the nucleus are ribosomal RNA precursors. It can be seen that both 75S RNA and rRNA appear in steep gradients in the cytoplasm as compared to the evenly distributed 4S RNA.

the light and heavy RSU, measured as 18S RNA and 28S RNA and/or 5S RNA, appear in steep gradients after their release from the nucleus (1) (Fig. 4). The light RSU can be seen in the cytoplasm about 30 min before the heavy RSU and the 18S RNA gradient is 30-50 μm ahead of the 28S RNA gradient showing that the rate of spread is of the order of 1 μm per min (1). Studies of the distribution of total ribosomal RNA between RER poor and RER rich cytoplasm shows an increased association to RER rich areas during the interval between 18 hrs and 6 days (2) suggesting that initially the RSU are predominantly in the cytosol (as free RSU and free polysomes) to become associated with the endoplasmic reticulum to an increasing extent over longer time intervals. How can this delay be explained? To some extent it is probably due to the relatively long translation time for giant polysomes (of the order of 30 to 60 min., ref. 2) as compared to that of other polysomes, which should delay the entry of RSU into the pool of RSU associated in the giant polysomes. This seems, however, not sufficient to explain that RSU are pulled from cytosol rich to RER rich regions during times, much longer than the 30-60 minutes needed for 75S RNA translation.

Other studies of the RSU gradients provide information that helps to understand this delayed association to the RER. After administration of puromycin the light RSU gradient was completely levelled out (1). However, the heavy RSU gradient was partially resistant to the drug (9). Furthermore, in untreated animals part of the heavy RSU gradient remained for at least two days after labelling whereas the light RSU gradient was usually extinguished already after six hrs (9). This stability of the heavy RSU gradient is not due to sustained export since the gradient can be observed also after all RNA synthesis has been blocked by actinomycin D (9). Probably the stable, drug resistant heavy RSU gradient is due to an association of heavy RSU to the RER independently of polysome formation. It is unlikely to be due to polysome formation as such in a reversible association to the RER because one would then have expected the two gradients to have a similar duration. Our conclusion that part of the heavy RSU bind as free units to the RER would also explain why after both 18 hrs and 6 days there is a marked excess of labelled heavy RSU in RER rich as compared to RER poor regions (2).

Further information on the behaviour of the heavy RSU has come from experiments in which the rate of growth of the gland cells was varied.

Under normal feeding conditions the cells grow continuously and increase protein and RNA content in parallel. If animals are starved the gland cells stop growing (10). The semi-permanent, puromycin resistant gradient of the heavy RSU did not form during growth inhibition by starvation (10). This could mean that heavy RSU only attach to the RER when newly formed membrane area is offered. Furthermore, if animals are starved after a heavy RSU gradient has formed this gradient becomes permanent, remaining as long as the animals can be kept alive (10).

These results do not necessarily mean that heavy RSU mainly attach to the RER in the central part of the cell. The semi-permanent gradient may form because the heavy RSU attaching to central parts of the RER have a higher specific radioactivity than the heavy RSU attaching to more peripheral parts, because they are still flowing from the nucleus towards the periphery. It is in fact more likely that heavy RSU attachment and new giant polysome formation occurs over the whole cytoplasm, since 75S RNA probably spreads over the whole cytoplasm before it attaches to the RER (2).

If newly labelled heavy RSU attach as free units to the RER forming radioactivity gradients why then do gradients disappear with time? One possibility is that the irreversibility is not absolute during feeding conditions and/or that loss of heavy RSU occurs through decay. Another possibility is that there is sufficient membrane mobility in growing cells to level out gradients over periods of a couple of days. The mobility in the cytoplasm of the gland cells can be well demonstrated in experiments om mitochondrial migration (11).

Giant polysome formation

Our results that the spread of 75S mRNA in the cytoplasm is a faster process than its association to the RER suggest that new giant polysome formation occurs over the whole cytoplasm and that it is not restricted to e.g. a perinuclear zone. This excludes the necessity of postulating a net peripheral flow of RER membranes (12).

Obviously the formation and attachment of a new giant polysome to the RER must offer a considerable problem of organization. About 80 ribosomes have to be placed on the reticulum within the range of one mRNA molecule and the polysome has to find room between preexisting polysomes. The RER is unusual in being tubular (13, 14), an architecture that it has in common with another cell dominated by giant polysomes, the silk gland cell of Bombyx mori (15) which could be of significance in this context. Thus stereo-EM observations (14) suggest that polysomes are arranged in spirals encircling the RER tubules. It is unlikely that 75S mRNA dissociates from ribosomes between rounds of translation since Miller spread polysomes do not show intermediate states of ribosome dressing, i.e. ribosomes only at the 5' or 3' end of the mRNA (6). The mRNA, therefore, has to continue spiralling around the tubules in the path left by the advancing preceding mRNA. Such a path would consist of ribosome binding sites (16) to which the newly formed translation complexes at the 5' end of the advancing mRNA could bind through their heavy RSU. The mechanics of such a continuous process are perhaps easier to visualize for a tubular reticulum than for a cisternal arrangement.

It is more difficult to see, however, how new complete giant polysomes are able to insert in an expanding reticulum unless there is considerable mobility of the RER (17) so that the 80 sites required for a new polysome can be brought into a continuous array together with sufficient membrane area. The gross relative stability of the heavy RSU gradients suggests that normally such a mobility has effects only over a short range.

The evidence we have that free heavy RSU bind to the RER independently of polysome formation and are in a relative excess over light RSU is of course in agreement with the fact that the RER contains binding sites for the heavy RSU and that such independent binding has been experimentally demonstrated (18). Furthermore our data indicate that the binding can occur very soon after export and is, therefore, unlikely to be due only to subunits remaining on the RER after translation. What functional role such an association might have is enigmatic. Perhaps it is a necessity in an expanding RER that ribosome binding sites become filled also when such a demand cannot be completely fulfilled by functional polysomes.

Acknowledgements: I am indebted to Dr. C. Grond for critical comments. The work described here was supported by the Swedish Cancer Society.

References

1. Edström, J.-E. and Lönn, U. J. Cell Biol. 70, 562-572 (1976).

2. Edström, J.-E., Rydlander, L. and Thyberg, J. Eur. J. Cell Biol. 29, 281-287 (1983).

3. Daneholt, B. and Hosick, H. Proc. Nat. Acad. Sci. USA 70, 442-446 (1973).

4. Rydlander, L., Pigon, A. and Edström, J.-E. Chromosoma 81, 101-113 (1980).

5. Daneholt, B., Anderson, K. and Fagerlind, M. J. Cell Biol. 73, 149-160 (1977).

6. Francke, C., Edström, J.-E., McDowall, A. and Miller, O. L. Jr. EMBO J. 1, 59-62 (1982).

7. Stevens, B. J. and Swift, H. J. Cell Biol. 31, 55-77 (1966).

8. Daneholt, B., Case, S. T., Hyde, J., Nelson, L. and Wieslander, L. Progr. Nucleic Acid Res. Mol. Biol. 19, 319-334 (1976).

9. Lönn, U. and Edström, J.-E. J. Cell Biol. 70, 573-580 (1976).

10. Lönn, U. and Edström, J.-E. J. Cell Biol. 73, 696-704 (1977).

11. Thyberg, J., Sierakowska, H., Edström, J.-E., Burvall, K. and Pigon, A. Devel. Biol. 90, 31-42 (1982).

12. Lönn, U. Cell 13, 727-733 (1978).

13. Kloetzel, J. A. and Laufer, H. J. Ultrastruct. Res. 29, 15-36 (1969).

14. Olins, A. L., Olins, D. E. and Franke, W. W. Eur. J. Cell Biol. 22, 714-723 (1980).

15. Blaes, N., Couble, P., Prudhomme, J. C. Cell Tissue Res. 312, 311-324 (1980).

16. Kreibich, G., Czako-Graham, M., Grebenau, R., Mole, W., Rodriquez-Boulan, E. and Sabatini, D. J. Supramol. Struct. 8, 279-302 (1978).

17. Ojakian, G., Kreibich, D. and Sabatini, D. J. Cell Biol. 72, 530-551 (1977).

18. Borgese, D., Blobel, G. and Sabatini, D. J. Mol. Biol. 74, 415-438 (1973).

Recent Improvements in Microscopy Towards Analysis of Transcriptionally Active Genes and Translocation of RNP-Complexes*

M. Trendelenburg[1,2], R. D. Allen[3], H. Gundlach[4], B. Meissner[1,2], H. Tröster[1,2], and H. Spring[2]

[1]Institute of Cell and Tumor Biology,
[2]Institute exp. Pathology, German Cancer Research Center, D-6900 Heidelberg, FRG
[3]Department of Biological Sciences, Dartmouth College, Hanover, NH, USA
[4]Division of Applied Microscopy, Carl Zeiss, D-7082 Oberkochen, FRG

Whereas insight into the structural and functional organization of defined cytoplasmic compartments has increased dramatically in the past years (see,e.g. Cold,Spring Harbor Symp. Vol.XLVI, 1982), predominantly by application of antibody labelling techniques, insight into nuclear chromatin organization is still rather limited. Among the obvious reasons for these limitations are (i) the general absence of defined - e.g. membrane surrounded - intranuclear domains, and (ii) the high compaction of most transcriptionally active genes to very dense, small complexes. As a consequence most high magnification analyses of nuclear structures and RNP transport phenomena did require up to now - the application of electron microscopical (EM) techniques. This generally does not allow a direct visualization of in vivo occurring transport phenomena.

In this contribution (i) a short outline of the different microscopical techniques and their application to analysis of nuclear chromatin structures is given. (ii) As indicated above, the dimensions of most gene domains and their associated RNP transcripts are below the range of conventional light microscopy and thus had to be analysed predominantly by EM techniques including Miller-type spread preparations and EM-autoradiography. (iii) Recently, due to the advent of AVEC light microscopy (Allen Video Enhanced Contrast LM; c.f. Allen and Allen, 1983), optical microscopy can now be used for the visualization of defined nuclear chromatin configurations as well as for the analysis of specific RNP complexes in unfixed nuclei and/or in low speed sedimented nuclear chromatin preparations.

---

*This article is dedicated to the memory of Professor Robert Day Allen.

Nucleocytoplasmic Transport
Edited by R. Peters and M. Trendelenburg

1. Microscopic approaches to study nuclear chromatin organization and RNP translocation.

The use of light microscopy to study the nucleo-cytoplasmic organization in unfixed specimens was generally restricted to observations at relatively low magnification such as the analysis of gross morphological changes of nuclear structures (for review see Trendelenburg, 1983). More specific questions to be analysed with direct LM observation include for example the documentation of the interesting phenomenon of nuclear rotation (Bard et al., 1985). Phase contrast microscopy had in particular been used as an important control (in many cases as the only possible fast control) for a variety of nuclear isolation procedures (for review see Agutter, 1982). Since direct observation was restricted - in most cases - by the known limitations of the light microscope a more detailed information as to structural aspects of specific nuclear domains and also the architecture of the juxtanuclear cytoplasmic zone had to be obtained via electron microscopic (EM) investigations (see below).

As it is evident from the concept of EM investigations, this type of structural investigation provides predominantly static information but allows to obtain structural data at high magnification. It is from EM studies, that most of our present knowledge concerning the overall structure of the nuclear envelope (NE) as well as the fine structure of the pore complexes is derived (see Franke, 1974, Maul 1977, Franke et al., 1981, Unwin and Milligan, 1982). Interestingly, it had not only been possible to obtain structural information on the nuclear envelope proper, but also it was possible to see the structural correlate of what can be interpreted as translocation of large RNP complexes via nuclear pores. As it is evident from the respective electronmicrogaphs, the large RNP complexes which were seen to traverse the NE (c.f. Franke, 1974, Franke and Scheer, 1974, Maul, 1977), are strikingly larger as compared to the average width of a nuclear pore complex, especially of one considers the narrow space of 9 nm which is thought to be open for the transfer of macromolecules (Clawson and Smuckler, 1982, Maul, 1982, c.f. also Feldherr, 1980). Certainly, it remains one of the most challenging research areas in the field of nucleocytoplasmic transport studies to determine the precise molecular characteristics of such specific translocation processes.

Representative micrographs of the type of information as provided by conventional microscopy are shown in Fig.1. In both figures, sections through large amphibian oocytes are shown with focus on the region close to the NE. Differential interference contrast (DIC)-LM according to Nomarski (see Allen et al. 1969, for details) offers the advantage

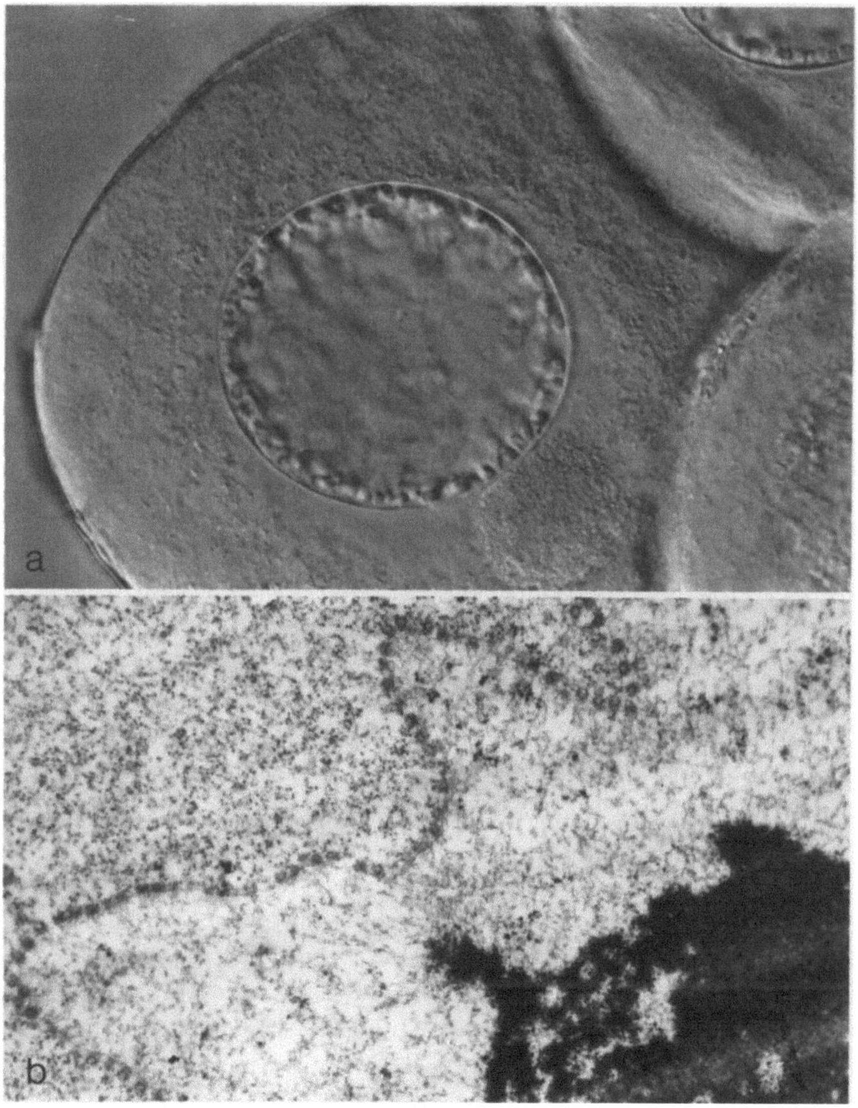

Fig.1. Nucleo-cytoplasmic compartimentalization of an amphibian oocyte as seen by conventional light microscopy (a, Xenopus laevis) and electron microscopy (b, Triturus alpestris. a, X 460; b, X 17.000.

of the visualization of unfixed and unstained cellular and in particular nuclear structures within an intact living cell. In particular, the arrangement of the extrachromosomal nucleoli close to the nuclear periphery as well as the position of small segments of the lampbrush chromosomes at the nuclear center can be seen (Fig.1a, Xenopus laevis, for additional explanations, see next section). However, up to now, this LM-method could only be used for low magnification studies (see Fig.1a). The reasons for this are (i) lack of contrast of unstained specimens and (ii) straylight and other light diffraction phenomena which interfere with the quality of the image.

Structural analysis at high magnification was restricted to EM analysis of thin sections of fixed and stained material. A typical example is shown in Fig.1b (Triturus alpestris). The close position of an extrachromosomal nucleolus to the NE can be easily seen. As it is also evident from this micrograph, it still requires further studies to elucidate the likely, direct, structural interrelationship between parts of the marginal zones of such nucleoli and the NE, as well as the cytoplasmic zone close to the outer margins of the NE (for a recent investigation using EM methods, see Arechaga and Bahr, 1985).

In order to obtain some kinetic data from EM investigations, suitable labelling experiments have to be carried out. Visualization of translocation of ribonucleoprotein (RNP) transcripts in EM sections can result in a distinction of the major types of RNP transcripts for a given cell type (see Fakan and Puvion, 1980, for review). In the case of nucleolar ribosomal RNA transcripts, EM autoradiography has to be preferentially carried out on dispersed nucleolar chromatin after a Miller-Type spread preparation (Miller and Beatty, 1969; Franke et al., 1979; Miller, 1981). Due to the high compaction of nucleolar chromatin (c.f. Fig.1b) it is difficult to localize exactly the site of actively transcribing rRNA genes in thin sectioned material. If chromatin spread preparations are made from early vitellogenic Triturus oocytes (Fig.2a, T. alpestris, diameter of the oocyte 350 µm), a suitable modification of the basic spreading technique can be developed, allowing the visualization of intact complexes which consist of NE-fragments and directly associated clusters of actively transcribing nucleolar rRNA genes (Fig.2b). In addition, EM autoradiography can be applied to this kind of spread preparations (c.f. Angelier et al., 1976). In the example shown in Fig.2c, EM autoradiography was performed using a spread preparation from a labelled Triturus cristatus oocyte: As shown on the micrograph, labelled nucleolar precursor (pre-rRNA) transcript fibrils can be identified (i) at the level of actively transcribing genes and (ii) as labelled structures associated to several parts of the NE.

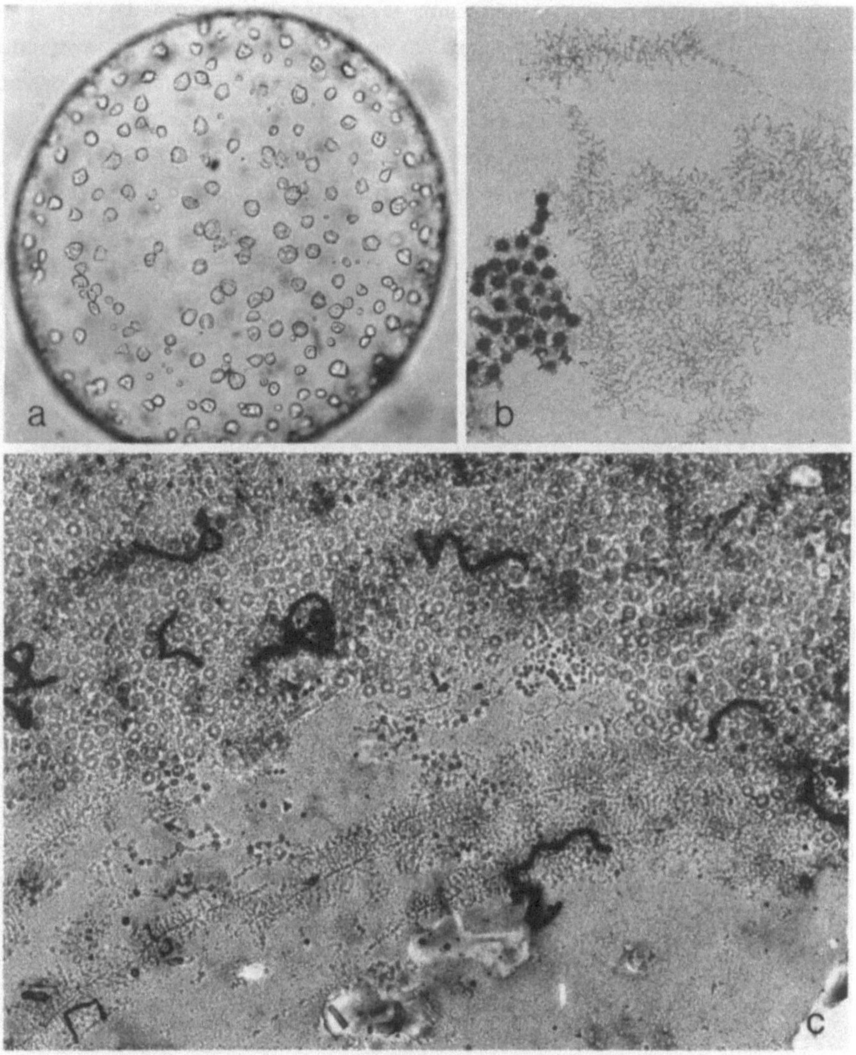

Fig.2. Light and electron microscopy of the close structural association of extrachromosomal nucleoli and the nuclear envelope in amphibian oocyte nuclei. (a) Phase contrast light micrograph of an isolated nucleus of a Triturus oocyte. (b) Miller-type chromatin spread preparation of a NE - nucleolus complex. (c) EM-autoradiography of a spread preparation as shown in (b). a, X 200; b, X 13.500; c, X 22.000

2. Light Microscopy of Unfixed Chromatin

An essential prerequisite for the direct analysis of transcription and transcript transport processes is the observation of unfixed cells and, in particular, the analysis of fine structural changes in the nuclear interior. Theoretically, a variety of optical contrast methods can be used for this purpose. However, for the direct visualization of nuclear structures, predominantly two optical contrast methods are used (i) phase contrast (according to Zernicke, c.f. Ross, 1967) and (ii) differential interference contrast (DIC) (according to Nomarski, c.f. Allen et al., 1969).

In our experience it is almost always rewarding to apply both of these optical contrast methods for the same specimen (c.f. Tröster et al., 1985), since the imaging principles are widely different for the two methods (for details see refs. quoted above). The precise location of the cell nucleus can be analysed using DIC optical contrast (Fig.1a). It is clear from the overall aspect of this micrograph, that structures surrounded by membranes are particularly well visible using DIC light microscopy. This method does allow photography of 'optical sections' through relatively thick cells such as previtellogenic oocytes of Xenopus laevis. The example shown represents an optical section through the center of the cell. Note that this is an unfixed, unstained cell (diameter approx. 180 µm, 40 x Planobjective). Using these optical conditions, fine structural details can be recognized throughout the cytoplasm, in particular the conspicuous mass of tightly packed mitochondria (denoted by bar in Fig.1a) a characteristic component for this early stage of oocyte differentiation. In conclusion valuable information on the nucleo-cytoplasmic compartimentalization can be obtained using DIC light microscopy of unfixed and unstained specimens. However, fine structural details of small chromosome segments located in the central nuclear area cannot be sufficiently visualized using this type of conventional transmitted light microscopy of whole cells with thick diameters.

Far better optical conditions are obtained, if nuclei are manually isolated from the oocyte (Gall, 1954; Callan and Lloyd, 1960; Gall, 1966). The reason for this is, that the thick layer of cytoplasmic yolk platelets (present in the large vitellogenic oocyte) is removed from the light path by the nuclear isolation step, thus, resulting in much less beam diffraction above the specimen. An example for the image quality obtained from a large, isolated nucleus (diameter 600 µm) is shown in Fig.3. In this example, phase contrast is used as optical contrast method.

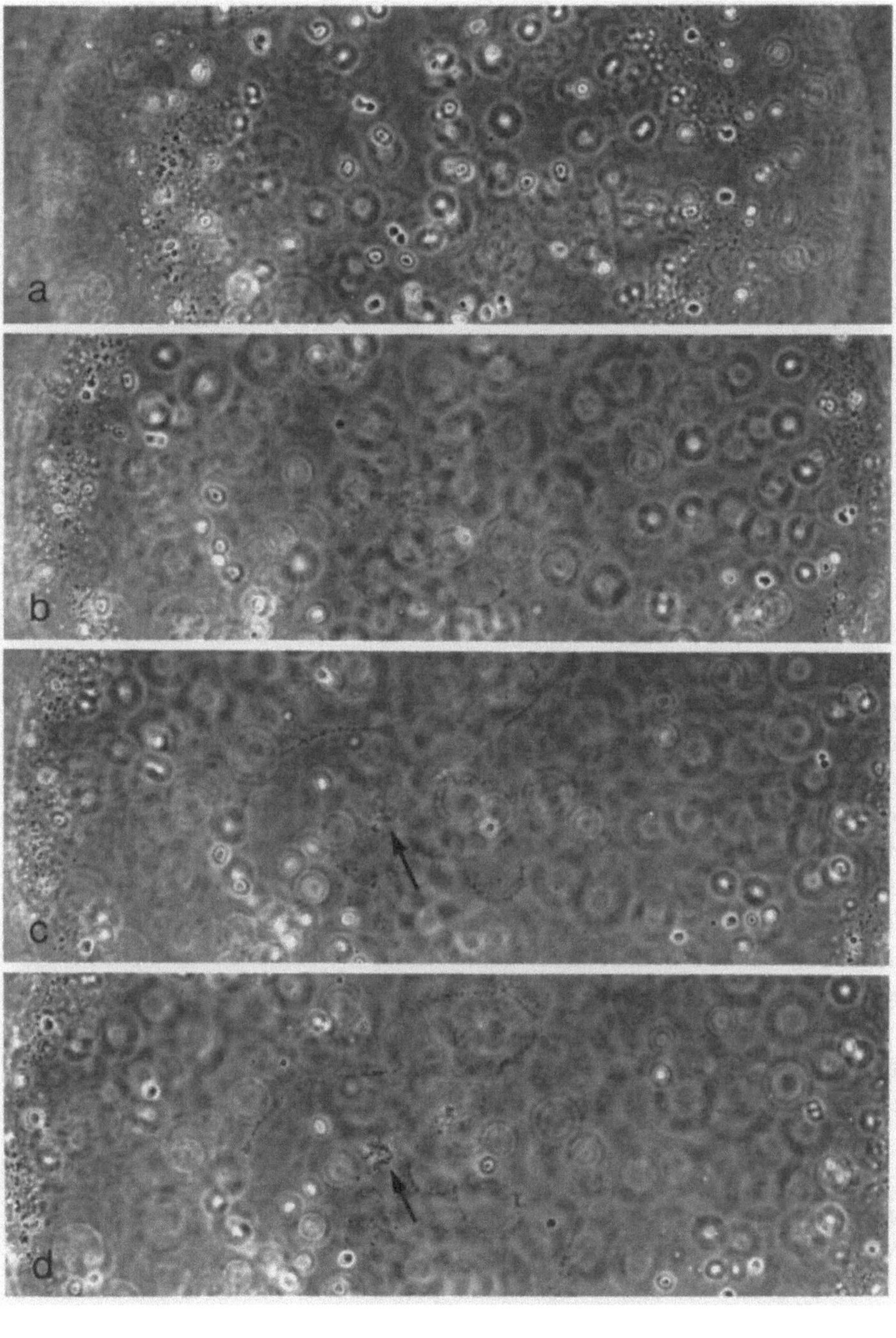

Fig.3. Examples of phase contrast light microscopy to visualize intranuclear chromatin structures (a-e, 40 X obj.) in an isolated nucleus from a large Pleurodeles waltlii oocyte. a-e, X 200. Arrowheads denote a conspicuous loop structure in c and d.

In this case, phase contrast can well be used for investigations on the three-dimensional architecture of the predominant nuclear chromatin components. As shown in Figs.3a-d, a series of consecutive images can be obtained for the nuclear periphery, e.g. 40 μm inside the nuclear membrane (Fig.3a), 70 μm (Fig.3b), 90 μm (Fig.3c) and 110 μm (Fig.3d). In the example of the focus series shown, the intranuclear position of a conspicuous lampbrush chromosome loop structure can be detected, which is located approximately 110 μm below the nuclear membrane. Due to the particularly high optical density of the chromatin segment shown, the recognition (e.g. by using a 40x objective) is facilitated in phase contrast microscopy. It should be kept in mind, however, that the result shown cannot be generalized since in many other cases, intranuclear focus levels cannot be obtained in sufficient quality due to the presence of very high numbers of amplified nucleoli. The close lateral association of these dense, large spherical structures causes a high amount of diffraction so that smaller structures within the nuclear interior cannot be visualized using phase contrast microscopy of thick, intact oocyte nuclei. On the other hand, the possibility for visualization of the in situ structure and arrangement using phase contrast clearly facilitates an eventual subsequent correlation of the structure seen in situ (Fig.3d, 4a) with a loop structure identified in a lampbrush chromosome preparation from the same type of oocyte nuclei (Fig.4b).
In summary, conventional light microscopy of unfixed, unstained specimens can yield a high amount of information as to position and structure of the large, major chromatin components in amphibian oocyte nuclei and other cells. For isolated nuclei up to a diameter of 600 μm predominantly phase contrast and/or differential interference contrast (DIC) transmitted light microscopy can be used. Chromatin structures visualized in phase contrast generally exhibit a higher contrast, e.g. when compared to interference contrast microscopy of the same specimen.

3. Light microscopic recognition of actively transcribing genes by video enhanced contrast microscopy

As already outlined in detail in the previous sections, a direct light microscopic observation of gene transcription and of the subsequent translocation of the RNP transcripts had not yet been possible, due to the obvious technical limitations.

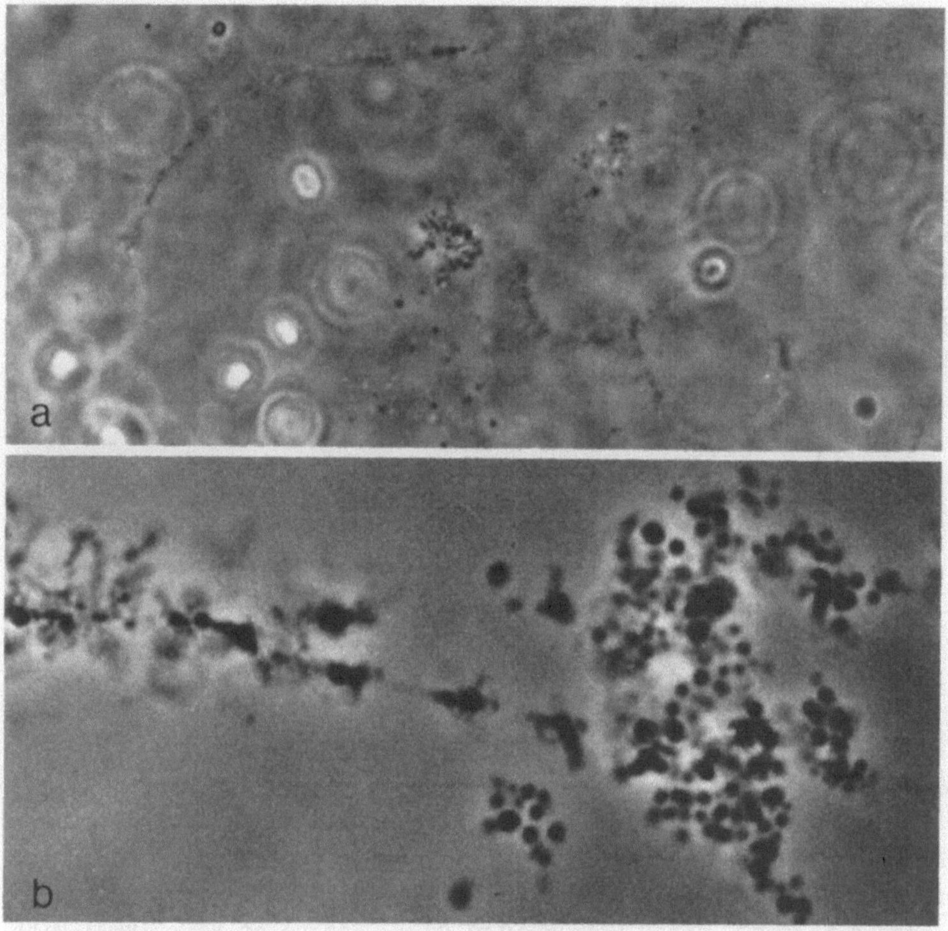

Fig.4. Possibility for a correlation of an in situ identified chromatin structure (a, c.f. Fig. 3d) with similar loop structures after lampbrush chromosome preparation (b). a, X 450; b, X 1.500

However, an appropriate instrumentation, using light microscopy, is a stringent prerequisite especially of one wishes to investigate the dynamic aspects of RNP translocation processes. As an example for the intricate complexity which one has to take into account for the coordinate expression of nucleolar rDNA genes, a diagrammatic flow-chart is shown (Fig.5; for detials see Ford, 1979; c.f. also Hadjiolov, 1985). As it is evident from this scheme, our knowledge has increased rapidly in particular in regard to the level of rDNA sequence elements as well as to the level of transcription (1 and 2 in Fig.5). Much less, however, is known for the structural aspects of transcript

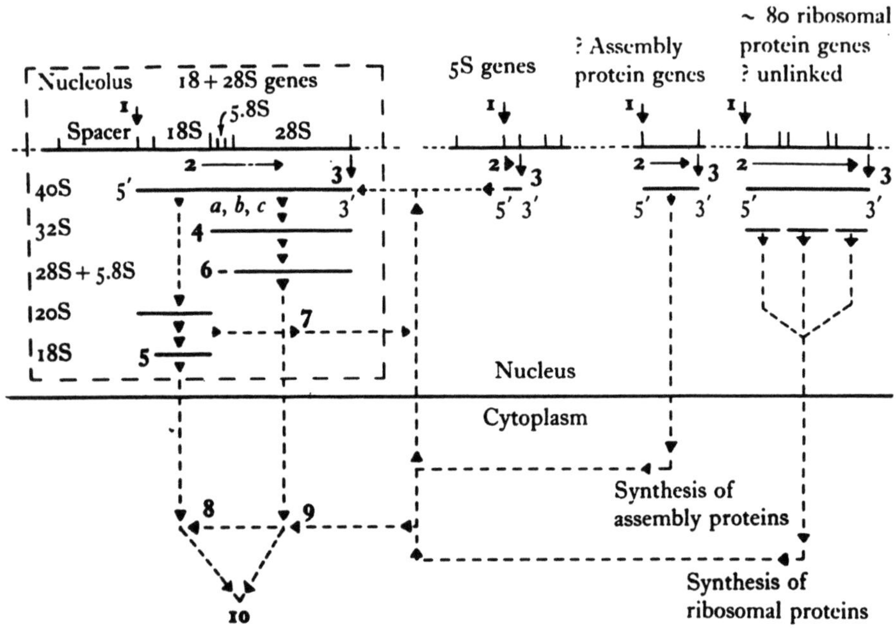

Fig.5. Flow chart of the complexity of ribosome biosynthesis (from Ford, 1979)

processing as well as for the protein assembly reactions (3 to 7 in Fig.5). In an earlier investigation (Moreno Diaz de la Espina et al., 1982) evidence was presented, that some of the rRNP processing intermediates and products may be associated to small, spherical units ("medusoid bodies", sensu Moreno Diaz de la Espina et al., 1982) with mean diameters of 0.2-2 µm. These microspheres were shown to be located preferentially at the nucleolar periphery as well as within the narrow nucleoplasmic zone between the extrachromosomal nucleoli and the NE. This cytological situation is different from the situation which was observed for the large RNP transcripts of the Balbiani ring genes of Chironomus salivary gland cell nuclei, which had been extensively characterized both by biochemcal as well as structural investigations (see Beermann, 1963; Stevens and Swift, 1966); Olins et al., 1980; Edström et al., 1983; Skoglund et al., 1983; and the contributions by Edström et al. and Skoglund et al., this volume). In the case of the giant Balbiani ring transcripts, small, singular particles of diameters of 30 nm could be identified in EM of thin sections and traced up to the passage through the nuclear pore complexes.

In conclusion, most identified RNP transcripts are of relatively small size in the range of 20-40 nm (Beermann, 1963; Stevens and Swift, 1966 and refs. cited above; c.f. also Dubochet et al., 1973; Matsuura et al., 1974). In order to trace directly the translocation of a specific RNP transcript from the site of the actively transcribing gene up to its passage through the nuclear pore complex, light microscopic observation at high mangification is required.
Recently, a major progress in light microscopic image quality had been achieved by the development of videomicroscopy instrumentation for the AVEC method (Allen Video Enhanced Contrast, c.f. Allen and Allen,1983; Allen, 1985). The necessary instrumentation basically consists of a high quality light microscope connected to a video camera and a special image processing unit. The essential characteristics of the AVEC method are: (i) Electronic contrast enhancement by adjustments of the videocamera, (ii) continuous subtractions of background noise, (iii) electronic contrast modulation of the digitized, cleared images. As it had been shown by Allen and Allen (1983), the AVEC principle is applicable for a variety of light optical contrast methods, in particular for Nomarski DIC, as well as for phase contrast (PH) microscopy. Originally, the parameters for the AVEC method had been developed for very flat microscopic specimens like the extruded axoplasm of the squid giant axon (Allen and Allen, 1983). However, in the course of the present study, it was possible to apply the method also for observations of isolated oocyte nuclei of insects and amphibians as well as for observations on dispersed, nucleolar and lampbrush-type chromatin from such oocyte nuclei. Thus, basically, a direct continuation could be reached, linking the results and also the specific specimen preparation methods which had been developed for the conventional LM (see section 2; and Gundlach, 1979; Spring and Franke, 1982; Trendelenburg, 1983) to investigations using optical microscopy at high magnification (Allen, 1985).
In a first series of experiments we have focussed our attention on the visualization of transcriptionally active rDNA genes in order to obtain information on the <u>in vivo</u> compaction of these genes in the nucleus. As a suitable object we choose nucleolar chromatin contained in oocyte nuclei of the house cricket, Acheta domesticus (Insecta, Gryllidae). From previous experience (Tröster et al., 1985) it was known, that two, morphologically different, types of rDNA chromatin-configurations can be observed in isolated Acheta oocyte nuclei: rDNA chromatin is present (i) in numerous, small extrachromosomal micronucleoli, and (ii) in form of a larger, puff-like structure as a typical

nucleolus organizer region (NOR) of the oocyte's chromosome complement. The experimental strategy for a light microscopic gene visualization is outlined in Fig.6. The first step of the preparation is

---

LIGHT MICROSCOPIC VISUALIZATION OF ACTIVE GENES

Opening of an isolated nucleus at physiological ionic strength

Rapid screening of the extruded unfixed chromatin (PH-LM, DIC-LM)

Addition of low salt medium to induce slight dispersal

Observation with electronic contrast enhancement at high magnification (Axiomat, Hamamatsu) including (i) on-line substraction of background and (ii) electronic contrast modulation of the digitized, cleared images

---

Fig.6. Diagram of experimental protocol for light microscopic visualization of transcriptionally active genes.

carried out using a conventional inverted light microscope. The preparation chamber containing the sedimented chromatin is then transferred to the stage of an inverted version of the Axiomat microscope and observed at low magnification in order to screen for the presence of suited chromatin areas and to record rapidly the positions of interest for each specimen. Only at this point, the preparation is observed at high magnification using video-enhanced contrast microscopy. Following the recording of images of the rDNA chromatin compacting at physiological ionic strength, the recording is then briefly interrupted in order to allow a stepwise, controlled, addition of low salt medium for induction of a slight, slow dispersal process of the compact rDNA chromatin. The sequential unravelling of the two types of Acheta rDNA chromatin can then continuously be observed by videomicroscopy up to the moment, when the first small, individual, active genes become visible after the gentle unravelling of the dense rDNA chromatin masses. A typical example of this type of experiment is shown in Fig.7. The isolated Acheta oocyte nucleus, as seen in DIC microscopy, contains the dense, small, extrachromosomal micronucleoli as well as an transcriptionally active, puff-like NOR structure (Fig.7a). Following the opening of the NE, the NOR region and the micronucleoli can be identified in a flattened chromatin preparation. The structural connection of the NOR to the very thin chromosome strand becomes visible (phase contrast microscopy, Fig.7b). This is about the maximal information as obtainable from conventional LM. Note, that in both micrographs the individual rDNA genes are not visible due to the compaction of rDNA chromatin and to the association of the gene chromatin with a large proportion of nucleolar proteins.

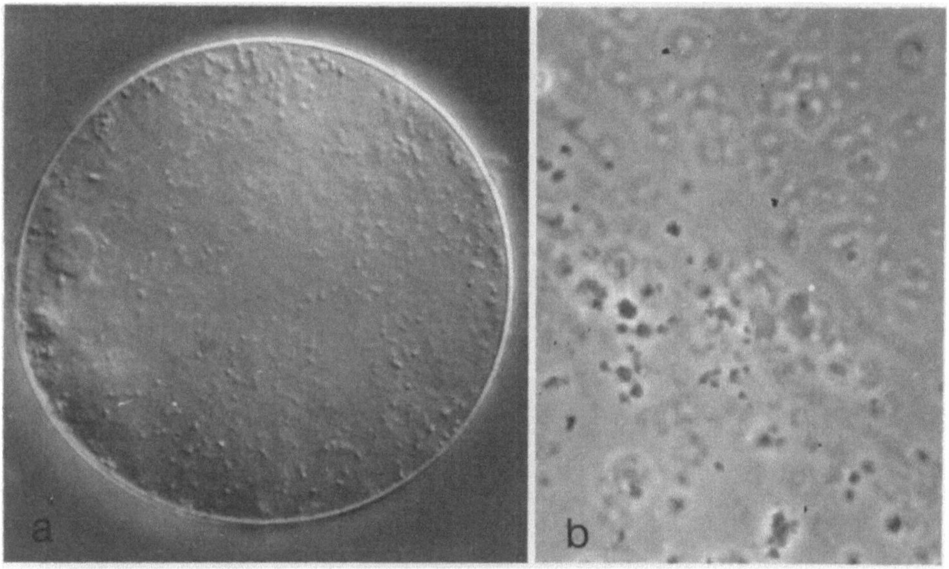

Fig.7. Survey light micrograph of the main rDNA chromatin forms present in an Acheta oocyte nucleus (a). Sedimented chromatin from an opend Acheta oocyte nucleus. Note connection of the NOR complex to a fine chromatin strand. a, X 650; b, X 1.300.

The light microscopic visualization of individual genes becomes possible after low salt treatment of a chromatin preparation of the type shown in Fig.7b. Following the reduction of the ionic strength from 100 mM (Figs.7a,b) to 70 mM salt, the dense rDNA chromatin masses start to disperse. In all preparations analysed so far, the first occurrence of individual rDNA genes was noted for the NOR regions (Fig.8a). This observation is in good agreement with data from thin section analysis of the same material which showed, that the rDNA genes of the NOR region are much less associated with nucleolar proteins as the genes of the micronucleoli (Tröster et al., 1985). Only much later during the dispersal process, the extrachromosomal rDNA genes of the micronucleoli can be visualized (Fig.8c). In Figs.8b and 8d a correlation of the LM data with EM observation as obtained from Miller type chromatin spread preparation of Acheta nucleolar chromatin

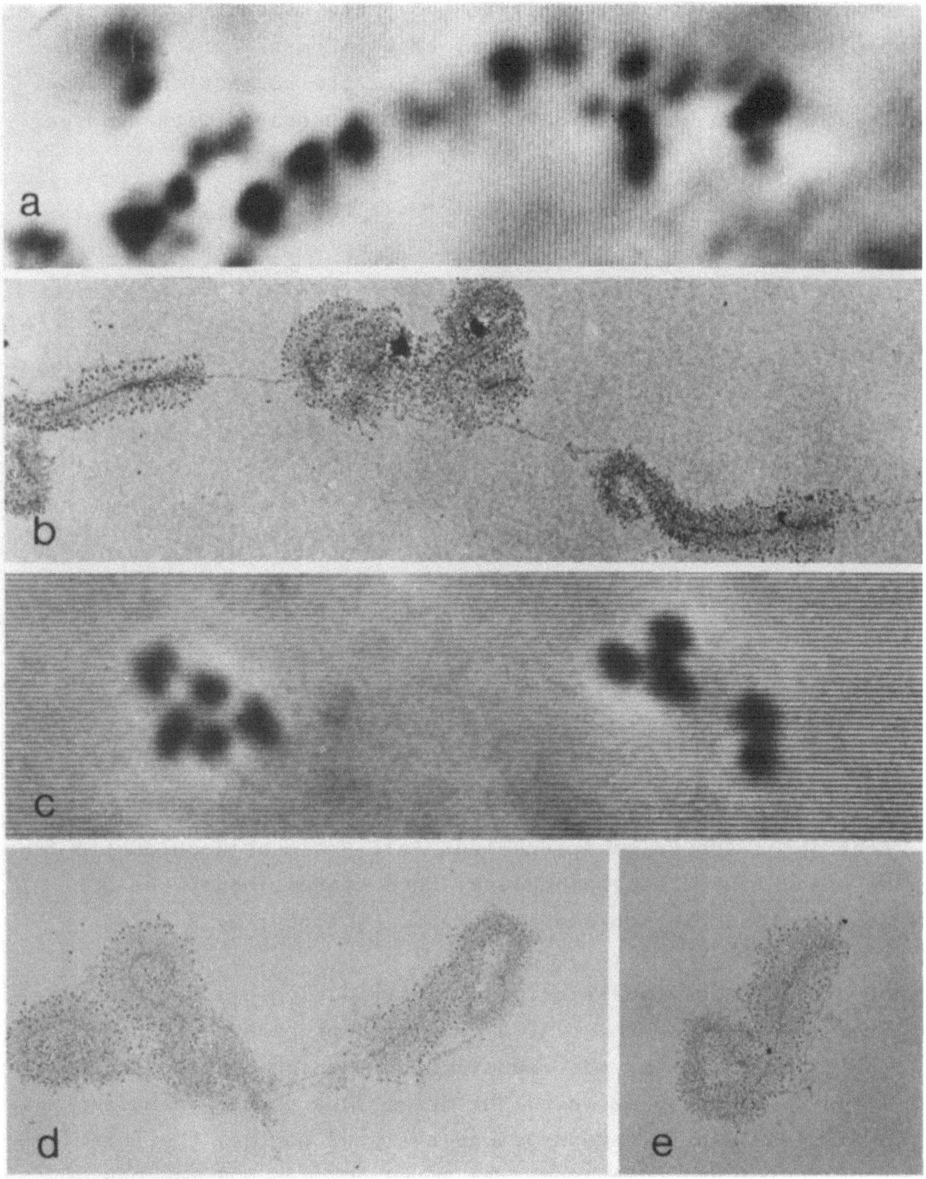

Fig.8. Correlation of light microscopic and electronmicroscopic data on Acheta rDNA chromatin organization. (a,b) rDNA genes present as large linear arrays within an NOR complex. (c,d) Individual rDNA genes visualized from compact, extrachromosomal micronucleoli. a,c X 7.800; b,d X 8.200.

is given. It is shown, that the rDNA genes of the NOR complex are present as long linear chromatin strands containing large numbers of rDNA genes in tandem array (Figs.8a,b). By contrast, the rDNA genes of the micronucleoli are visualized as circular chromatin units consisting of one up to 10 or more genes per circle (for detailed description see Trendelenburg et al., 1973; 1976; 1977; Tröster et al., 1985). Compared to the dimension of the rDNA transcription units (TUs) after Miller-type spread preparation (Figs.8b,d), e.g. after dispersal at very low ionic strength (Miller and Beatty, 1969; Franke et al., 1979; Miller, 1981; Hofmann et al., 1985) it is evident, that the genes visualized after gentle dispersal at 70 mM salt have still retained most of their 'native' in vivo compaction. This holds for the transcribed parts of the genes ('matrix units') as well as for the nontranscribed rDNA spacer segments. From thin section analysis a minimum estimate of a 4-5 fold compaction of 'native' genes as compared to fully extended genes after Miller type spread preparation had been determined (Tröster et al., 1985), a factor which is in agreement with similar determinations for the Balbiani ring gene TUs (Daneholt et al., 1982; for review, see Mathis et al., 1980).
It is clear, that these results can only be considered as essential initial steps towards a light microscopic documentation of gene transcription and RNP translocation.

Acknowledgement

We thank M. Ii, H. Kleinmeyer, and Dr. H. Waldvogel for support. Part of this study was presented at the 3rd Int. Congress Cell Biology, Tokyo, Sept. 3-7, 1984. M.F.T. is a recipient of a Heisenberg-Fellowship from the Deutsche Forschungsgemeinschaft (DFG). Part of this study was financially supported by the DFG (grant Tr 147/6-1).

REFERENCES

Agutter, P.S. (1982) In: The Nuclear Envelope and the Nuclear Matrix, pp. 91-109, Alan Liss, New York

Allen, R.D. (1985) Ann. Rev. Biophys. Biophys. Chem. 14, 265-290

Allen, R.D., David, G.B. and Normarski, G. (1969) Z. wiss. Mikrosk. Mikrotechn. 69, 193-221

Allen, R.D. and Allen, N.S. (1983) J. Microscopy 129, 3-17

Angelier, N., Hemon, D. and Bouteille, M. (1976) Exp. Cell. Res. 100, 389-393

Arechaga, J. and Bahr, G.F. (1985) In: Nuclear Envelope Structure and RNA maturation, pp. 23-50, Alan Liss, New York

Bard, F., Bourgeois, C.A., Costagliola, D. and Bouteille, M. (1985) Biol. Cell 54, 135-142

Beermann, W. (1963) In: 13. Mosbacher Kolloquium, pp. 64-100, Springer Berlin

Callan, H.G. and Lloyd, L. (1960) Phil. Trans. B 243, 135-219

Clawson, G.A. and Smuckler, E.A. (1982) In: The Nuclear Envelope and the Nuclear Matrix, pp. 271-278, Alan Liss, New York

Cold Spring Harbor Syp. Quant. Biol. (1982) Organization of the Cytoplasm, vol. XLVI

Daneholt, B., Andersson, K., Björkroth, B. and Lamb, M.M. (1982) Eur. J. Cell Biol. 26, 325-332

Dubochet, J., Morel, C., Lebleu, B. and Herzberg, M. (1973) Eur. J. Biochem. 36, 465-472

Edström, J.E., Rydlander, L. and Thyberg, J. (1983) Eur. J. Cell Biol. 29, 281-287

Fakan, S. and Puvion, E. (1980) Int. Rev. Cytol. 65, 255-299

Feldherr, K.M. (1980) Cell Tissue Res. 205, 157-162

Ford, P.J. (1979) In: Maternal Effects in Development (D.R. Newth & M. Balls, eds.), pp. 81-110, Cambridge University Press, Cambridge

Franke, W.W. (1974) Int. Rev. Cytol. 39, 71-236

Franke, W.W. and Scheer, U. (1974) Symp. Soc. Exp. Biol. XXVIII, 249-278

Franke, W.W., Scheer, U., Spring, H., Trendelenburg, M.F. and Zentgraf, H. (1979) In: The Cell Nucleus (H. Busch, ed.), vol. 7, pp. 49-95, Academic Press, New York

Franke, W.W., Scheer, U., Krohne, G. and Jarasch, E.D. (1981) J.Cell Biol. 91, 39-50

Gall, J.G. (1954) J. Morphol. 94, 283-351

Gall, J.G. (1966) In: Methods in Cell Physiology (D.M. Prescott, ed.) vol. II, pp. 37-60, Academic Press, New York

Gundlach, H. (1979) In: Moderne Untersuchungsmethoden in der Zytologie (S. Witte, R. Ruch, eds.) pp. 27-35, Witzstrock, Baden-Baden

Hadjiolov, A. (1985) Cell Biol. Monogr. Vol. 12, pp. 1-268, Springer Wien

Hofmann, A., Laier, A. and Trendelenburg, M.F. (1985) In: Molekular- und Zellbiologie (N. Blin, M.F. Trendelenburg & R.E. Schmidt, eds.) pp. 144-158, Springer, Berlin

Mathis, D., Oudet, P. and Chambon, P. (1980) Progr. Nucleic Acid Res. Mol. Biol. 24, 1-55

Matsuura, S., Morimoto, T., Tashiro, Y., Higashinakagawa, T. and Muramatsu, M. (1974) J. Cell Biol. 63, 629-640

Maul, G.G. (1977) Int. Rev. Cytol. Suppl. 6, 75-86

Maul, G.G. (1982) In: The Nuclear Envelope and the Nuclear Matrix, pp. 1-11, Alan Liss, New York

Miller, O.L. (1981) J. Cell. Biol. 91, 15-27

Miller, O.L. and Beatty, B.R. (1969) Science 164, 955-957

Moreno Diaz de la Espina, S., Franke, W.W., Krohne, G., Trendelenburg, M.F., Grund, C. and Scheer, U. (1982) Eur. J. Cell Biol. 27, 141-150

Olins, A.L., Olins, D.E. and Franke, W.W. (1980) Eur. J. Cell Biol. 22, 714-723

Ross, K.F.A. (1967) In: Phase Contrast and Interference Contrast Microscopy for Cell Biologists, Arnold, London

Skoglund, U., Andersson, K., Björkroth, B., Lamb, M.M. and Daneholt, B. (1983) Cell 34, 847-855

Spring, H. and Franke, W.W. (1981) Eur. J. Cell Biol. 24, 298-308

Stevens, B.J. and Swift, H. (1966) J. Cell Biol. 31, 55-77

Trendelenburg, M.F. (1983) Hum. Genet. 63, 197-215

Trendelenburg, M.F., Scheer, U. and Franke, W.W. (1973) Nature New Biol. 245, 167-170

Trendelenburg, M.F., Scheer, U., Zentgraf, H. and Franke, W.W. (1976) J. Mol. Biol. 108, 453-470

Trendelenburg, M.F., Franke, W.W. and Scheer, U. (1977) Differentiation 7, 133-158

Tröster, H., Spring, H., Meissner, B., Schultz, P., Oudet, P. and Trendelenburg, M.F. (1985) Chromosoma 91, 151-163

Unwin, P.N.T. and Milligan, R.A. (1982) J. Cell Biol. 93, 63-75

A Structural Model for the Nuclear Pore Complex

R. A. MILLIGAN
Department of Cell Biology, Sherman Fairchild Building, Stanford University
School of Medicine, Stanford, CA 94305, USA

The nuclear pore complex (NPC) occupies a unique position in the architecture of the eucaryotic cell in that it forms the gateway between two environments, the nucleus and the cytoplasm, which are topologically equivalent but which differ in their macromolecular composition. Strictly speaking, the nucleus is not a membrane-bound organelle but rather can be thought of as an area of the cell surrounded by incompletely fused membrane vesicles with NPCs occupying the pores. Undoubtedly, one of the more fundamental functions of the NPC is simply to keep these pores open and to prevent the vesicles from fusing completely and isolating the nucleus from the cytoplasm. The existence of such pores in the nuclear envelope appears to be a simple way to allow the free passage of ions and small molecules between the nucleus and cytoplasm [14,15] without the necessity for a sophisticated system of channels which is required with a truly membrane bound organelle. Nevertheless, the NPC does carry out a barrier function since molecules larger than ~90 Å in diameter do not traverse it freely [14]. Large molecules are selectively imported into the nucleus and evidence is accumulating that such proteins contain a nuclear localization signal - a short sequence containing positively charged amino acid residues [3,5,10,16]. Macromolecular assemblies such as ribosomal subunits and RNPs containing mature mRNA molecules also appear to be selectively exported or released from the nucleus [13,19]. The papers contained in this volume describe the present state of knowledge of these processes of nucleocytoplasmic transport.

Despite the importance of the NPC in nucleocytoplasmic transport [4,6] and therefore as a potential point of regulation in the pathway of gene expression, relatively little is known about its detailed structure [9] and composition [8]. In this paper I will summarize our work on the structure of the NPC and speculate on how the structural components we have described might be involved in both the active and passive transport processes.

Nucleocytoplasmic Transport
Edited by R. Peters and M. Trendelenburg

PROCEDURES

The oocytes of Xenopus laevis contain a macronucleus or germinal vesicle which can be isolated manually by extruding it through a small hole punched in the animal pole. We used this simple isolation procedure for obtaining nuclei and then spread the envelopes on electron microscope grids by splaying them out using fine glass needles. This approach has the advantage that the envelopes can be fixed and stained within about a minute of their removal from the oocyte, greatly increasing the chance of observing components of the NPC which are weakly or transiently attached to it. Annulate lamellae, cytoplasmic membrane stacks studded with pore complexes [11], are also found close to the animal pole of these oocytes and, on occasions, these structures are extruded with the nucleus and adhere to the EM grid. It is thought that annulate lamellae are formed as a result of overproduction of NPC components in these cells and in embryonic and neoplastic tissues [18]. We incubated the spread envelopes in various solutions and, after staining them with uranyl acetate or gold thio-glucose, examined them in an electron microscope. Minimal irradiation procedures were used in recording images. We processed the images by electron crystallographic methods to obtain two-dimensional projection maps and a three-dimensional map of the NPCs. A complete description of the materials and methods used can be found in reference 20; where necessary for an understanding of this text, some details have been included.

ISOLATED PORE COMPLEXES

Incubating nuclear envelopes spread on grids in solutions containing low concentrations of detergent resulted in the release of some intact pore complexes from the nuclear envelope and from annulate lamellae. These NPC's were found attached to the grid in three orientations termed en face (fig. 1a), edge-on (fig. 1a, arrows) and oblique (fig. 1, inset). En face views show that the NPC is composed of eight elongated subunits extending radially from a 300-350Å diameter central granule. Computer processing and averaging of many NPC images reveals the reproducible structural details of the assembly (fig. 1b). Each of the eight subunits of the NPC consists of three regions; an inner region at a radius of ~350Å, a large peak of material at a radius of ~450Å and a small peak at the perimeter at a radius of ~550Å. The inner regions or spokes are connected circumferentially and encircle the central granule. Density variations in

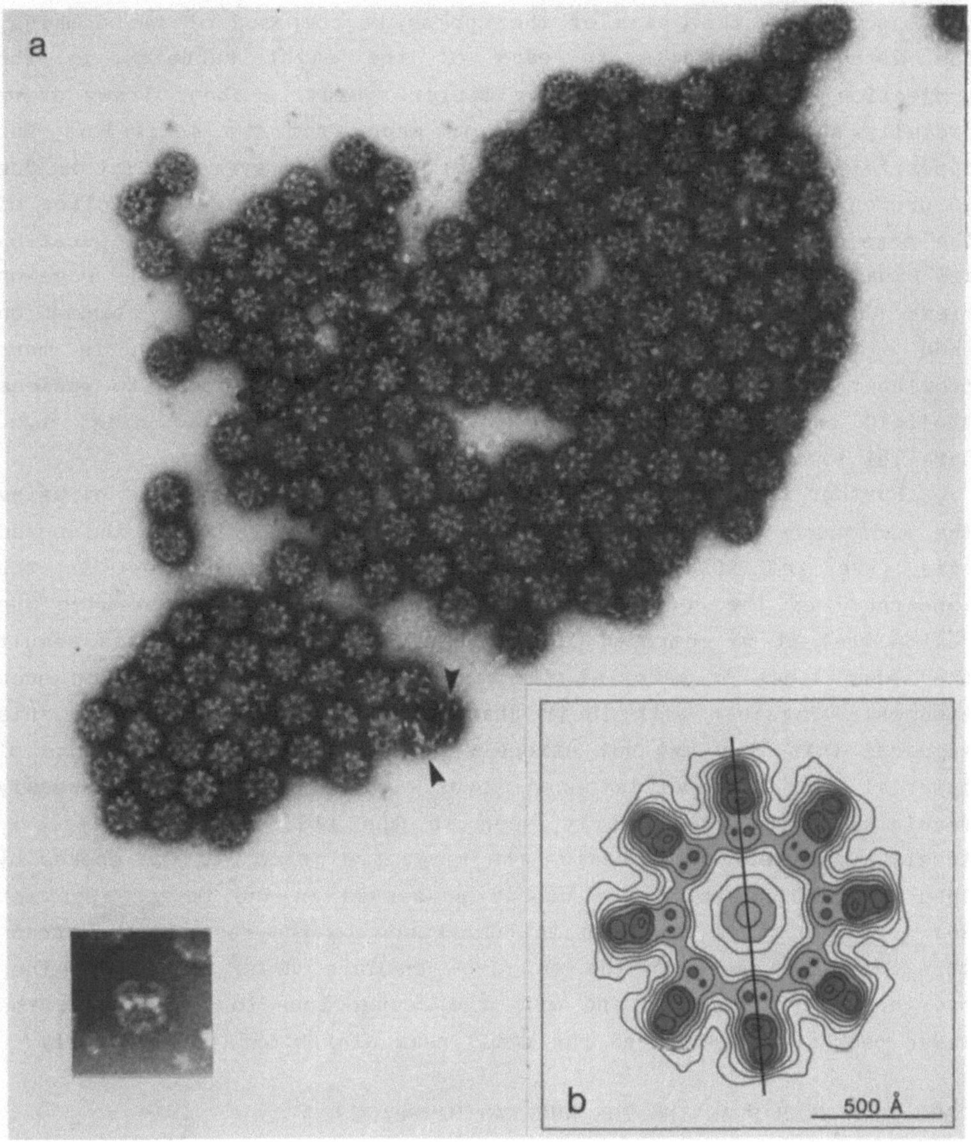

Figure 1: **a.** Image of a cluster of isolated pore complexes (40,000X) released from the nuclear envelope or from annulate lamellae by incubation in 0.1% Triton X100. En face views are more commonly seen and are easily identified by their eight-fold rotational symmetry. Also found are edge-on (arrows) and oblique (inset, 63,000X) views. **b.** Projection map obtained by averaging ~30 NPCs. Averaging of NPCs with strong eight-fold symmetry was carried out using the SEMPER suite of programs [17]. The shaded regions represent parts of the structure where biological material is concentrated. There is approximate mirror symmetry about lines drawn radially (one shown). Deviations from perfect mirror symmetry are due to unequal staining of the two sides of the structure, a phenomenum commonly found when a single carbon support film is used. The resolution of the map is ~65Å.

the map suggest that each of the spokes is composed of two domains. The structural details in each of the eight subunits in the projection map are approximately mirror symmetric about lines drawn radially, suggesting that the NPC is made from two equivalent but oppositely-facing halves. Lack of perfect mirror symmetry may be due to uneven staining of the two sides of the structure. Inspection of the edge-on views (fig. 1, arrows), where the two sides in question are equally exposed to stain, confirms this arrangement. The edge-on views show a central broad zone of matter, the spokes, flanked on each side by a parallel faint line of material which is more prominent at each end. Thus these isolated NPCs appear to possess two-fold axes of symmetry perpendicular to the octad axis, i.e. parallel to the plane of the membrane if it were present.

Further information about the structure can be obtained by using the microscope goniometer to tilt the edge-on NPCs about the octad axis (see ref 20). During tilting there are changes in the appearance of the central broad zone of matter consistent with the notion that it is composed of subunits. In contrast to this result the faint lines of material flanking the central zone show no such changes. Together with their characteristic staining pattern this suggests that they are not strongly modulated but are thin rings of material each like a flattened torus. Such continuous and weakly modulated rings are clearly seen in the oblique views (fig. 1, inset). At their outer radius the rings are joined to the spokes by thin connections which are poorly preserved in the images (see ref 20). As the rings are thin in comparison to the rest of the structure, they are seen in the map from en face views only where they overlap with the spokes and with the connections to the spokes, the large peak at r = 450Å and the small peak at r = 550Å respectively.

PORE COMPLEXES SPANNING THE NUCLEAR MEMBRANES

To obtain equivalent staining of both sides of the NPC, we spread nuclear envelopes on holey carbon films and stained them by immersing the grid in gold thio-glucose solution (fig. 2a). In situ, the ~1200Å diameter NPCs are tightly packed with an average center to center spacing of ~1600Å. Particles identifiable as ribosomes can be seen on the membrane stretches between the NPCs. As with isolated NPCs those in the nuclear envelope exhibit eight-fold symmetry. However, additional features are present which are seen most clearly after computer processing of the images. One new feature seen in the

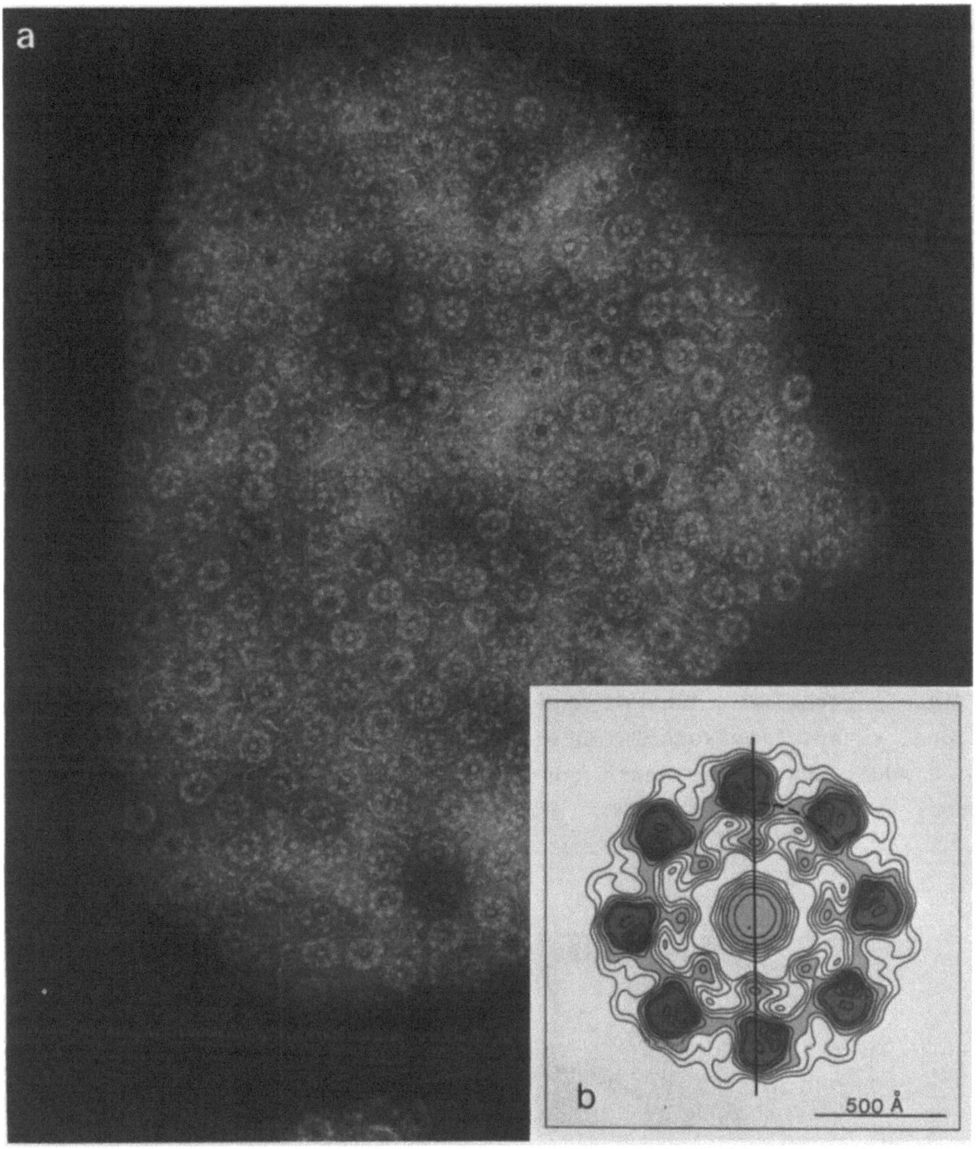

Figure 2: **a.** Image of pore complexes in situ (40,000X). Nuclear envelopes were spread over a holey carbon film, fixed and stained on both sides of the grid with 1% gold thio-glucose solution. The isolation procedure is rapid and therefore predisposes to the retention of loosely or transiently associated parts of the NPC. **b.** Projection map obtained from a single well preserved NPC by rotational filtering [2]. Images of NPCs which have strong peripheral eightfold components give very similar maps. The resolution of the map is ~90 Å. The dotted line shows the approximate location of the membrane border of the pore, i.e. where the inner and outer nuclear membranes come together. There is a pronounced lack of mirror symmetry at the periphery of the structure about lines drawn radially.

projection map (fig. 2b), is a circle of density at a radius of ~420Å. This lies just outside the spokes and shows the location of the membrane border of the pore, i.e. where the inner and outer nuclear membranes come together. Beyond the membrane border the perimeter of these in situ NPCs is characterized by a strong eight-fold density modulation. In place of the large and small peaks at r = 450 and 500Å found in isolated NPCs we now find a single peak of density, ~220Å in diameter. In addition, this feature is no longer disposed symmetrically about lines drawn radially implying the absence of two-fold axes of symmetry in the plane of the membranes. These features of the map suggest that, in situ, the NPCs have additional material associated with either the nuclear or cytoplasmic rings.

To resolve this question we carried out a three-dimensional analysis of the NPC in situ. As before, we used envelopes spread over holey films to minimize artifacts due to unequal staining of the two sides. Because the evelopes were spread manually we were able to deduce which side originally faced the cytoplasm and which the nucleus. Sections through the map perpendicular to the octad axis (fig. 3) reveal the lack of two-fold symmetry and show that the additional material is associated with the cytoplasmic side of the NPC.

Additional experiments confirm our conclusions about the structural components of the NPC. Small numbers of the components can be

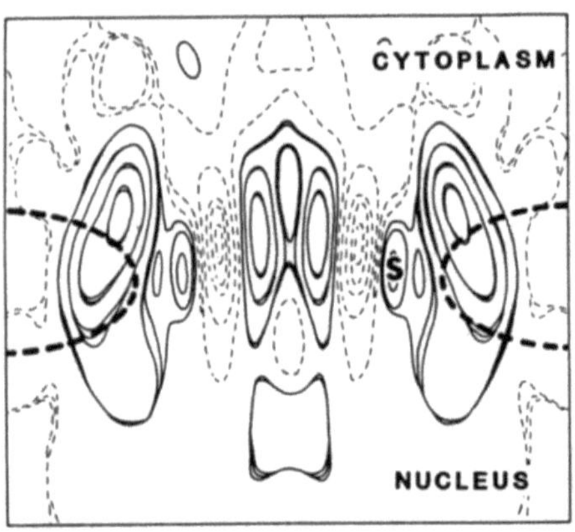

Figure 3: Details of the low resolution three-dimensional map of a pore complex spanning the nuclear membrane. Shown is a 25Å-thick projection of sections perpendicular to the plane of the membranes and passing through the peaks of the eight-fold density modulation. Positive contours (solid lines) show the location of the regions where biological material is concentrated. Thin broken lines are negative contours. The spokes (S), which lie in the central plane in isolated (detached) NPCs, are flanked at high radius by material unequally distributed between nuclear and cytoplasmic sides. There is clearly more material on the cytoplasmic side, (compare with edge-on views of isolated NPCs in fig. 1). The estimated location of the nuclear membranes is indicated by the heavier broken lines. Because of the variable preservation of the central granule we attach no significance to details in that region of the map.

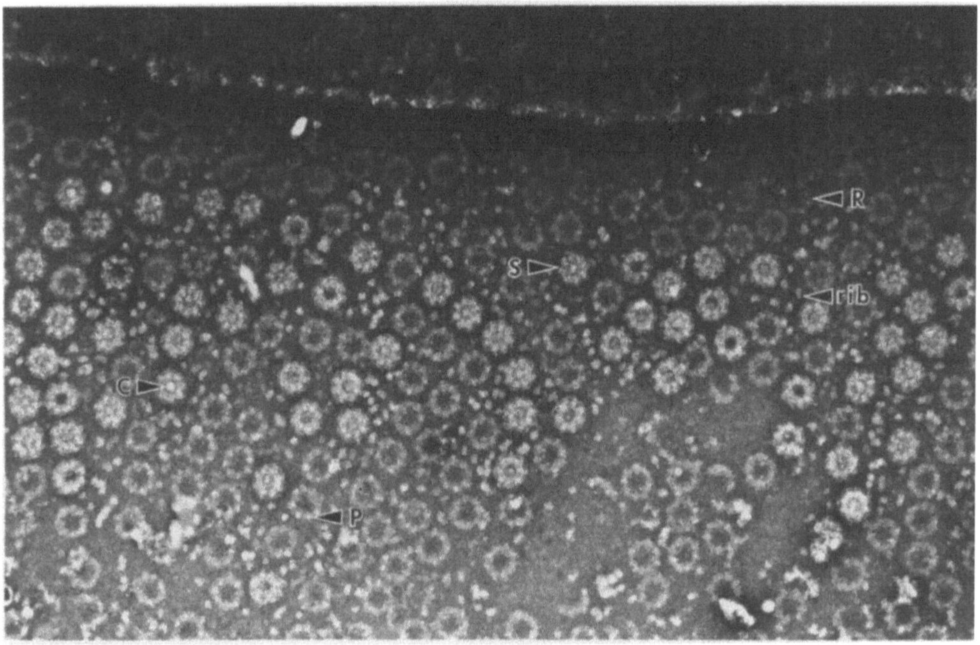

Figure 4: "Finger-print" of material released from nuclear envelopes (38,000X). Such preparations were obtained by spreading a nuclear envelope on a poly-lysine-coated carbon film and incubating in high salt and/or detergent before staining. In places the envelope skeleton (seen at top) detaches leaving various components of the NPC attached to the "sticky" carbon support film. Identifiable are rings (R), spokes (S), central granules (C), particles associated with the rings (P) and ribosomes (rib) which occupy sites on the membrane between the pores.

visualized separately on the grid by preparing "finger-prints" of the nuclear envelope. For this technique envelopes were spread on poly-lysine-coated carbon support films and incubated in high salt and/or detergent solutions before staining. In some parts of the grid the envelope skeleton detaches leaving the released components in their original positions on the "sticky" carbon support. In such preparations (fig. 4) components of the NPC identifiable as rings (R), spokes (S), and central granules (C) can be distinguished. On occasions large particles (P) are associated with the rings. These particles are similar in size and appearance to the ribosomes found occupying sites on the membrane between the NPCs. These observations together with the evidence for pronounced asymmetry of the NPC in the envelope, suggest that the additional material associated with the cytoplasmic side of the NPC may be ribosomes.

DISCUSSION

Our study of the isolated pore complexes reveals some interesting principles of design for the framework of this large assembly (fig. 5). There are obvious similarities between the framework on the cytoplasmic side and on the nuclear side, with equivalent but oppositely-facing rings of material on each side. These are linked in similar fashion to a more massive structure lying in the central plane of the assembly - the circle of spokes. Given the two-fold symmetry of the arrangement of rings, it seems likely that the spokes themselves are also composed of material arranged with the same symmetry. This would require a total of sixteen subunits making up the circle of spokes in the central plane, eight connected to each ring. Such an arrangement would fit the observed densities in the higher resolution projection map (fig. 1b) where there appear to be two peaks of material making up each of the eight spokes. The unequal prominence of the two peaks, which is due to unequal staining of the two sides of the NPC, suggests that all sixteen subunits do not lie in exactly the same plane but are staggered, one towards one side, one towards the other, this arrangement being repeated around the circle. The two subunits making up each spoke are poorly resolved suggesting that they are intimately associated. This association may be responsible primarily for holding the two symmetry related halves of the NPC together. Adjacent spokes are well resolved implying that their association is less intimate; presumably a strong association here is not as important since the task of holding the NPC together circumferentially is shared with the rings.

Free passage of ions and molecules up to ~90Å in diameter could be accomodated by the model described. Not only might this take place between the membrane borders and the spokes, but might occur also in the region between the spokes and the central granule (asterisks in fig. 5b). In the projection maps both regions contain low concentrations of biological material and are of approximately the required dimensions (see figs. 1b, 2b).

On the other hand, the transport of large molecules or assemblies through the NPC poses a greater problem as no obvious large channel exists. It has been suggested that large molecules are unfolded and translocated linearly through small channels in the NPC [1], however, the elegant experiments of Feldherr and associates [6] show unequivocally that particles up to 200Å in diameter pass through the NPC without observable deformation. Moreover, while in transit

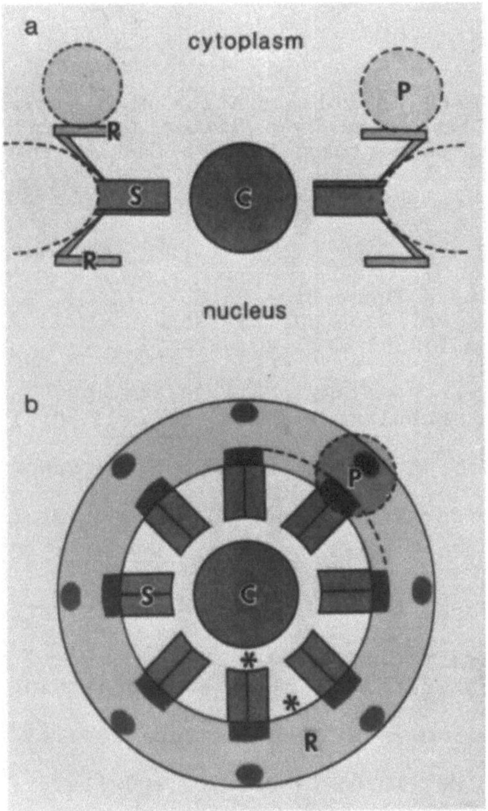

Figure 5: Schematic representation of the nuclear pore complex in central cross-section (**a**) and in projection down the octad axis (**b**). The parts making up the framework include the spokes (S), each composed of two domains or subunits, and the rings (R). The central granule (C) is represented as a sphere ~350Å in diameter. Poorly resolved connections exist between the outer parts of the spokes and the rings. When viewed down the octad axis (**b**) superposition of the rings and the spokes gives rise to the characteristic pattern of stain exclusion at high radius. Dashed lines show the position of the additional features seen in pore complexes in the nuclear envelope i.e., the membrane border and the large particles (P) which resemble ribosomes. The particles are easily detached and are not always present.

the particles are located in the region of the central granule. It appears, therefore, that the central granule seen in electron micrographs either contains a large channel or is itself composed of molecules in transit. As yet there is no compelling evidence for either possibility.

Although it seems likely that the pore complex has a framework embodying the design principles of the model described, the NPC spanning the nuclear envelope is functionally asymmetric (nuclear proteins pass in one direction and mRNAs in the opposite direction) and it is unlikely that a structurally symmetric assembly can be responsible for this. One possibility is that the functional NPC in the envelope is composed of the framework described, which is further embellished by side-specific molecules or modifications which confer asymmetry of function. The large particles which we have found associated with the cytoplasmic rings probably fall into this category. Another example might be the nuclear lamina [7] to which the NPCs are attached on the nuclear side. Such additional side-specific material may or may not be arranged with eight-fold symmetry and may indeed be tissue or cell type specific [12].

ACKNOWLEDGEMENTS

Most of the experimental work was done in collaboration with Nigel Unwin [20]. I also thank Jean Mullenax and Owen Saxton for their help, The Royal Society for financial support and Barbara Buckley for preparing the manuscript.

REFERENCES

1. Clawson GA and Smuckler EA (1982) J Theor Biol 95:607-613.
2. Crowther RA and Amos LA (1971) J Mol Biol 60:123-130.
3. Davey J, Dimmock NJ and Colman A (1985) Cell 40:667-675.
4. Dingwall C (1985) TIBS 10:64-66.
5. Dingwall C, Sharnick SV and Laskey RA (1982) Cell 30:449-458.
6. Feldherr CM, Kallenbach E and Schultz N (1984) J Cell Biol 99:2216-2222.
7. Gerace L, Comeau C and Benson M (1984) J Cell Sci, Suppl 1:137-160.
8. Gerace L, Ottaviano Y and Kondor-Koch C (1982) J Cell Biol 95:826-837.
9. Harris, JR and Marshall P (1981) in Electron Microscopy of Proteins, Vol 1 (New York: Academic Press).
10. Kalderon D, Roberts BL, Richardson WD and Smith AE (1984) Cell 39:499-509.
11. Kessel RG (1983) Int Rev Cytol 82:181-303.
12. Krohne G, Debus E, Osborn M, Weber K and Franke WW (1984) Expl Cell Res 150:47-59.
13. Melton DA, DeRobertis EM and Cortese R (1980) Nature 284:143-148.
14. Paine PL, Moore LC and Horowitz SB (1975) Nature 254:109-114.
15. Peters R (1984) EMBO J 3:1831-1836.
16. Richardson WD, Roberts BL and Smith AE (1986) Cell 44:77-85.
17. Saxton WO, Pitt TJ and Horner M (1979) Ultramicroscopy 4:343-354.
18. Stafstrom JP and Staehelin LA (1984) J Cell Biol 98:699-708.
19. Wickens MP and Gurdon JB (1983) J Mol Biol 163:1-26.
20. Unwin PNT and Milligan RA (1982) J Cell Biol 93:63-75.

Nuclear Lamins During Gametogenesis, Fertilization and Early Development

G. G. MAUL[1] and G. SCHATTEN[2]
[1]The Wistar Institute, Philadelphia, PA 19104, USA
[2]Department of Biological Science, Florida State University, Tallahassee, FL 32306, USA

INTRODUCTION

Lamins have been described as major components of the nuclear envelope (Gerace et al., 1978; Gerace and Blobel, 1982). These proteins of approximately 70 to 60 kD seem to be present in nuclei of all species investigated albeit in different sizes and numbers. Somatic cells like rat liver have three lamins named A, B, and C according to size (Gerace et al., 1978); Xenopus red blood cells have two lamins (Krohne et al., 1981), where lamin I has properties of the mammalian lamin B and the amphibian lamin II is similar to the mammalian lamin C in location, size and pI.

In contrast to somatic cells, the nuclei participating in gametogenesis fertilization and embryogenesis appear to be developmentally regulated (Stick and Schwarz, 1983; Benavente et al., 1985; Schatten et al., 1985; Stick and Hausen, 1985). The oocyte nucleus, (germinal vesicle) from Xenopus (Stick and Krohne, 1982), and the surf clam S. solidissima has only one lamin (Maul and Avdalovic, 1980). The diversity of lamins is, however, increasing and a new species has been found in Xenopus spermatids (Benavente and Krohne, 1985) and mouse spermatids (Maul et al., 1986). The presence of different lamins in nuclei that undergo or have undergone extreme structural and functional changes may well mean that these proteins are instrumental in these changes. Here we report on the distribution of lamins in several species in order to determine whether any differences detected can be correlated with differences in function of cells in spermatogenesis, the germinal vesicle stage of oogenesis and early development. We investigated the mammal Mus musculus, the echinoderm, Lytechinus variegatus,

Nucleocytoplasmic Transport
Edited by R. Peters and M. Trendelenburg

and the surf clam, Spisula solidissima. Fertilization in these species occurs at different stages: the germinal vesicle stage for clam, second meiotic metaphase for the mouse, and the pronuclear stage for the sea urchin. The speed with which the conceptus begins to divide, the amount of storage proteins present and the mRNA synthesis is also variable in these species.

RESULTS AND DISCUSSION

Antibodies used in our investigation came from autoimmune sera of patients with Scleroderma and Lupus erythematosus. Those directed against mammalian lamin AC have been characterized by McKeon et al. (1984), and those against lamin B, by Maul et al. (1986). As shown in Fig. 1, human serum 44 (MS44) reacts with the nuclear periphery of 3T6 cells and little if any staining is present inside the nucleus. If tissue culture cells are treated in situ with 1% Triton, digested with DNase and RNase, and extracted with 1 M NaCl, a treatment where the lamina remains intact, nearly the same image is seen (Fig. 1b).

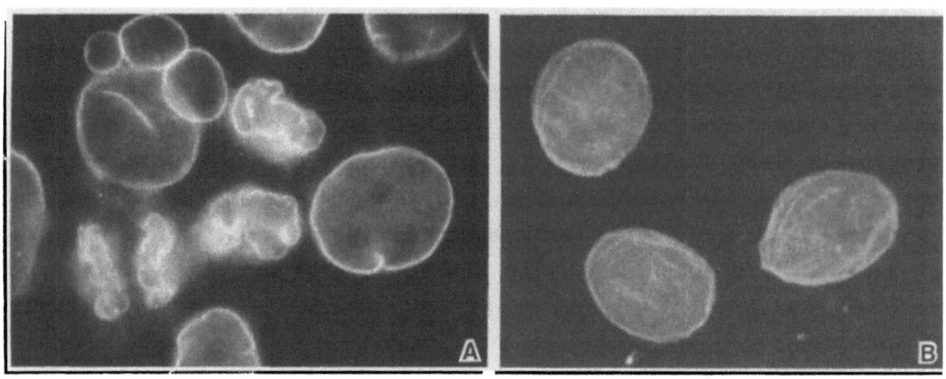

Fig. 1. a) Human serum 44 reacts with the nuclear periphery and little if any staining is present inside the nucleus of 3T6 cells. b) This staining remains if the nucleus is digested sequentially with DNase, RNase and 1 M NaCl in situ.

At mitosis, staining is diffuse throughout the cell. Nuclear envelope staining was also demonstrated on liver sections of mouse and Xenopus. In the clam, we used the hepatopancreas and in the sea urchin, coelomocytes. The somatic cells of these species all showed nuclear rim staining, indicating that the epitopes recognized belonged to a nuclear envelope protein.

Immunoblotting of 2D-SDS gel electrophoretically separated proteins from rat liver and Xenopus red blood cells showed that HS44 reacts with lamin B and its amphibian equivalent lamin I (see Fig. 2c, lane RNE for a 1-dimensional blot of rat liver nuclear envelope). A third antibody, which detects the clam germinal vesicle lamin (lamin G), was used in these studies. Clam germinal vesicle and nuclear envelope (Fig. 2a, lanes CNU and CNE, respectively) are shown immunoblotted in Fig. 2b. Only a single 67 kD protein is labeled. These three antibodies (anti-lamin AC, Fig.

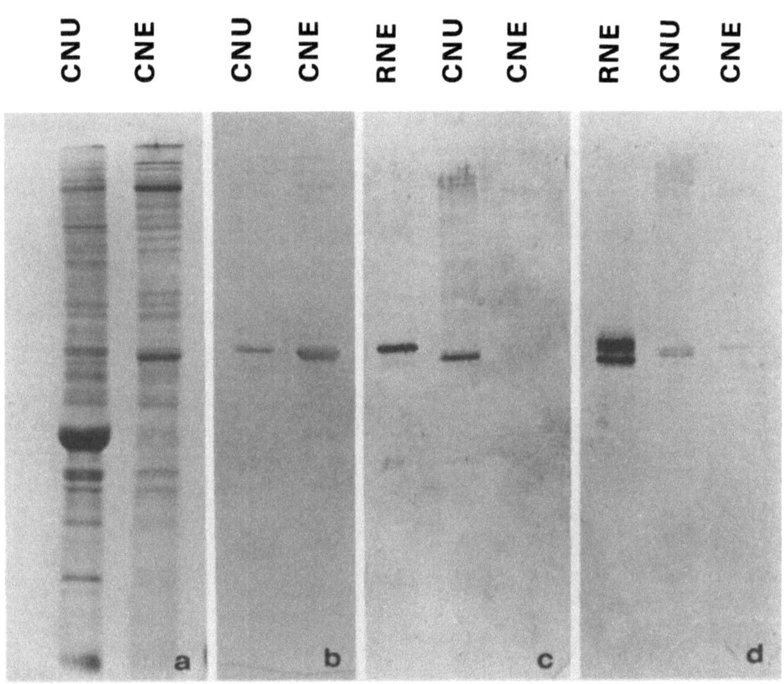

Fig. 2. a) Coomassie blue-stained clam germinal vesicle (CNU) and clam germinal vesicle nuclear envelope (CNE). b) Same as (a) but blotted and reacted with clam anti-lamin G antibodies. In both lanes a single protein of 67 kD reacts. c) Rat liver nuclear envelope (RNE), CNU and CNE reacted with HS44. Lane RNE shows reaction with rat lamin B, and lane CNU, with the clam lamin B equivalent. No reaction is seen in the isolated clam nuclear envelope. d) The same preparations as in (c) reacted with LS1 serum. The antibody reacts with lamins AC and their proteolytic products in rat (RNU). In both lanes with clam proteins, only a faint reaction can be recognized. Proteins were separated on 7-15% SDS polyacrylamide gels transblotted and assayed with the Vector stain biotin-avidin peroxidase system.

d, lane RNE; anti-lamin B, Fig. 2c, lane RNA; and anti-lamin G, Fig. 2b, lane CNE) were used to localize the lamin equivalent in mouse, sea urchin and clam nuclei at different stages of their differentiation.

Lamins During Spermatogenesis

We have identified the presence of lamin epitopes in most phases of spermatogenesis of the mouse. We have found that while some lamins disappear in spermatogenesis, others appear or become specifically localized to one or more regions of the sperm head.

Incubation of mouse testes sections (ethanol-fixed and wax-embedded) with anti-lamin B antibodies revealed reaction at the nuclear periphery of some nuclei. Sertoli cells are most strongly

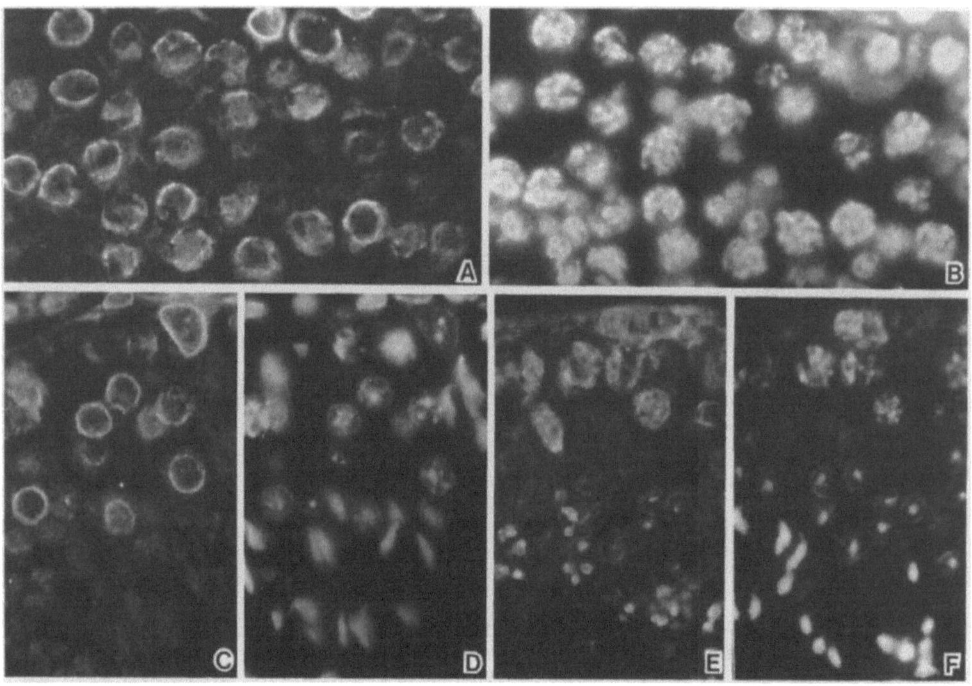

Fig. 3. Cross-sections through mouse seminiferous tubules (ethanol-fixed and wax-embedded). (a) Deparaffinized sections were incubated with HS44 and labeled with biotinilated anti-human IgG which was visualized by fluorescinated avidin (b) Same section as (a) but the DNA was stained with Hoechst 33258. Pachytene nuclei show reaction at the nuclear envelope. c) Secondary spermatocytes also stain strongly but less that Sertoli cells. e) Late stage round spermatids still show faint staining. Stronger reaction is present in the residual bodies (arrow points to cytoplasmic lobes). f) Same as (c) but Hoechst 33258 stained.

stained. Less intense staining was seen on the periphery of spermatocytes (Fig. 3a) at pachytene; see the same section Hoechst stained for DNA distribution in Fig. 3b. Secondary spermatocytes are also strongly reactive at the nuclear envelope (Fig. 3c). Staining in the round spermatids becomes faint. In later stages, but before nuclear elongation, one can find the antigen in the cytoplasmic lobes that remove the excess cytoplasm during spermiogenesis (Fig. 3c, arrow). No fluorescence could be recognized in elongating spermatid nuclei. However, when epididymal sperm were fixed in ethanol on glass slides and incubated with different sera, a number of specific locations could be seen. In Fig. 4a and b, we used a monoclonal antibody that reacts both with a lamin B and the lamin AC epitopes. Staining can be recognized in the post-acrosomal area of the spermatozoa, indicating that lamins are present in the competely condensed sperm nucleus.

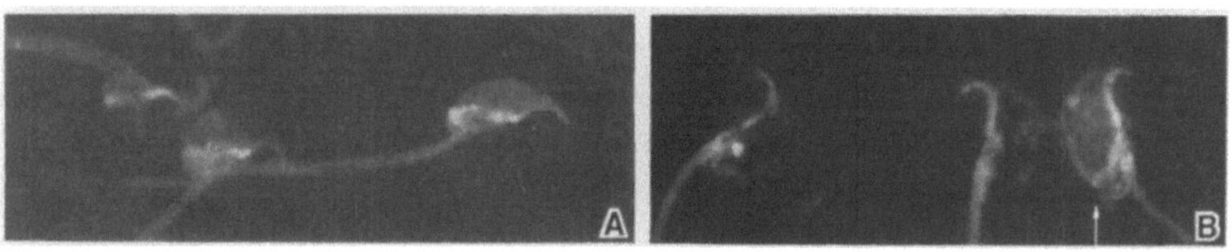

Fig. 4. Epididymal sperm fixed in ethanol on glass slides and incubated with a monoclonal antibody reacting with lamins AC and B. a) The post-acrosomal area of the sperm head reacts. b) After treatment as in Fig. 1b, most of the antigen is removed but some remains.

Anti-lamin AC staining after formalin fixation is very weak

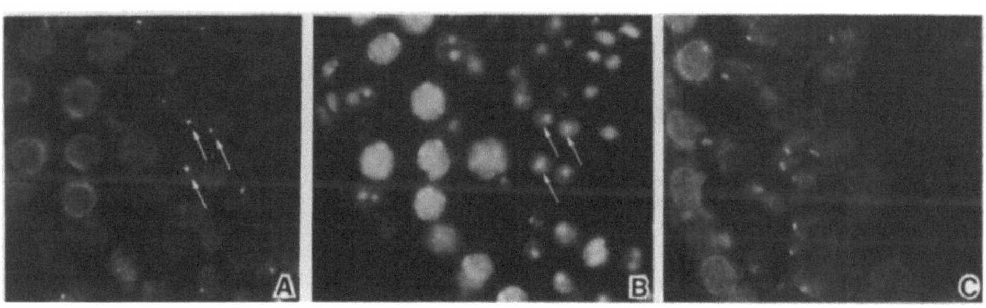

Fig. 5. Cross section of mouse seminiferous tubules formalin-fixed, wax-embedded. Deparaffinized sections were incubated with LS1 serum recognizing lamin AC. Early stage round spermatids show acrosomally located small dots (a) or short lines (c). (b) is the same as (a) but stained with Hoechst 33258. Arrows point to the location of the immunoreactive spots which lie acentrically and are not situated over areas of DNA concentration.

but recognizable on spermatogonia, spermatocytes and Sertoli cells, but cannot be seen on spermatids except as precise dots or short lines in a small number of seminiferous tubule cross-sections (Fig. 5a and b). From their location and counterstaining with an acrosomal antibody, the staining dots are identified as having an acrosomal location during very early stages (stage 3-4, according to Oakberg, 1956). In later stages, this antigen is not recognizable. Instead, a new lamin epitope appears which is recognized by the anti-clam lamin G. Like the anti-lamin AC epitope, this epitope is seen as dots in some tubules and as crescent or ring shapes in others (see Fig. 6a) later stages of the round spermatids. From this observation, we conclude that lamin AC is exchanged for a new lamin, one that has epitopes in common with another germ cell lamin, i.e., that of the evolutionarily distant clam. The appearance of the new lamin (lamin G) in mouse spermatogenesis parallels that of Xenopus oocytes where a new lamin (lamin III) appears after pachytene, and the two lamins previously present are lost (Stick and Schwarz, 1983). It seems likely, therefore, that during gametogenesis in general, an exchange to different lamins takes place.

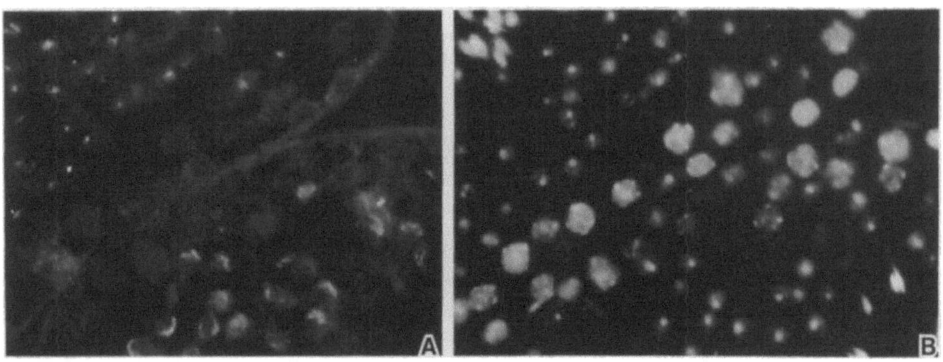

Fig. 6. Cross-sectioned mouse seminiferous tubule (ethanol-fixed and wax-embedded). Deparaffinized sections were incubated with clam anti-lamin G antibodies. Immunoreactive spots like in Fig. 5a are found in round spermatids of somewhat later stages but no reaction is seen in Sertoli cells and spermatocytes. b) Same as (a) but stained with Hoechst 33258.

During our investigation, we became aware that the fixation protocol, wax embedding, the use of heat for wax infiltration, the type of antibodies, as well as the sensitivity of the second antibody system used may have effects on the observability of the antibody reaction. The 10- to 30-fold higher sensitivity of the biotin-avidin system over the directly fluorescinated second antibody played a major part in the recognition of antigen at the very low levels present during the later stages of spermatogenesis. We have been able to repeat the results on mouse with Xenopus and sea urchin, and the recent finding of Benavente and Krohne (1985) of a new lamin-like protein in Xenopus spermatogenesis supports these findings.

Lamins in the Germinal Vesicle Stage Oocyte

The germinal vesicle stage of late oocytes in different species varies as to whether the chromatin is attached to the nuclear envelope. Of the species investigated previously, Xenopus and clam have no chromatin attached and only one lamin-like protein. Somatic cells have chromatin attached to the nuclear envelope and more than one type of lamin. Since, mouse and sea urchin germinal vesicles (unlike Xenopus and clam) have chromatin attached, we tested them for the possible presence of more than one lamin. The density of the cytoplasm and the possibility that lamins are stored in the cytoplasm made it necessary to perform the *in situ* assay of the lamins present in extracted oocytes.

When we used anti-lamins AC and B antibodies, we found that both detect determinants in the germinal vesicle envelope of mouse (Fig. 7) and sea urchins (not shown). In clam, a reaction was found with the total nucleus but not with the nuclear envelope. Any lamin B present in these oocytes must therefore either be stored in the germinal vesicle or in the cytoplasm. We were able to test this possibility in the clam by isolation of these compartments, separation of the proteins, and immunoblotting since techniques for mass isolation of germinal vesicles are available (Maul, 1980). When probed with the lamin B binding HS44 serum, no reaction is found in the germinal vesicle envelope (Fig. 2c, lane CNE). Total nuclei, however, contain a strongly binding protein (Fig. 2c, lane CNU). For comparison, rat nuclear envelope was added in lane RNE, localizing the position of the mammalian lamin B. We had found no structures of the germinal vesicle that bind lamin B antibodies using immunofluorescence. The possibility that a lamin B equivalent might be present in soluble form was tested by rapidly isolating germinal vesicles, rupturing them in a syringe and pelleting all particulate material. The 100,000 g super-

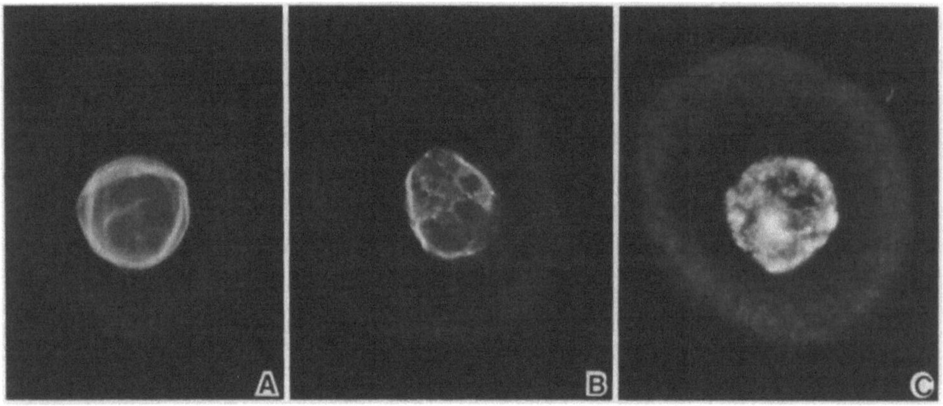

Fig. 7. Extracted mouse oocytes reacted with (a) LS1 serum (lamin AC) and (b) HS44 (lamin B). Both antigens are present.

natant of the germinal vesicle, but not that of cytoplasm, contained the lamin B binding antigen (data not shown). Anti-lamin AC antibodies showed only faint binding to a single protein in the nuclear envelope (Fig. 2d, lane CND), an observation that is in line with our previous observation that there is only minor cross-reactivity of polyclonal anti-AC antibodies with lamin G and, conversely, minor cross-reactivity of anti-lamin G with mammalian lamin AC.

Based on our observations (Maul et al., 1986) and the previous findings on Xenopus (Stick and Krohne, 1982), we suggest that germinal vesicles with chromatin attached to the nuclear envelope as in mouse and sea urchin have a lamin B equivalent stably integrated into the nuclear envelope in addition to a lamin G or AC equivalent. Xenopus and clam, however, with condensed chromosomes in the center of the germinal vesicle or attached to the nuclear envelope, respectively, have only one integrated lamin. Despite the limited number of observed cases, one might therefore correlate lamin B with binding of chromatin to the nuclear envelope.

Lamins During Early Development

Fertilization requires several dramatic changes in nuclear organization. The three systems investigated represent extremes in the fertilization mechanism. In clam, the egg is spawned in the germinal vesicle stage and the meiotic maturation and reduction division take place with the male genome unenveloped in the egg. The ovulated mouse oocyte is arrested at second meiotic metaphase; fertilization is only complete at first mitosis when the parental chromosomes align at metaphase. The sea urchin egg is spawned as a mature egg with a female pronucleus, and pronuclear fusion or syngamy occurs shortly after sperm incorporation.

Clam. Lamins do not appear on any recognizable structure after germinal vesicle breakdown about 7-10 min after sperm penetration until the formation of the pronuclei. This indicates that no nuclear envelope is made between first and second polar body formulation. During these two divisions, the chromosomal mass of the male genome does not acquire lamins but may acquire other proteins. The signals that maintain the lamins in solution (or phosphorylated) are therefore maintained through two successive divisions. As shown with anti-lamin G antibodies, all nuclei, i.e., those of the macromere and micromeres, react strongly. No lamin B reaction on the nuclear envelope was found up to 48 hr after fertilization (trochophore larvae).

Mouse. Both anti-lamin AC and B antibodies react with the two pronuclei in mouse. The pronuclei acquire the lamins immediately after fertilization during pronuclear formation (Fig. 8a). The polar body nucleus, however, must have obtained only very small amounts of lamins since its outline is faint relative to that of the pronuclei despite the presence of a complete nuclear envelope. The availability of lamins due to a restricted amount of cytoplasm may account for the reduced lamin incorporation but some regulated mechanism cannot be excluded. During subsequent development, lamin B is present in all early developmental stages. Lamin epitopes as recognized by the human anti-lamin AC autoantibodies, however, diminish during the morula stage (Fig. 8b) (Schatten et al., 1985), but reappear in the early blastocyst nuclei (Fig. 8c).

Xenopus and clam germinal vesicles have a single lamin incorporated in their nuclear envelope. Whether mouse has an equivalent lamin, i.e., one distinct from the somatic lamins AC, cannot be determined because of the lack of sufficient material for immunoblotting. However, the disappearance of a protein recognized by lamin AC antibodies at the morula stage and the reappearance of anti-lamin AC recognizable antigen in the blastocyst, might be analogous to what occurs in early *Xenopus* development. The lamin

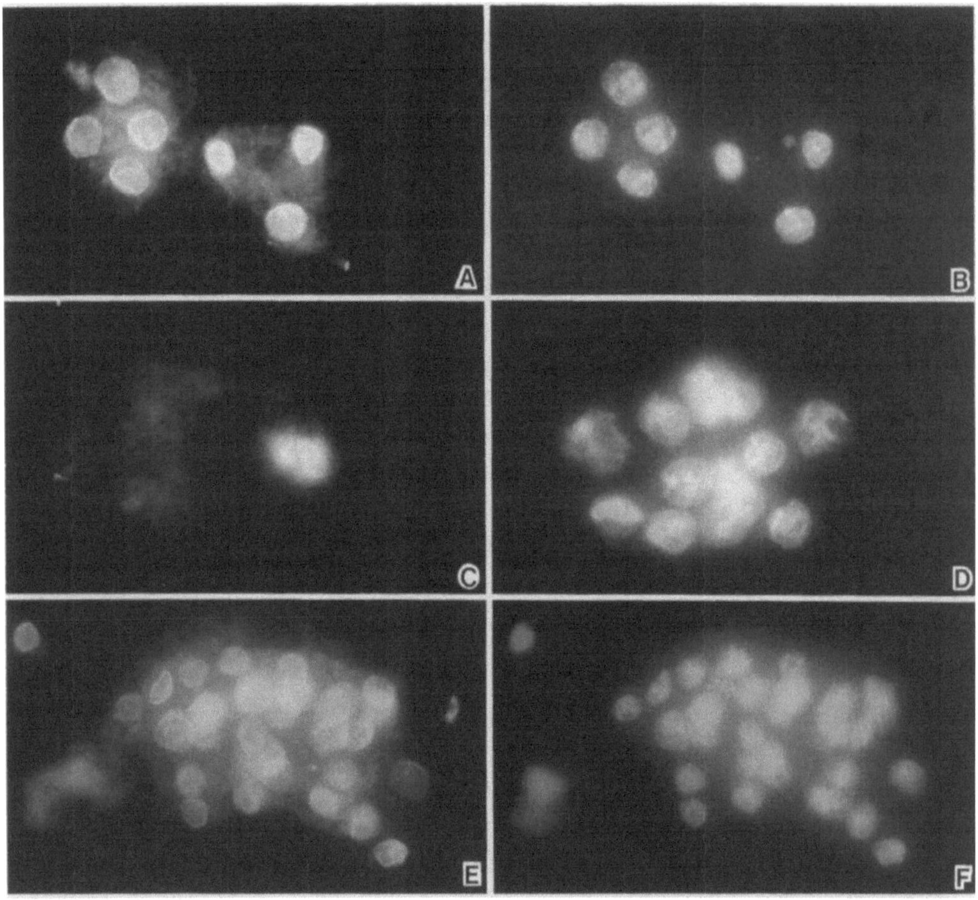

Fig. 8. Detergent-extracted mouse early embryos reacted with serum LS1 (lamin AC) show that the antigen recognized is: a) present before the morula stage; one nucleus has been lost in this extracted embryo; b) absent during the morula stage; and c) again present in an early blastocyst.

III (which is antigenically related to lamin II and AC in mammals) diminishes during early development and lamin II is newly synthesized at the neurula stage (Stick and Hausen, 1985). For this system, monoclonal antibodies are available but polyclonal antibodies would recognize both. In the case of mouse early develop ment, we have only a polyclonal antibody that may recognize both lamins, a specific lamin G and the somatic lamins AC. The observation of a strong decrease of antibody binding at the morula stage and the reappearance could then, in light of the findings in Xenopus, be interpreted as showing the decline of lamin G and the new synthesis of the somatic lamin AC. Proof of this interpretation awaits development of the appropriate monoclonal antibodies.

<u>Sea urchins</u>. The female pronucleus of the unfertilized sea urchin egg is spawned with an intact nuclear envelope and incorporated lamins as recognized by anti-lamin AC and B antibodies. The sea urchin sperm, like the mouse sperm, brings only minimal amounts of lamins recognized most strongly at the acrosomal and centriolar fossa. After sperm incorporation, the decondensing male pronucleus acquires the lamins immediately. From this observation, it is clear that nuclear envelope incorporated lamins can co-exist with unincorporated or soluble lamins over extended periods of time in the same cell.

The lamins as recognized by anti-AC and anti-B antibodies behave in an unusual fashion during mitosis in sea urchin. Unlike in mouse zygotes and somatic cells (Gerace and Blobel, 1980), where lamins are never observed on chromosomes, lamins delineate most of the sea urchin mitotic chromosomes even during metaphase. This system may give us an early clue to the question of whether the lamins bind first to the envelope or to the chromatin. It may also be a natural system to study this question further, and particularly, to identify the protein (or DNA) to which the lamins bind on chromosomes.

During early development in sea urchin, both the epitopes recognized by human anti-lamin AC and B were lost or unrecognizable by the blastula stage. Epitopes recognized by an anti-lamin AC monoclonal antibody, however, appeared. Again, there seem to be certain similarities between the observations on Xenopus and mouse early embryogenesis. From the extended absence of the lamin epitopes (lamin B and AC) in sea urchin and lamin B equivalents in clam, we anticipate that synthesis may only be initiated after the larval stages and possibly at the switch-over to the adult stage. There are various larval and adult tissues simultaneously present during the switch-over. It will be interesting to determine whether the lamins remain differentiated between those tissues.

The presence of lamin B in nuclear envelopes of germinal vesicles of some species but not in others might reflect functional differences. If lamin B is essential in chromatin attachment to the nuclear envelope in other species, as Burke and Gerace's (1985) evidence indicates for somatic mammalian cells, then Xenopus lamin III must be able to substitute for it until the reemergence of lamin I (the lamin B equivalent) during the midblastula transition. The presence of lamin B in nuclear envelopes on only those nuclei where chromatin is attached and where RNA synthesis takes place might be fortuitous or might have functional significance, the latter possibility being more intriguing than the notion of lamin B as a mere storage product.

ACKNOWLEDGEMENTS

This work was supported by the National Institutes of Health grants GM-21615 and CA-10815 to G.G.M. and HD-12913 and HD363 and NSF PCM 83-15900 to G.S.

REFERENCES

Benavente, R. and Krohne, G. 1985. Proc. Natl. Acad. Sci. U.S.A. 82:6176-6180.
Benvente, R., Krohne, G., and Franke, W.W. 1985. Cell 41:177-190.
Burke, B. and Gerace, L. 1985. J. Cell Biol. 101:143a.
Gerace, L., Blum, A., and Blobel., G. 1978. J. Cell Biol. 79:546-566.
Gerace, L., and Blobel, G. 1982. Cold Spring Harbour Symp. Quant. Biol. 46:967-978.
Krohne, G., Debauvalle, M.C., and Franke, W.W. 1981. J. Mol. Biol. 151:121-141.
Maul, G.G. 1980. Exp. Cell Res. 129:431-438.
Maul, G.G., and Avdalovic, N. 1980. Exp. Cell Biol. 130:229-240.
Maul, G.G., French, B.T., and Bechtol, K.B. 1986. Dev. Biol., in press.
McKeon, F.D., Tuffanelli, D.L., Fukujama, K., and Kirscher, M.W. 1983. Proc. Natl. Acad. Sci. U.S.A. 80:4347 -4378.
Oakberg, E.F. 1956. Am. J. Anat. 99:391-413.
Schatten, G., Maul, G.G., Schatten, H., Chaly, N., Simerly, C., Balaczon, R., and Brown, D.L. 1985. Proc. Natl. Acad. Sci. U.S.A. 82:4727-4731.
Stick, R., and Krohne, G. 1982. Exp. Cell Res. 138:319-330.
Stick, R., and Schwarz, H. 1982. Cell Diff. 11:235-243.
Stick, R., and Schwarz, H. 1983. Cell 33:949-958.
Stick, R., and Hausen, P. 1985. Cell 41:191-200.

A Pool of Soluble Nuclear Lamins in Eggs and Embryos of *Xenopus laevis*

G. KROHNE and R. BENAVENTE
Division of Membrane Biology and Biochemistry, Institute of Cell and Tumor Biology, German Cancer Research Center, D-6900 Heidelberg, FRG

The growing oocyte of the amphibian Xenopus laevis accumulates and stores large pools of various RNAs, ribonucleoproteins, and proteins (1-6). These stored molecules are sufficient to meet the needs of the embryo during the first 12 cleavage cycles until RNA synthesis starts at the midblastula stage ca. 4 hours after fertilization (7). Recent experiments indicate that Xenopus eggs also contain stored nuclear envelope components (8, 9). When chromatin or purified DNA are injected into eggs or incubated in extracts of activated eggs nucleus-like structures are assembled possessing nuclear pores and a nuclear lamina.

In this report we describe the pool of soluble nuclear lamins in Xenopus and discuss it in light of nuclear envelope assembly.

The nuclear lamins of Xenopus

In Xenopus laevis four different lamins have been identified (L_I, L_{II}, L_{III}, and L_{IV}; 10-13; for review see 14). These karyoskeletal polypeptides have apparent molecular weights of 68,000

Nucleocytoplasmic Transport
Edited by R. Peters and M. Trendelenburg

(L_{II}, L_{III}), 72,000 (L_I), and 75,000 (L_{IV}) and display cell type specificity in their expression (15). The majority of somatic cells contain L_I and L_{II} whereas a few specialized somatic cells (neurons, myocytes, Sertoli cells) express lamin L_{III} in addition to the two somatic lamins (L_I, L_{II}). Only one lamin is present in diplotene oocytes and eggs (L_{III}) and in spermatids and sperms (L_{IV}). The four lamins are closely related proteins as shown by immunological methods (14).

Lamin L_{III} is structurally bound in the Xenopus oocyte

The nuclear lamina of fully grown oocytes (diplotene stage of meiosis) is a proteinaceous meshwork of fibrils located subjacent to the inner nuclear membrane. It interconnects the nuclear pore complexes, and both components are present as structural entities of the final karyoskeletal residue which is referred to as the "pore complex-lamina fraction" (10, 16, 17). This karyoskeletal residue contains one major polypeptide, lamin L_{III}, which makes up more than 80% of the total protein present (for minor components see 10, 11, 18).

Lamin L_{III} is soluble in the egg

The fully grown oocyte is arrested in the diplotene stage of meiotic prophase until hormonal signals induce the resumption of meiosis. The nuclear envelope is disintegrated, a meiotic spindle is formed that gives rise to the first polar body and the egg is arrested after ovulation in the second metaphase of meiosis.

Almost all L_{III} detectable in the egg is present in a soluble form and can be recovered from the 100,000 x g supernatant in form of a 9 S particle (15). The exact protein composition of the particle is yet not known.

We compared soluble lamin L_{III} of eggs with the polymeric form of the oocyte by isoelectric focusing and two-dimensional gel electrophoresis and found that lamin L_{III} of the egg was more acidic than the oocyte lamin (Fig. 1a and b). When the soluble lamin L_{III} of the egg is treated with alkaline phosphatase its mobility in two-dimensional gels is nearly identical with that of L_{III} prepared from oocytes (Fig. 1c). These results indicate that the charge differences of lamin L_{III} of the egg are primarily due to phosphorylation.

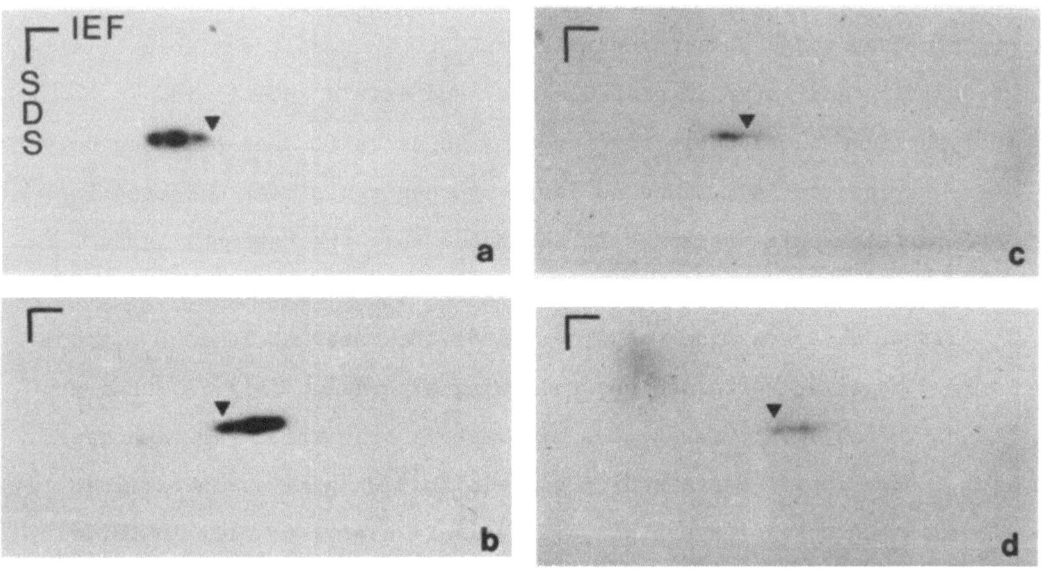

Fig. 1: Comparison of polymeric and soluble lamin L_{III} by two-dimensional gel electrophoresis and immunoblotting with antibodies specific for L_{III} (for technical details see 15). The arrowheads mark the position of the reference protein bovine serum albumin subjected to coelectrophoresis. a: Nuclear pore complex-lamina fraction of oocytes. b: Fraction of eggs enriched in soluble lamin L_{III}. c: Soluble L_{III} of eggs incubated for 1 hour at 25°C with alkaline phosphatase. d: Control of the experiment shown in c; same conditions without enzyme. IEF, isoelectric focussing in the first dimension; SDS, second dimension.

The data shown here indicate that phosphorylation may be involed in lamina depolymerization during amphibian meiosis in a manner similar to that previously described for mitosis of cultured mammalian cells (19). The factors involved in lamina phosphorylation and nuclear envelope breakdown appear to be conserved during evolution since extracts of Xenopus eggs are capable to use mammalian lamins as substrate (for details see 20).

L_{III} is the only nuclear lamin of Xenopus embryos up to midblastula transition

The composition of the nuclear lamina of Xenopus embryos during development has been described (15, 21). Embryonic cells up to the midblastula stage possess only lamin L_{III}. Then at the midblastula transition and later in gastrula the synthesis of new lamin polypeptides characterisitic for somatic cells (L_I and L_{II}) begins (for review see 14). Since no lamin synthesis has been detected in earlier embryonic stages it is suggested that the maternal lamin L_{III} is recycled during early embryogenesis.

In order to examine the dynamics of the maternal lamin L_{III} pool more closely we performed the following experiment (Fig. 2). Embryos at different ages prior to the midblastula transition were incubated for 1 hour in 10% Barth medium containing 400 µg/ml cycloheximide. At this concentration protein synthesis is nearly totally inhibited and blastomeres are arrested in interphase (20). Six embryos of a defined stage were homogenized and after pelleting of the nuclei, the proteins present in the 8,000 x g supernatant were analyzed by one-dimensional gel electrophoresis and immunoblotting with lamin antibodies specific for L_{III}. The experiment revealed that a considerable amount of lamin L_{III} was not structurally bound to nuclei and, therefore, could be recovered in the supernatant. In addition the amount of soluble L_{III} in interphase cells of embryos decreased during development (Fig. 2), concomitant with the increase in the number of nuclei.

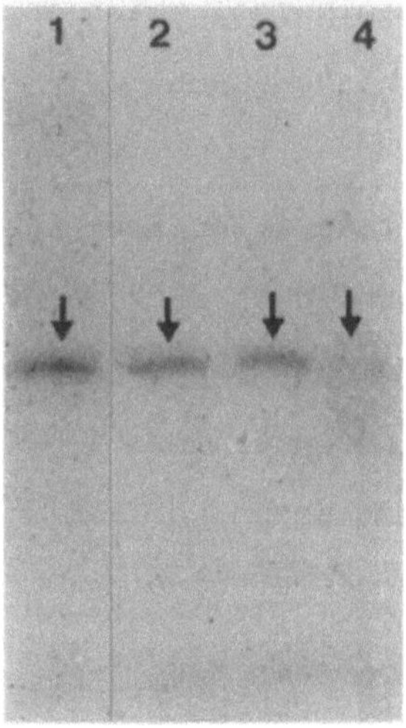

Fig. 2: Analysis of soluble lamin L_{III} in early embryogenesis by one-dimensional gel electrophoresis and immunoblotting with lamin antibodies (for details see 15). Embryos of various stages (for description of stages see 22) were incubated for 1 hour in 10% Barth medium containing 400 μg/ml cycloheximide to arrest all embryonic cells in interphase. Total soluble proteins present in the 8,000 x g supernatant of homogenates from six eggs or embryos were loaded on each lane. 1: Unfertilized eggs. 2: Stage 5. 3: Stage 6 1/2. 4: Early stage 8. Lamin L_{III} is marked by arrows.

Lamina assembly and the soluble lamin pool

Gerace and colleagues have proposed a model for the organization of the nuclear lamina and its dynamics during the cell cycle (for review see 23). According to this model extensive phosphorylation of lamins during mitosis is involved in nuclear lamina disassembly and nuclear envelope breakdown whereas dephosphorylation of lamins at telophase allows for the repolymerization of lamins.

The dynamics of lamin L_{III} in early embryos of Xenopus can be only partially explained by this model. In contrast to the situation in cultured cells, the early embryo contains a pool of nuclear lamins of which a considerable proportion remains in a soluble state during interphase. Our experiments (Fig. 2) and recently published data (8, 9) indicate that the proportion of the lamin pool which is assembled into embryonic nuclei is correlated with the amount of chromatin present in the embryo.

These results allow us to speculate that the transition from the soluble and phosphorylated form to the polymeric and dephosphorylated lamin occurs not "in solution" but in direct contact with the chromatin along the surface of the reforming nucleus. This suggests that in the amphibian embryo a significant amount of lamin remains in the soluble (mitotic) form when the number of binding sites is low. Hopefully, further studies of nuclear lamina dynamics in early embryonic development will contribute to our knowledge about the mechanisms regulating the assembly-disassembly cycle of the nuclear envelope and should permit the localization of these reactions in the cell.

Acknowledgements

We thank Dr. W.W. Franke for stimulating discussions and Dr. F.J. Longo for reading and correcting the manuscript. This work received financial support from the Deutsche Forschungsgemeinschaft (grant Kr 758/2-1).

References

1. Pestell, R.Q.W. (1975) Biochem. J. 145: 527-534.
2. Laskey, R.A., Mills, A.D. & Morris, N.R. (1977) Cell 10: 237-243.
3. Woodland, H.R. & Adamson, E.D. (1977) Dev. Biol. 57: 118-135.
4. DeRobertis, E.M., Lienhard, S. & Parisot, R.F. (1982) Nature 295: 572-577.
5. Kleinschmidt, J.A., Scheer, U., Dabauvalle, M.C., Bustin, M. & Franke, W.W. (1983) J. Cell Biol. 97: 838-848.
6. Zeller, R., Nyffenegger, T. & DeRobertis, E.M. (1983) Cell 32: 425-434.
7. Newport, J.W. & Kirschner, M. (1982) Cell 30: 675-686.
8. Lohka, M.J. & Masui, Y. (1983) Science 220: 719-721.
9. Forbes, D.J., Kirschner, M. & Newport, J.W. (1983) Cell 34: 13-23.
10. Krohne, G., Franke, W.W. & Scheer, U. (1978) Exp. Cell Res. 116: 85-102.
11. Krohne, G., Dabauvalle, M.C. & Franke, W.W. (1981) J. Mol. Biol. 151: 121-141.
12. Krohne, G., Debus, E., Osborn, M., Weber, K. & Franke, W.W. (1984) Exp. Cell Res. 150: 47-59.
13. Benavente, R. & Krohne, G. (1985) Proc.Natl.Acad.Sci.USA 82: 6176-6180.
14. Krohne, G. & Benavente, R. (1986) Exp. Cell Res. 162: 1-10.
15. Benavente, R., Krohne, G. & Franke, W.W. (1985) Cell 41: 177-190.
16. Scheer, U., Kartenbeck, J., Trendelenburg, M., Stadler, J. & Franke, W.W. (1976) J. Cell Biol. 69: 1-18.
17. Aaronson, R.P. & Blobel, G. (1975) Proc.Natl.Acad.Sci.USA 72: 1007-1011.
18. Benavente, R., Krohne, G., Schmidt-Zachmann, M.S., Hügle, B. & Franke, W.W. (1984) J. Cell Sci., suppl. 1: 161-186.
19. Gerace, L. & Blobel, G. (1982) Cold Spring Harbor Symp. Quant. Biol. 46: 967-978.
20. Miake-Lye, R. & Kirschner, M. (1985) Cell 41: 165-175.
21. Stick, R. & Hausen, P. (1985) Cell 41: 191-200.
22. Nieuwcoop, P.D. & Faber, J. (1967) Normal Table of Xenopus laevis (Daudin). North-Holland Publishing Co., Amsterdam.
23. Gerace, L., Comeau, C. & Benson, M. (1984) J. Cell Sci., suppl. 1: 137-160.

Uptake of Oocyte Nuclear Proteins by Nuclei of *Xenopus* Embryos

CHR. DREYER, R. STICK, and P. HAUSEN
Max-Planck-Institut für Entwicklungsbiologie, Spemannstrasse 35/V, D-7400 Tübingen, FRG

Introduction

The Xenopus oocyte nucleus, or germinal vesicle, accumulates a variety of proteins during the several months of oocyte growth. Many of these nuclear proteins are probably stored for later use in embryogenesis (1-2).When the first meiotic division is completed during egg maturation, germinal vesicle breakdown leads to the loss of a compartment for the nuclear proteins (3, 4). They are distributed in the cytoplasm of the mature egg, mainly in the animal hemisphere where later in development most of the nuclei of the embryo will appear (3, 4). After fertilization, these proteins are reaccumulated by the newly generated nuclei of the embryo.

Since the sum of the volumes of all nuclei during cleavage and early blastula stages is smaller than the volume of the original germinal vesicle, the nuclei of the early embryo cannot be expected to harbour all of the proteins inherited from the oocyte nucleus. It is not before midblastula that the total volumes of all cell nuclei of an embryo add up to that of one germinal vesicle (5).

Making use of monoclonal antibodies against germinal vesicle proteins (6, 7), we observed that not all nuclear antigens partition between the nuclear and the cytoplasmic compartments in the same way. During the first hours of development, at least two classes of nuclear antigens may be distinguished: Early migrating antigens, whose accumulation into nuclei starts upon pronucleus formation, and late migrating antigens that appear to be excluded from the nuclei during cleavage. Later, these antigens are shifted into the nuclei too, each at a characteristic stage of development (3, 8). Recently, comparable observations have been made on Drosophila embryos (9).

Nucleocytoplasmic Transport
Edited by R. Peters and M. Trendelenburg

In this report, the selectivity of nuclear antigen uptake at different developmental stages in vivo is compared to the selectivity of protein accumulation by nuclei that have been injected into unfertilized eggs. Moreover, an in vitro system is described in which the selective uptake of specific proteins can be studied in extracts from activated eggs.

The biological meaning and possible mechanisms of the well scheduled shift of different nuclear antigens from the cytoplasm to the nuclei at characteristic developmental stages are discussed.

Results

I. Observations in early development

As the pronuclei are formed after fertilization, they selectively accumulate certain proteins from the egg cytoplasm, whereas others are excluded. This has been studied by indirect immunofluorescent staining of embryo sections making use of monoclonal antibodies against germinal vesicle proteins (3, 8). Of twelve Xenopus germinal vesicle proteins studied, six start being accumulated by the pronuclei and are highly concentrated in the nuclei at subsequent stages. Nucleoplasmin, N_1, N_2 and N_4 fall into this class of early migrating nuclear antigens (table 1). The germinal vesicle derived lamin L_{III} is assembled into the lamina of pronuclei and cleavage nuclei (10). Of the other nuclear antigens, two are shifted into the nuclei at blastula stages, one during early gastrula, and three between late gastrula and neurula (table 1). The shift of late migrating proteins into the nuclei is a gradual process. For example, the first signs of accumulation of antigens b6-6E7 and 37-1A9 into interphase nuclei can be observed as early as after the sixth division, but the most striking change is seen between the twelfth division and stage 9, as cell cycle length is gradually extending (11, 12,). In several cases, uptake of a nuclear antigen seems to be complete in endodermal cells earlier than in the other germ layers (3, 8).

Table 1: Oocyte nuclear antigens in early development

mAB	antigen	MW (kDa)	IP	$\frac{N}{C}$ in oocyte[1)]	phase of transition from cytoplasm to nuclei of embryo
b7-1D1	nucleoplasmin	30	4.9-5.5	600	
b6-3B7	N_1[2)]	110	4.8	190	
b2-2B10	N_1[3)]	110	4.8	190	
b3-5E8	N_2	95	4.8	210	cleavage[5)]
b7-2H4	N_4	34	4.9	210	
32-4A1		33	5.3-5.6		
L6-8A7	lamin L_{III}[4)]	68	6.2		
37-1A9		80	5.5	50	blastula st 8-9[6)]
b6-6E7	nucleolar	86	5.6		
35-1A7		90	5.5		blastula to gastrula st 9-12
b7-1B4		68	6.4		gastrula st 10-12
37-1B2		68	6.6		gastrula to neurula st 12-14
32-5B6		46	6.3	15	gastrula to neurula st 12-14

1) Ratio of nuclear concentration to cytoplasmic concentration according to Paine (26).
2) Specific of X. borealis (3, 7). 3) Reacts with N_1 of X. laevis and with antigen from heterochromatin of X. borealis. 4) Stick and Hausen (10).
5) First signs of accumulation seen in pronuclei. 6) First signs of accumulation seen after 6th division.

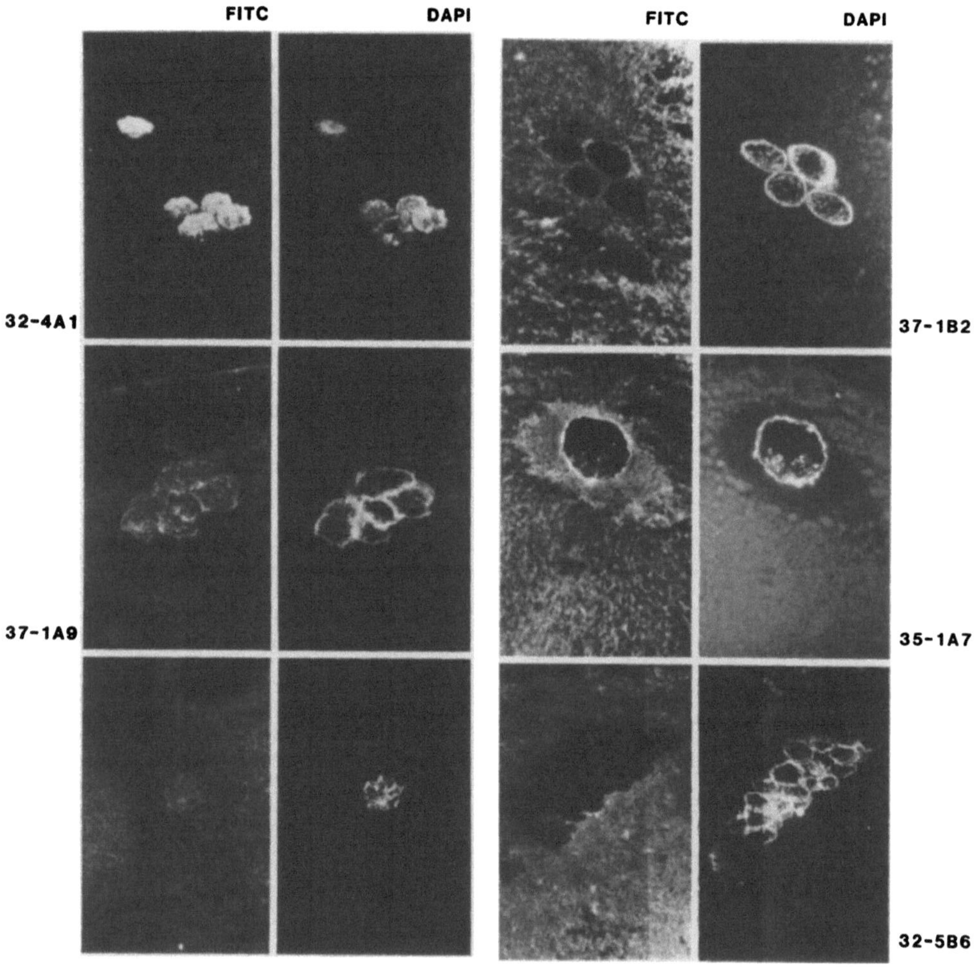

Figure 1: Sperm nuclei 2 hours after injection into unfertilized eggs
Sperm nuclei were isolated from *X. laevis* essentially as described by Lohka and Masui (15). Twenty nl of a suspension containing 10^7 nuclei per ml were injected into each unfertilized egg of *X. laevis*. After 2 h incubation specimens were fixed with TCA and embedded in polyester wax. Serial sections were stained with antibody as indicated and counterstained with DAPI.

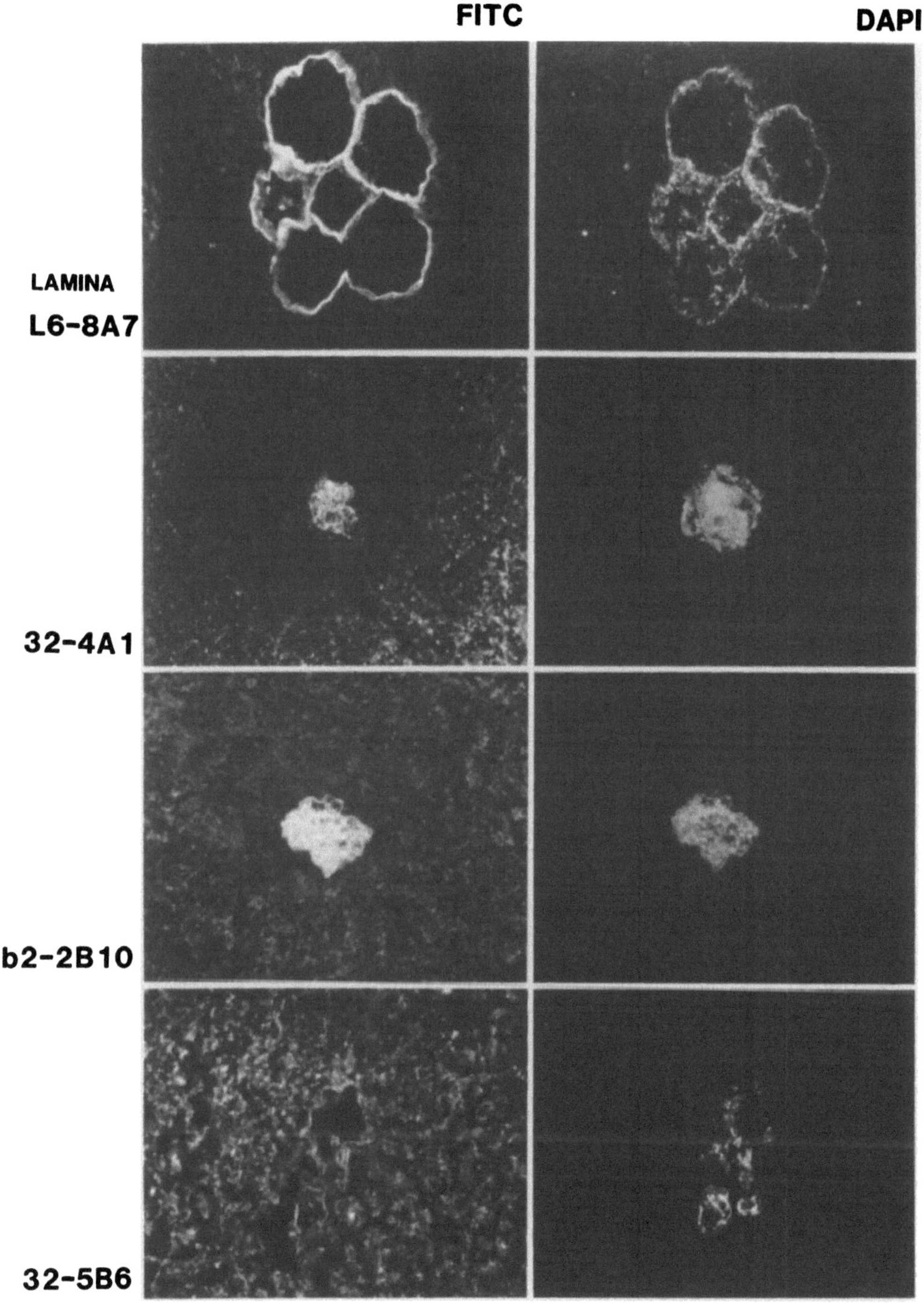

Figure 2: Erythrocyte nuclei after injection into unfertilized eggs
X. laevis erythrocyte nuclei were injected into unfertilized eggs and the specimens were processed for antibody staining as described in Fig.2.

The nucleolar antigen b6-6E7 is first accumulated into the nucleoplasm at blastula, before it is enriched in the newly generated nucleoli during gastrulation (data not shown). During mitosis, these antigens are shed into the cytoplasm (8).

II. Uptake of nuclear proteins into injected nuclei

When sperm nuclei are injected into an unfertilized egg of Xenopus, they round up and swell like pronuclei *in vivo* (13). We have assayed such additional "pronuclei" for their uptake of nuclear antigens. The injected sperm nuclei assemble a lamina and highly accumulate such antigens that are normally found in pronuclei and cleavage nuclei. Late migrating antigens are poorly taken up or even excluded (Fig. 1). Assembly of lamin L_{III} and accumulation of antigens b2-2B10 and 32-4A1 could be observed as early as 30 min after injection of the nuclei. Occasionally, some sperm nuclei did not change their shape and did not swell as usual. These did not assemble a nuclear lamina, nor did they accumulate any of the antigens tested (data not shown). If *Xenopus* erythrocyte nuclei are injected, they also swell and take up the same subset of nuclear antigens as the sperm nuclei (Fig. 2, table 2). Neither sperm nor erythrocyte nuclei contained any of these germinal vesicle antigens before swelling.

As was recently described, pronucleus-like bodies can also be formed after injection of pure DNA into unfertilized eggs (14). We have injected λ-DNA into unfertilized eggs and have then identified the DNA on fixed sections by staining with DAPI. Immunofluorescent staining of the same sections revealed that a nuclear lamina is assembled around the λ-DNA and that the resulting bodies accumulate the same complement of antigens that have been found in sperm-derived pronuclei whereas late migrating nuclear antigens are excluded (Fig. 3, table 2).

In conclusion, the unfertilized egg can provide all components to form multiple pronuclei from injected sperm nuclei or even naked DNA. The source of the nuclei and the sequence of the DNA appear to be irrelevant for the selective uptake of early migrating nuclear antigens.

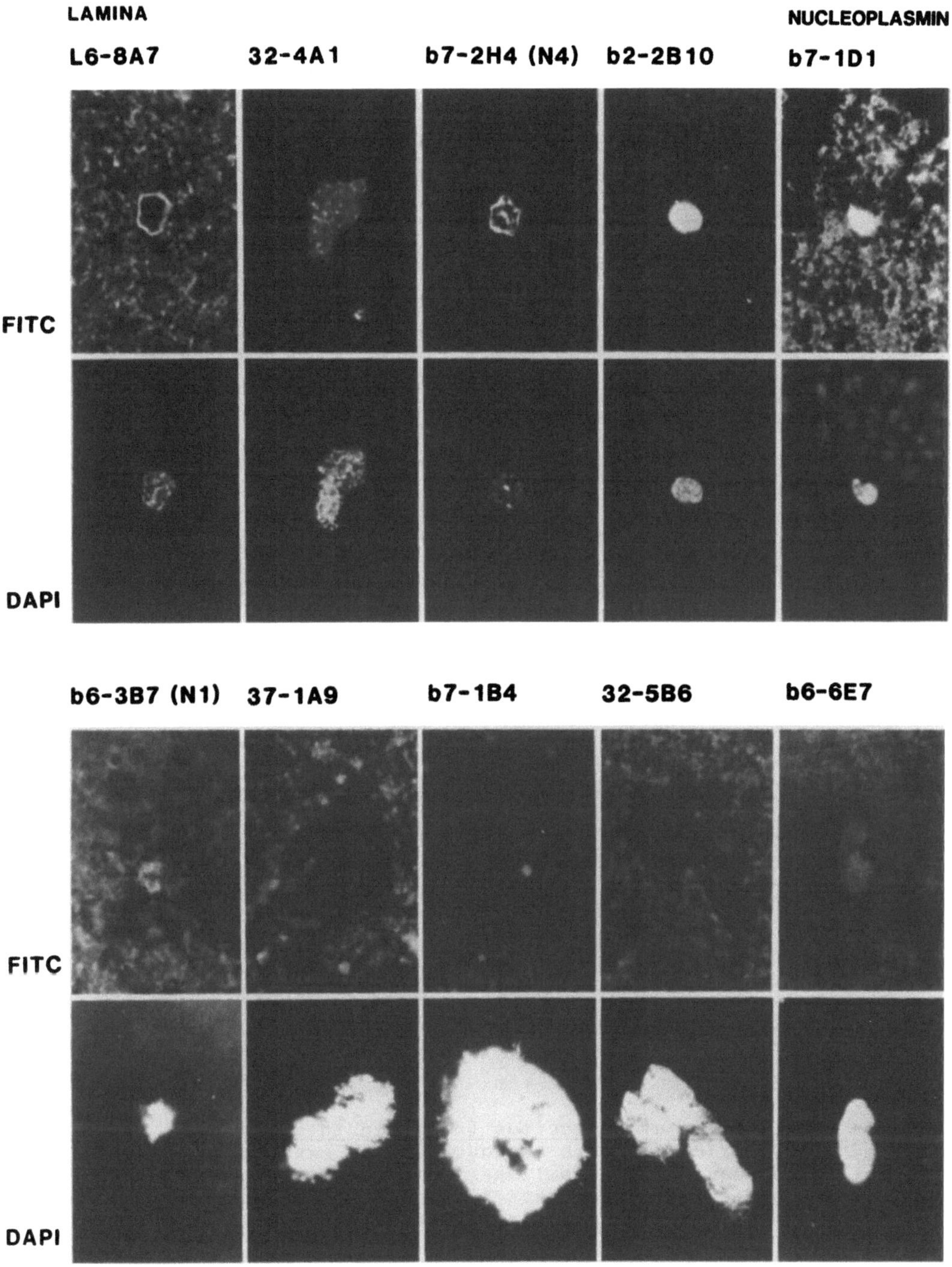

Figure 3: Pronuclei 2,5 hours after injection of λ-DNA into unfertilized eggs
Five to ten ng of λ-DNA were injected into each egg of *X. borealis*. Upper panels: antibody staining as indicated. Lower panels: DAPI counterstaining.

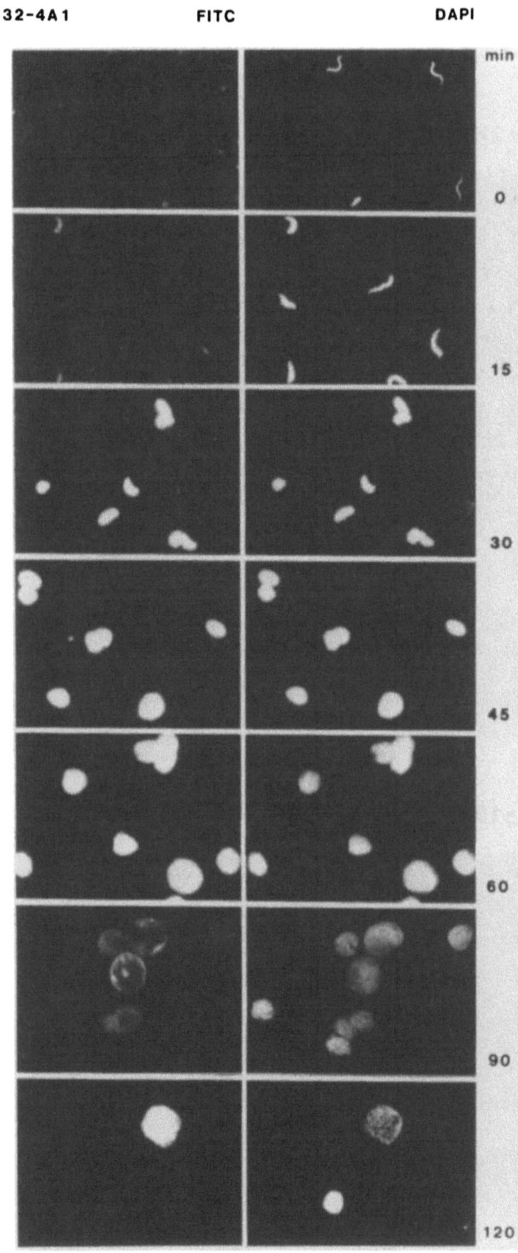

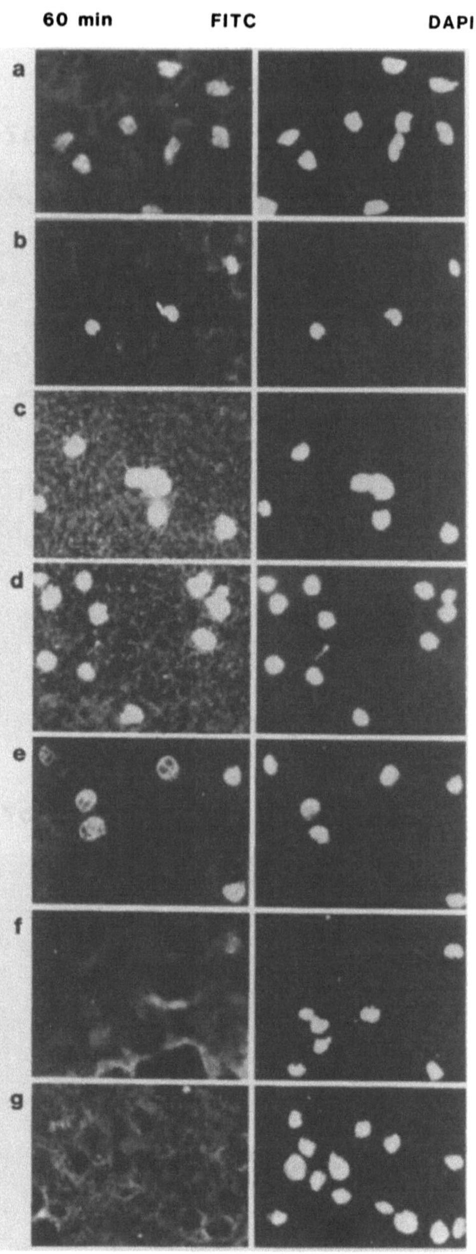

Figure 4: Kinetics of pronucleus formation in vitro
Sperm nuclei (1x10^6/ml) were incubated in an extract from electrically activated eggs. After the times indicated, samples were smeared on slides and air-dried. These were fixed with formaldehyde and stained with mAB 32-4A1 (left panels) and counterstained with DAPI (right panels).

Figure 5: Selective uptake of oocyte nuclear antigens by pronuclei in vitro
Sperm nuclei were incubated in activated egg extract as described in Fig. 5. After 60 min, samples were assayed for antigens b2-2B10 (a), L6-8A7 (b), , 32-4A1 (c), b7-1D1 (d), b7-2H4 (e), 37-1A9 (f), and 32-5B6 (g).

III. Accumulation of nuclear proteins in vitro

When sperm nuclei are incubated in a cell-free extract from electrically activated eggs, they round up and swell like pronuclei *in vivo* (15). Under favourable conditions, they even undergo chromosome condensation and mitosis *in vitro* (16).

Sperm nuclei reproducibly round up within the first 30 min of incubation in the extract. They then swell up to 1000 times their original volume (Fig. 4). In several experiments, chromosome condensation and nuclear envelope breakdown were seen after about 90 min. Thereafter new nuclei of heterogeneous sizes are formed (Fig. 4). The first signs of assembly of a nuclear lamina and of accumulation of early migrating proteins are seen after 15 min, while the sperm nuclei are still rod-shaped. The accumulation of some nucleoplasmic antigens seems to commence concomitant with or even before nuclear lamina assembly. Upon swelling of the nuclei, early migrating proteins are highly accumulated, whereas late migrating proteins are excluded (Fig. 5). Antigens 37-1A9, 35-1A7 and b6-6E7 that are shifted into the nuclei at blastula *in vivo*, are taken up to a moderate degree (Fig. 6, Table 2). When mitotic figures are observed, antigens are dissipated again, but newly accumulated thereafter (Fig. 4).

The nuclei formed *in vitro* could be isolated and probed for their antigen content by immunoblotting. There was good agreement between immunohistological and biochemical results: lamin L_{III}, nucleoplasmin, and N_4 are highly concentrated, 37-1A9 and 35-1A7 are moderately enriched, and 32-5B6 and 37-1B2 (and actin) are hardly found in the nuclear fraction. The observations *in vitro* have been variable with respect to N_1 and N_2, most probably because these proteins are prone to degradation in egg extracts.

In some experiments, erythrocyte nuclei have been used instead of sperm nuclei. They swell to a lesser degree, but their selectivity is comparable (Table 2).

In conclusion, the *in vitro* system described has very much the same selectivity of nuclear protein uptake as pronuclei or cleavage nuclei *in vivo*.

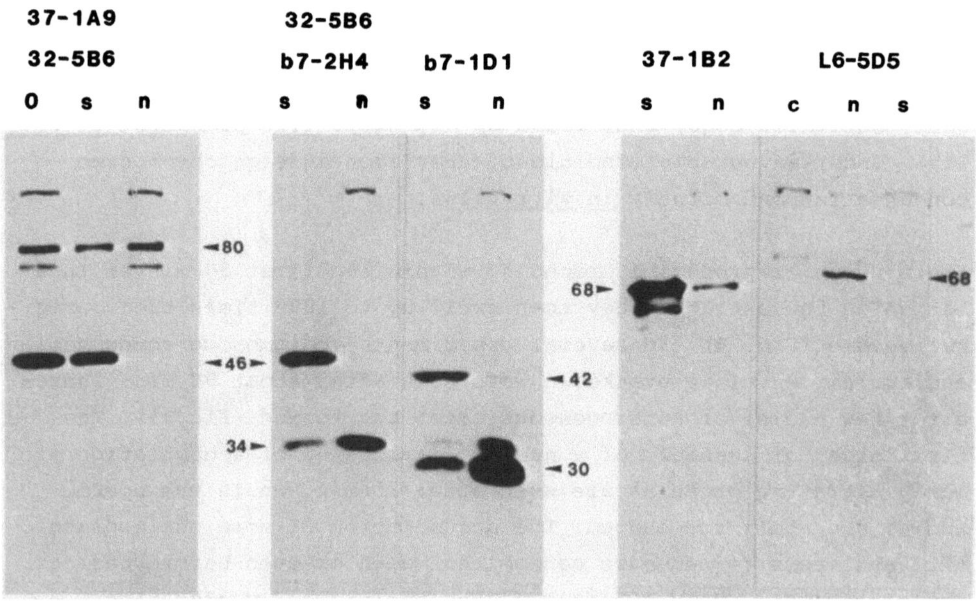

Figure 6: Pronuclei formed in vitro assayed by immunoblotting
After one hour of incubation in egg extract pronuclei were isolated as an interface over 60% Percoll. Equal amounts of protein from egg extract at time zero (0), nuclei (n), supernatant (s), or interface from controls without sperm nuclei (c) were separated on SDS-polyacrylamide gels and subjected to immunoblotting with mABs 32-5B6 (46 kD antigen), 37-1A9 (80 kD); b7-2H4 (34 kD), 37-1B2 (68 kD); b7-1D1 (nucleoplasmin, 30 kD, and actin, 42 kD), L6-5D5 (lamin L_{III}, 68 kD. The M_R of L_{III} is dependent on the ge system (10). All Antibodies were binding to two irrelevant protein bands of 145 and 75 kD.

Table 2: Uptake of oocyte nuclear antigens by artificial pronuclei

Antigen to mAB[1)]	Time of transition to nuclei in vivo[1)]	Accumulation by injected			Accumulation in egg extracts by	
		sperm nuclei	erythrocytes nuclei	λ-DNA	sperm nuclei	erythrocyte nuclei
b7-1D1	cleavage	n.d.	n.d.	+++	+++	+++
b2-2B10		+++	++	++	++/±[2)]	n.d.
b7-2H4		++	n.d.	++	++	++
32-4A1		+++	++	++	+++	++
L6-8A7		+++	+++	++	+++	++
37-1A9	blastula	+	n.d.	±	+	+
b6-6E7		n.d.	n.d.	+	+	n.d.
35-1A7	blastula to gastrula	+	+	n.d.	+	n.d.
b7-1B4	gastrula	±	-	±	±	-
37-1B2	gastrula to neurula	-	-	-	±	-
32-5B6	gastrula to neurula	-	-	-	-	-

1) See table 1. 2) Results variable; N_1 and N_2 are degraded in egg extract.

Discussion

The concept of signal sequences that are contained in nuclear proteins and are required for their guidance into the nuclear compartment (17, 18) has recently attained much support from sequence analysis and genetic engineering experiments (19-24).

In early development, the sequential shift of nuclear proteins from the cytoplasm to the nuclei might be achieved by covalent addition of such signal to a nuclear protein, before it can be reaccumulated into the nuclei. Several antigens have previously been analyzed on two-dimensional gels. However, no indication of structural changes with developmental time has been found for late migrating antigens (2). This is in agreement with observations of Dabauvalle and Franke, who showed that proteins translated *in vitro* can migrate into nuclei (25).

Alternatively, different nuclear proteins could contain different signal sequences that allow accumulation into nuclei with different kinetic parameters: The most potent (or multiple) signal sequences might be contained in early migrating proteins such as nucleoplasmin. Proteins with less efficient signal sequences may require a longer interphase period for the process of accumulation and may therefore not be shifted into the nuclei before the midblastula transition, when the cell cycle length increases (11, 12). This model could explain why complete localization of some antigens into the nuclei is observed earlier in the endoderm than in mesoderm and ectoderm, where more antigen is located and the cell cycles are shorter. Signals of different efficiencies could also explain why early migrating proteins tend to be most highly concentrated in the oocyte nucleus as compared to the cytoplasm (Table 1). Of the late migrating proteins, only two could be identified in the work of Paine, and these tend to partition to a greater extent between nucleus and cytoplasm of the oocyte (26).

To assess the role of kinetic parameters, we have tried to extend the cell cycle length at early blastula by injection of cycloheximide and of aphidicolin. No change of the selectivity was observed after 90 min (data not shown).

Artificial pronuclei *in vitro* could be kept for four hours and longer under conditions where nuclear swelling was retarded and maturation promoting factor was absent. Nevertheless late migrating nuclear antigens such as 32-5B6 were not accumulated. We therefore suppose that kinetic parameters alone cannot account for the observations in normal development.

Changes in selectivity might also be induced by an alteration of the nuclear structure with development. Several, probably gradual changes would have to be postulated to explain the observations made *in vivo*. The gradual replacement of the germinal vesicle specific lamin L_{III} by the somatic lamins L_I and L_{II} provides an example that has recently been studied in detail (10). Nevertheless, a causal relationship between selectivity and nuclear lamina composition is rather unlikely, since all the nucleoplasmic proteins studied are accumulated by the germinal vesicle, which does not contain lamins L_I and L_{II} (10).

Transport of late migrating nuclear proteins might be dependent on *de novo* synthesis of a protein or a nucleic acid that is not present at early stages but required for transport of certain proteins (18, 27). The *in vitro* system described here can be manipulated in its composition and might therefore be useful for studying this question.

If observations on embryos are compared with those on injected nuclei and on artificial pronuclei, the avidity with which different antigens migrate into nuclei is strikingly comparable *in vivo* and *in vitro* (Tables 1 and 2). *In vitro*, later migrating antigens are accumulated with falling tendencies the later in development they are shifted into the nuclei. For this selectivity the source of the nuclei or the sequence of their DNA seem to be irrelevant(Fig.3). Moreover, the three-dimensional architecture of the cell seems to be dispensible at least for the accumulation of early migrating antigens (Fig. 4 and 5).

Antigens accumulated during cleavage are believed to be required either for the architecture of the nuclei e.g. the lamin, or for DNA replication and chromosome formation. The latter is probably the case for nucleoplasmin and for N_1 and N_2, which are believed to complex histones and aid nucleosome assembly (28-30). Late migrating nuclear proteins are possibly required for nuclear events pertinent to gene activity after the midblastula transition.

We wish to thank Brigitte Gläser for expert technical assistance throughout this work, and Ulrike Gossweiler for excellent tissue culture work. Roswitha Groemke-Lutz is thanked for help with preparing the figures, and Brigitte Hieber and Ingrid Baxivanelis for carefully typing the manuscript.

References

1) Laskey, R.A., Gurdon, J.B. and Trendelenburg, M. (1979). In: "Maternal effects in development", D.R. Newth and M. Balls, eds. Symp. Brit. Soc. Develop. Biol. 4, pp. 65-80.

2) Dreyer, C. and Hausen P. (1983). Dev. Biol. 100, 412-425.

3) Dreyer, C., Wang, Y.H., Wedlich, D. and Hausen, P. (1983). In: "Current problems in germ cell differentiation", A. McLaren and C.C. Wylie, eds. Brit. Soc. Develop. Biol. Symp., pp. 329-351.

4) Hausen, P., Wang, Y.H., Dreyer, C. and Stick, R. (1985). J. Expl. Embryol. Morphol., in press.

5) Gerhart, J. (1980). In: "Biological regulation and development", Z. Goldberger, ed., Plenum Press, pp 133-316.

6) Dreyer, C., Singer, H. and Hausen, P. (1981). W. Roux's Arch. 190, 197-207.

7) Dreyer, C., Wang, Y.H. and Hausen P. (1985). Develop. Biol. 108, 210-219.

8) Dreyer, C., Scholz, E. and Hausen P. (1982). W. Roux's Arch. 191, 228-233.

9) Dequin, R., Saumweber, H., Sedat, J.W. (1984). Dev. Biol. 104, 37-48.

10) Stick, R. and Hausen P. (1985). Cell 41, 191-200.

11) Boterenbrood, E.C., Narraway, J.M. and Hara, K. (1983). W. Roux's Arch. 192, 216-221.

12) Newport, J. and Kirschner, M. (1982). Cell 30, 675686.

13) Graham, C.F. (1966). J. Cell Sci. 1, 363-374.

14) Forbes, D.J., Kirschner, M. and Newport, J. (1983). Cell 34, 13-23.

15) Lohka, M.J. and Masui, Y. (1983). Science 220, 719-721.

16) Lohka, M.J. and Maller, J.L. (1985). J. Cell Biol. 101, 518-523.

17) De Robertis, E.M., Longthorne, R.F. and Gurdon, J.B. (1978). Nature 272, 254-256.

18) De Robertis, E.M. (1983). Cell 32, 1021-1025.

19) Dingwall, C. Sharnick, S.V. and Laskey, R.A. (1982). Cell 30, 449-458.

20) Dingwall, D. and Allan, J. (1984). EMBO J. 3, 1933-1937.

21) Hall, M.N., Hereford, L. and Herskowitz, I. (1984). Cell 36, 1057-1065.

22) Lanford, R.E. and Butel, J.S. (1984). Cell 37, 801-813.

23) Kalderon, D., Richardson, W.D., Markham, A.F. and Smith, A.E. (1984). Nature 311, 33-38.

24) Bürglin, T. This volume.

25) Dabauvalle, M.C. and Franke, W.W. (1982). Proc. Nat. Acad. Sci. 79, 5302-5306.

26) Paine, P.L. (1982). In: "The nuclear envelope and the nuclear matrix", G.G. Maul, ed., New York, pp. 75-83.

27) Mattaj, I.W. and De Robertis, E.M. (1985). Cell 40, 111-118.

28) Mills, A.D., Laskey, R.A., Bead, P. and De Robertis, E.M. (1980). J. Mol. Biol. 139, 561-568.

29) Kleinschmidt, J.A. and Franke, W.W. (1982). Cell 29, 799-809.

30) Kleinschmidt, J.A., Fortkamp, E., Krohne, G., Zentgraf, H. and Franke, W.W. (1985). J. Biol. Chem. 260, 1166-1176.

Sequence Features of the Nucleoplasmin Tail Region and Evidence for a Selective Entry Mechanism for Transport into the Cell Nucleus

C. DINGWALL, T. R. BÜRGLIN[1], S. E. KEARSEY, S. DILWORTH, and R. A. LASKEY
CRC Molecular Embryology Group, Department of Zoology, University of Cambridge, Downing Street, Cambridge CB2 3EJ, United Kingdom
[1] Department of Cell Biology, Biozentrum, Klingelbergstrasse 70, 4056 Basel, Switzerland

Specific transport processes exist within the cell to ensure that molecules are directed to their correct subcellular or extracellular location. The transport of macromolecules into and out of the nucleus is arguably the most important of these processes because of its involvement in the regulation of gene activity.

Leaving aside the transport of macromolecules from the nucleus to the cytoplasm and considering only the entry of proteins into the nucleus, one might expect that other macromolecular transport processes such as protein secretion or the entry of proteins into chloroplasts and mitochondria would serve as a paradigm for nuclear protein uptake. However there is a major difference between these processes and the entry of proteins into the nucleus.

In general, proteins that are to be secreted or are to be incorporated into chloroplasts or mitochondria are synthesised as larger precursor polypeptides that bear an additional sequence of amino acids that acts as a "signal" to target the protein to its correct location. This "signal" is removed at some stage in the transport process to generate the mature protein which is now correctly located within an organelle or outside the cell (Kreil, 1981; Sabatini et al., 1981; Gething, 1985).

Nuclear proteins are not synthesised as precursors but contain within their mature structure all the information that is needed to ensure specific localization within the nucleus (Gurdon, 1970; Bonner, 1975; De Robertis et al., 1978). Therefore an examination and comparison of the structure of nuclear proteins should give some indication of the nature of the nuclear location "signal".

We have studied the selective accumulation of proteins in the nucleus using nucleoplasmin, an acidic protein which binds histones in vitro (Laskey et al., 1978; Earnshaw et al,. 1980) and in vivo (Kleinschmidt et al,. 1985).

Nucleocytoplasmic Transport
Edited by R. Peters and M. Trendelenburg

In this paper we shall review our studies of the transport of nucleoplasmin into the *Xenopus* oocyte nucleus which show that information for its accumulation in the nucleus is specified by a carboxyterminal domain, and that this domain specifies selective entry into the nucleus rather than selective retention after entry by diffusion (Dingwall *et al.*, 1982). We present a comparison of amino acid sequences of the "tail" region of nucleoplasmin with sequences from SV40 large T antigen (Kalderon *et al.*, 1984b; Smith *et al.*, 1985) and the yeast MAT α2 protein (Hall *et al.*, 1984) that have been shown to confer nuclear localisation upon non-nuclear proteins when linked to them at the gene level.

In addition we present calculations which indicate that the observed rate of nucleoplasmin accumulation in the nucleus is too rapid to be accounted for by simple diffusion.

The identification of a polypeptide domain that specifies entry into the nucleus.

Nucleoplasmin was first identified as a component of a cell-free chromatin assembly system produced from unfertilised eggs of *Xenopus laevis* (Laskey et. al., 1977; 1978). It is an acidic, pentameric, highly phosphorylated thermostable protein with subunits of 30,000Mr. It is the most abundant protein in the *Xenopus* oocyte nucleus where its concentration is of the order of 5 - 7 mg/ml. Like all other nuclear proteins studied to date it has the ability to accumulate rapidly in the nucleus after microinjection into the cytoplasm.

By the use of partial proteolysis of the nucleoplasmin pentamer we have been able to isolate a polypeptide domain that is both necessary and sufficient for entry into the nucleus (Dingwall *et al.*, 1982).

Digestion of nucleoplasmin with trypsin (or a variety of other proteases) generates a relatively protease resistant pentameric "core" molecule with subunits of 23,000 Mr. This core molecule, unlike intact nucleoplasmin is unable to enter the nucleus after microinjection into the cytoplasm (Fig. 1b). This result implies that the region of the subunits ("tail") removed by proteolysis contains a signal specifying nuclear entry. Digestion of the nucleoplasmin pentamer with pepsin cleaves each subunit at a specific site to liberate the "tail" region as an intact protein fragment. A partial pepsin digest of nucleoplasmin is therefore a mixture of partially cleaved pentamers and free, dissociated tail fragments. Microinjection of these digestion products into the cytoplasm of *Xenopus* oocytes reveals that the isolated "tail" fragment can accumulate rapidly in the nucleus (Fig. 1a) and at a rate comparable to intact nucleoplasmin (Dingwall *et al.*, 1982).These studies demonstrate that the ability of nucleoplasmin to accumulate in the nucleus is not a property of the whole molecule but is a property of a discrete polypeptide domain and this domain is both necessary and sufficient for nuclear entry. Additionally, partially proteolysed pentamers can enter the nucleus, even a pentamer which bears a single intact subunit (Fig. 1a).

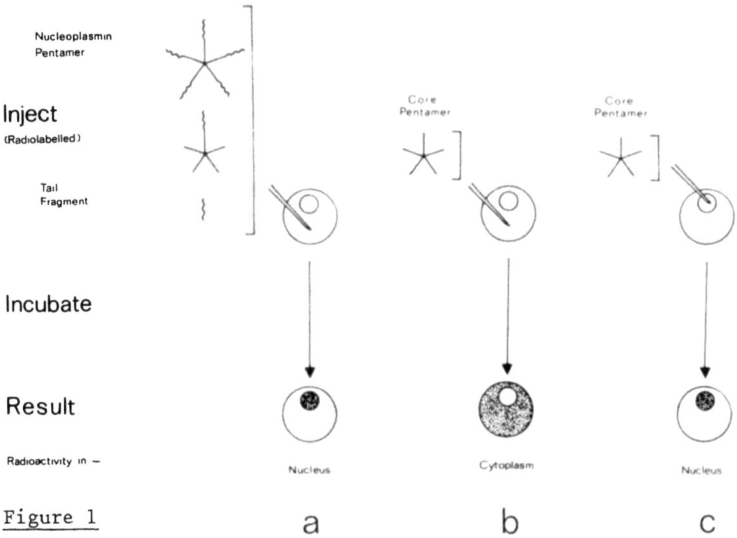

Figure 1

These data do not allow us to distinguish between the two possible mechanisms of nuclear accumulation namely selective entry of nuclear proteins or free diffusion followed by selective retention of nuclear proteins. Experimental results that favour selective entry are discussed further below.

Putative nuclear location signals in the nucleoplasmin tail region

We prepared a cDNA library from *Xenopus* oocyte mRNA in the expression vector λgt11 (Young and Davis, 1983a,b) We have isolated cDNA clones for nucleoplasmin from this libraray and seqeunced them using standard cloning and sequencing techniques (Messing 1983).

The "tail" region of nucleoplasmin as identified by proteolysis is probably about 100 amino acids long and lies at the carboxyterminus of the molecule (Dingwall et al,. 1982). Analysis of the cDNA clones reveals that the tail region has a polyglutamic acid tract containing approximately thirteen contiguous glutamic acid residues consistent with the known histone binding properties of nucleoplasmin (Earnshaw *et al.*, 1980). We have also examined the sequences of the "tail" region for sequences showing homology to postulated nuclear location signals that have been identified in other proteins.

Figure 2. Amino acid sequence of part of the carboxyterminal "tail" region of nucleoplasmin (deduced from the cDNA seqeunce).

```
                 1*                                10
... Glu(n) .... Ser Pro Pro Lys Ala Val Lys Arg Pro Ala Ala Thr Lys Lys
                                                - - - - - - - - -
                                                Sequence A

                                      20
                Ala Gly Gln Ala Lys Lys Lys Lys Leu Asp Lys Glu Asp Glu
                            ______________________
                            Sequence B

                    30                                      40
                Ser Ser Glu Glu Asp Ser Pro Thr Lys Lys Gly Lys Gly Ala
                                        ______________________
                                        Sequence C

                                                50
                Gly Arg Gly Arg Lys Pro Ala Ala Lys Lys - COOH
                                - - - - - - - -
                                Sequence D
```

* Residues are numbered from the serine residue at the end of the polyglutamic acid tract.

———— Sequences homologous to the SV40 nuclear location signal

— — — — Sequences homologous to the yeast nuclear location signal

In the case of SV40 large T antigen the mutation of a single lysine residue at position 128 abolishes the ability of the protein to become located in the nucleus (Lanford and Butel, 1984; Kalderon et al., 1984a). This lysine residue lies in a sequence of seven amino acids that have been shown to contain information for nuclear localisation by deletion mutagenesis (Kalderon et. al., 1984a). Also, this sequence has the ability to cause chicken pyruvate kinase to relocate to the cell nucleus after expression as a fusion protein (Kalderon et al., 1984b).

In yeast, E.coli β-galactosidase can be targeted to the nucleus when fused to the thirteen amino-terminal amino acids of the MAT α2 protein (Hall et al., 1984). Within this peptide is a sequence of four amino acids that is also present in a number of other nuclear proteins (e.g. MATα1, histone H2B) that Hall et al. (1984) suggest may be a nuclear location signal.

In the nucleoplasmin "tail" there are two regions showing homology to the SV40 nuclear location signal and two regions showing homology to the yeast nuclear location signal (Figure 2).

Figure 3. Comparison of sequences in the nucleoplasmin "tail" region with known nuclear location signals.

			128				
SV40 T antigen	Pro	Lys	Lys	Lys	Arg	Lys	Val
Tolerated as	Ser	Thr		Thr	Thr	Thr	Ile
single	Leu	Ile		Arg	Ile	Met	
substitutions	Ala			Met	Lys	Arg	
Nucleoplasmin	*	*	*	*	*		
Sequence B	Ala	Lys	Lys	Lys	Lys	Leu	Asp
	*	*	*	*		*	
Sequence C	Pro	Thr	Lys	Lys	Gly	Lys	Gly

* Denotes amino acids that match those in the wild type SV40 sequence or tolerated substitutions.

Yeast	3					
MAT α2 homology	Lys	Ile	Pro	Ile	Lys	
Nucleoplasmin Sequence A	Arg	Pro	Ala	Ala	Thr	Lys
Nucleoplasmin Sequence D	Lys	Pro	Ala	Ala	Lys	

Numbers denote the position of the relevant amino acid in the MAT α2 or SV40 large T antigen sequences.

The SV40 large T antigen sequence has been mutated at a number of locations (Smith et al., 1985) and the mutations that have no effect on nuclear location (tolerated mutations) are given in figure 3. This comparison reveals that in the nucleoplasmin "tail" region sequences labelled B and C bear a striking homology to the SV40 T antigen sequence, and the sequences labelled A and D show similar close homology to the yeast MATα2 sequence, having the common feature of a central core of hydrophobic residues one of which is proline flanked on either side by a lysine residue.

We have been unable to find any sequences homologous to the karyophilic sequence identified in influenza virus nucleoprotein (Davey et al., 1985).

This comparison reveals that the "tail" region of nucleoplasmin bears at least four regions that are strikingly homologous to postulated nuclear location signals and raises the possibility that this domain could be an extremely potent signal for nuclear entry, perhaps accounting for the ability of a single "tail" region to transport a partly cleaved nucleoplasmin pentamer (Mr 118,000) into the oocyte nucleus.

Nucleoplasmin entry into the nucleus is too rapid to be accounted for by free diffusion and binding

There is a considerable body of data (reviewed by Bonner, 1978) consistent with the idea that proteins can diffuse freely into the nucleus through pores in the nuclear envelope. The specific sequestration of nuclear proteins in the nucleus could be accounted for by their selective retention in the nucleus by binding to insoluble nuclear components (eg. the nuclear matrix).

The pores in the nuclear envelope have patent radii of approximately 45 Angsroms and exogenous (non-nuclear) proteins and dextrans with dimensions greater than this fail to enter the nucleus or do so only very slowly. This corresponds to a maximum molecular weight of 60,000 for proteins that can enter the nucleus by diffusion.

Nucleoplasmin (Mr 165,000) accumulates in the oocyte nucleus after microinjection into the cytoplasm much more rapidly than could apparently be accounted for on the basis of free diffusion (Dingwall et al., 1982).

In an attempt to address this point more systematically we have applied a mathematical analysis used by Feldherr et al., (1983) to the data for nucleoplasmin.

Nucleoplasmin appears as a disc in electron micrographs with a radius of approximately 37 Angstroms (Earnshaw et al., 1980). This measurement of radius is close to the radius of the largest dextran (35.5 Angstroms) used by Paine et al. (1975) to examine nuclear envelope permeability. Hence if we assume that nucleoplasmin behaves in solution as a sphere with a Stokes radius of 35.5 Angstroms, the mathematical analysis for the nuclear permeation by the dextran can be applied directly to the data for nucleoplasmin.

Table 1. Comparison of the observed and calculated rates of nucleoplasmin uptake into the *Xenopus* oocyte nucleus after microinjection into the cytoplasm.

		FASTEST OBSERVED UPTAKE		
		N/C obs. ($X_{n,c}$)	N/C calc.	calculated as % of observed
Time after injection (hours)	0.5	10	0.2	2
	1.5	13	0.58	4.5
	3.0	37	1.4	3.7
	($K_{n,c}$ = 139)			

		SLOWEST OBSERVED UPTAKE		
		N/C obs. ($X_{n,c}$)	N/C calc.	calculated as % of observed
Time after injection (hours)	0.5	1.0	0.039	0.1
	2.0	3.0	0.16	0.6
	4.0	6.2	0.32	1.1
	($K_{n,c}$ = 27)			

N/C. Nuclear to cytoplasmic concentration ratio assuming the nucleus represents 12% of the aqueous volume of the oocyte.

N/C (calculated) was determined using the formula

$$1 - \frac{Xn,c}{Kn,c} = e^{-k/t}$$

Where $X_{n,c}$ is the nuclear to cytoplasmic concentration ratio at time t.

$K_{n,c}$ is the nuclear to cytoplasmic concentration ratio at equilibrium (48 hours post injection, $K_{n,c}$ = 139 for the fastest observed rate and $K_{n,c}$ = 27 for the slowest observed rate).

k is the rate constant of nuclear permeation determined by plotting

$$\ln \; 1 - \frac{Xn,c}{Kn,c} \quad \text{against time}$$

For simple diffusion and no binding, and not taking water activity into account, the nuclear to cytoplasmic concentration ratio at equilibrium will be 1,

($K_{n,c}$ = 1). If nucleoplasmin diffused into the nucleus at the same rate as the 35.5 Angstrom dextran, then at 0.5, 1.5 and 3.5 hours after injection, the nuclear to cytoplasmic concentration ratios would be 0.0015, 0.004 and 0.01 respectively.

These values have been determined assuming that there is no binding in the nucleus, i.e. $K_{n,c}$= 1. With binding in the nucleus $K_{n,c}$ would be larger, hence if we assume that there is binding in the nucleus we can re-calculate the nuclear to cytoplasmic concentration ratios and compare these values with the observed values (Table 1). We have found great variability in the observed rates of nucleoplasmin accumulation between batches of oocytes, and therefore present the data for the fastest and slowest observed rates. Despite this variation, an examination of Table 1 indicates that the calculated accumulation based on the assumption of free diffusion and binding is no more than a few percent of the accumulation that was observed at these times.

These results are completely consistent with the findings of Feldherr et al., (1983) who applied the same mathematical analysis to their data for the uptake of the *Rana* oocyte nuclear protein R-N1 (Mr 148,000). They also concluded that the observed rate of accumulation of this protein following microinjection into the cytoplasm or the *in vivo* labelling of newly synthesised protein is too great to be accounted for by a simple "diffusion plus binding" mechanism.

In order to account for the observations one therefore needs to postulate the existence of a mediated transport process to transport nucleoplasmin and other large nuclear proteins rapidly into the nucleus.

One possibility is that the radiolabelled protein that is injected into the oocyte does not diffuse randomly throughout the cytoplasm in three dimensions but that some mechanism exists whereby the protein molecules are directed towards the nucleus or membranes or cytoskeletal elements. That is, a reduction in dimensionality of the diffusion process (Adam and Delbruck, 1968) would possibly bring about a substantial increase in the rate of arrival of molecules on the nuclear envelope and hence in the rate of accumulation.

The mathematical expression presented by Adam and Delbruck (1968) has been greatly simplified by Hardt (1980). Hardt defines a transit time or mean of the diffusion times for molecules arriving at the nucleus where

$$\text{Transit time in 2 dimensions} = \frac{L^2}{2D} \, . \ln\left(\frac{L}{a}\right)$$

$$\text{Transit time in 3 dimensions} = \frac{L^2}{2D} \, . \, \frac{2}{3} \, \frac{L}{a}$$

Where D = the Diffusion Coefficient

L = diameter of oocyte

a = diameter of nucleus

L^2 / 2D is a constant and so the increase in rate brought about by a reduction in dimensionality is represented by the ratio of

$$\frac{2}{3}\frac{L}{a} \quad \text{to} \quad \ln\left(\frac{L}{a}\right)$$

In the stage VI *Xenopus* oocyte L is approximately 900μ and a is approximately 150μ. Hence

$$\frac{2}{3}\frac{L}{a} = 4 \text{ (3 dimensions)}$$

$$\ln\left(\frac{L}{a}\right) = 1.8 \text{ (2 dimensions)}$$

giving a ratio of 2.2

Hence a reduction in dimensionality of the diffusion process from three to two would only increase the rate of accumulation by a factor of 2.2 Therefore it is not necessary to invoke targeting of molecules to the nuclear envelope as diffusion to the nucleus in three dimensions, even in a cell or large as the *Xenopus* oocyte, is probably not a rate limiting step. This is supported by the observations of Paine *et al*. (1975) who noted that the diffusion of the 35.5 Angstrom radius dextran throughout the oocyte cytoplasm is rapid but the nuclear envelope presents a permeability barrier to the dextran.

This indirect evidence for a selective entry mechanism is supported by more direct evidence obtained by microinjecting the protease resistant nucleoplasmin "core" molecule directly into the *Xenopus* oocyte nucleus.

If the tail region specifies selective retention of nucleoplasmin (binding) within the nucleus afetr entry by diffusion then the expected result would be that the injected core molecules would diffuse back into the cytoplasm as a function of time.

If, on the other hand, the tail region of the molecule specifies selective entry, then the core molecules might be expected to remain in the nucleus when placed there, since their failure to accumulate in the nucleus is caused only by the lack of the signal for entry.

The observed result was that core molecules remained in the nucleus (Fig. 1c and Dingwall et. al., 1982) and hence the tail must be acting as a signal that specifies selective entry rather than binding within the nucleus.

These data therefore strongly support the hypothesis that large nuclear proteins are accumulated in the nucleus by a transport process operating at the level of the nuclear envelope. Results from Feldherr's laboratory (Feldherr et al., 1984) show convincingly that the nuclear pore complex is involved in this process.

The ability of the amphibian oocyte nucleus to accumulate injected nucleoplasmin or N1 (Feldherr et al., 1983) against a substantial concentration gradient is by definition an "active transport" process but the energy requirement and the mechanistic details of the process are as yet unknown.

Discussion

The accumulated data reviewed here and that from other laboratories (Feldherr et al., 1983; Feldherr et al., 1984) make it possible for us to assert confidently that nucleoplasmin (and probably most large nuclear proteins) cannot accumulate in the nucleus by passive diffusion into the nucleus followed by retention by binding but that some mechanism of selective entry must be operating. It could involve active transport of the protein molecule through the pore complex with the concomitant hydrolysis of ATP. Alternatively it could involve a "facilitated diffusion" type of mechanism in which the nuclear protein interacts with the pore complex and causes it to increase in diameter thus permitting the protein molecule to enter by diffusion. Paine et al. (1975) noted that a 10 Angstrom change in pore orifice would effect a one-thousand-fold change in the rate of equilibration of large proteins (i.e. proteins with hydrodynamic radii greater than 31 Angstroms). Retention could then be the result of binding in the nucleus.

A selective mechanism for protein entry implies that some feature of the protein is recognised by the transportation machinery (i.e. nuclear pore complex). Such a feature would constitute a signal for nuclear entry. Sequences that are postulated to function as nuclear location signals have been identified in other nuclear proteins (see above) and the comparison of these amino acid sequences with the amino acid sequence of the nucleoplasmin "tail" indicates the presence of several regions of homology. It will be interesting to determine which of these homologies function in the transport of nucleoplasmin into the nucleus.

REFERENCES

Adam, G. and Delbruck, M. (1968). In Structural Chemistry and Molecular Biology. (Eds. A. Rich and N. Davidson) W.H. Freeman & Co. San Francisco pp 198-215.

Bonner, W.M. (1975). J. Cell Biol. 64, 431-437.

Bonner, W.M. (1978). In The Cell Nucleus 6 (Ed. H. Busch part C). pp 97-148.

Davey, J., Dimmock, N.J., and Colman, A. (9185) Cell 40, 667-675.

De Robertis, E.M., Longthorne, R.F., and Gurdon, J.B. (1978). Nature 272, 254-256.

Dingwall, C., Sharnick, S.V. and Laskey, R.A. (1982). Cell 30, 449-458.

Earnshaw, W.C., Honda, B.M., Laskey, R.A. and Thomas, J.O. (1980). Cell 21, 373-383.

Feldherr, C.M., Cohen, R.J. and Ogburn, J.A. (1983). J. Cell Biol. 96, 1486-1490.

Feldherr, C.M., Kallenbach, E. and Schultz, N. (1984). J. Cell Biol. 99, 2216-2222.

Gething, M-J. (1985). Current Communications in Molecular Biology. Protein Transport and Secretion. Cold Spring Harbor Laboratory.

Gurdon, J.B. (1970). Proc. Roy. Soc. Lond. B 176, 303-314.

Hall, M.N., Hereford, L., and Herskowitz, I. (1984). Cell 36, 1057-1065.

Hardt, S. (1980). In Mathematical Models in Molecular and Cell Biology. (Ed. L.E. Segel). Cambridge University Press pp 453-457.

Kalderon, D., Richardson, W.D., Markham, A.F. and Smith, A.E. (1984a). Nature 311, 33-38.

Kalderon, D., Roberts, B.L., Richardson, W.D. and Smith, A.E. (1984b). Cell 39, 499-509.

Kreil, G. (1981). Ann. Rev. Biochem. 50, 317-348.

Kleinschmidt, J.A., Fortkamp, E., Krohne. G.,Zentgraf. F., and Franke. W.W. (1985) J. Biol. Chem. 260, 1166-1176.

Lanford, R.E. and Butel, J.S. (1984). Cell 37, 801-813.

Laskey, R.A., Mills, A.D. and Morris, N.R. (1977). Cell 10, 237-243.

Laskey, R.A., Honda, B.M., Mills, A.D. and Finch, J.T. (1978). Nature 275, 416-420.

Messing, J. (1983) Methods in Enzymology 101, 20-78.

Paine, P.L., Moore, L.C. and Horowitz, S.B. (1975). Nature 254, 109-114.

Sabatini, D., Kreibach, G., Morimoto, T. and Adesmik, M. (1981). J. Cell Biol. 92, 1-22.

Smith, A.E., Kalderon, D., Roberts, B.L., Colledge, W.H., Edge, M., Gillet, P., Markham, A., Paucha, E. and Richardson, W.D. (1985). Proc. Roy. Soc. Lond. B. 226, 43-58.

Young, R.A. and Davis, R.W. (1983a). Proc. Nat. Acad. Sci. 80, 1194-1198.

Young, R.A. and Davis, R.W. (1983b). Science 222, 778-782.

Intracellular Transport of a Karyophilic Protein

B. SCHULZ and R. PETERS
Max-Planck-Institut für Biophysik, Kennedyallee 70, D-6000 Frankfurt 70, FRG

The term 'karyophilic' was suggested (Dabauvalle and Franke, 1982) for cellular proteins which accumulate in the nucleus without binding to chromatin or other nuclear structures. Absence of binding means that these proteins are recovered from the supernatant of centrifuged cell homogenates. The classification originated from studies of Xenopus oocytes (Bonner, 1975; ; Merriam and Hill, 1976; De Robertis et al., 1978). The exceptionally large size of these cells permits manipulations not easily applied to average-sized somatic cells. For instance, Xenopus oocytes can be easily loaded with macromolecules by microinjection, they can be fractioned by particularly fast and mild manual procedures, frequently the fractions of even single cells are large enough to be analyzed by centrifugation and gel electrophoresis. With these methods several karyophilic proteins of the Xenopus oocyte were characterized and designated as N1-N4 and nucleoplasmin. Karyophilic proteins are also present in oocytes from amphibians other than Xenopus (Dabauvalle and Franke, 1982). However, their occurance in somatic cells is largely unresolved.

Laskey et al. (1977) observed that the assembly of nucleosomes in vitro is facilitated by small amounts of Xenopus oocyte homogenate. This effect was shown (Laskey et al., 1978) to be mediated by a protein referred to as nucleoplasmin (Earnshaw et al., 1980). By now, nucleoplasmin is the best characterized karyophilic protein (for review, see Dingwall, 1985). It occurs as pentamer (Earnshaw et al., 1980) of 30-kD subunits (Laskey et al., 1978; Krohne and Franke, 1980a). By enzymatic digestion the monomer can be cleaved into a 'core' and a 'tail' (Dingwall et al., 1982). Recently, cDNA

Nucleocytoplasmic Transport
Edited by R. Peters and M. Trendelenburg

clones of nucleoplasmin have been obtained and sequenced (see, Dingwall et al. in this volume). Nucleoplasmin is the most abundant karyophilic protein of the Xenopus oocyte (Mills et al., 1980) and presumably (Laskey and Earnshaw, 1980) plays a major role in the organization of chromatin in vivo. Immunological studies (Krohne and Franke, 1980a,b) suggest that nucleoplasmin occurs also in oocytes of other amphibians and in a variety of tissues and cultured cells of other vertebrates.

In the past, nucleocytoplasmic transport of karyophilic proteins was studied in amphibian oocytes employing radiotracer methods and electron microscopy. Purified radiolabeled preparations of nucleoplasmin were injected into Xenopus oocytes (Dingwall et al., 1982). The subcellular distribution of injected material was assessed by autoradiography of cell sections and by quantitation of radioactivity in isolated nuclei and cytoplasms. For intact nucleoplasmin, a rapid accumulation in the nucleus was observed after cytoplasmic injection. Nucleocytoplasmic concentration ratios ($C_{c/n}$) of larger than 100 were established within 24 h after cytoplasmic injection. In striking contrast, the core pentamer of nucleoplasmin remained in the cytoplasm after cytoplasmic injection and in the nucleus after nuclear injection. Colloidal gold particles with diameters of 110 Å or larger not normally permeate the nuclear envelope. However, if coated with nucleoplasmin gold particles of that size rapidly permeated the nuclear envelope via the pore complex (Feldherr et al., 1984). Nuclear uptake of karyophilic proteins other than nucleoplasmin was also studied (Bonner, 1975; De Robertis et al., 1978; Feldherr et al., 1983) suggesting that similar mechanisms are involved.

We have studied mobility and transport of nucleoplasmin in somatic cells. Nucleoplasmin and its tryptic core were fluorescently labelled and injected into hepatoma tissue culture (HTC) cells and various other transformed and non-transformed mammalian cells. Employing fluorescence microphotolysis the diffusional mobility of nucleoplasmin was measured in both cytoplasm and nucleus. The kinetics of nuclear accumulation was determined by the same method. Analog studies were performed using a covalent conjugate of nucleoplasmin with the intensely fluorescent macromolecule phycoerythrin. This communication shortly summarizes results and implications. At first, a short introduction to fluorescence microphotolysis is given. An account of intracellular transport of nucleoplasmin follows. Finally, nucleoplasmin transport in oocytes and somatic cells will be compared and potential mechanisms of nuclear accumulation discussed.

FLUORESCENCE MICROPHOTOLYSIS: MEASUREMENT OF NUCLEOCYTOPLASMIC TRANSPORT AND INTRACELLULAR MOBILITY IN AVERAGE-SIZED CELLS

Fluorescence microphotolysis (FM) was conceived as a method for measuring translational diffusion in the plasma membrane of single cells (Peters et al., 1974). Since then, the scope of the method has broadened and now FM is used to measure, at the spatial resolution of the light microscope, translational and rotational molecular motion and membrane transport in a variety of experimental systems (for review, see Cherry, 1979; Peters, 1981; Axelrod, 1983; Ware, 1984; Garland and Johnson, 1985; Peters, 1985, 1986). With regard to intracellular transport it may be helpful i) that transport coefficients of defined molecular species can be measured in highly heterogeneous media because specificity is achieved by fluorescent labeling, ii) that fluorescent labelling implies high sensitivity and time resolution, and iii) that molecular transport can be quantitated with discrimination between different mobility classes.

The concept of the FM method is illustrated in Fig.1 which discriminates between diffusion and membrane transport. Although both processes can be measured by discontinuous (flash) and continuous photolysis the present discussion, for simplicity, is restricted to flash methods.

In order to measure molecular mobility (Fig.1a) a thin layer containing fluorescently labelled macromolecules is illuminated by a cylindrical, largely attenuated laser beam and fluorescence is measured. Then, the power of the laser beam is increased by several orders of magnitude for a few milliseconds to photolyse a substantial fraction of fluorophores contained in the illuminated volume. After photolysis fluorescence measurement is continued employing again the initial low beam power. Fresh fluorescently labelled macromolecules will now diffuse into the illuminated volume and (partially) restore fluorescence. Diffusion coefficients can be derived from the time course of fluorescence recovery. The procedure

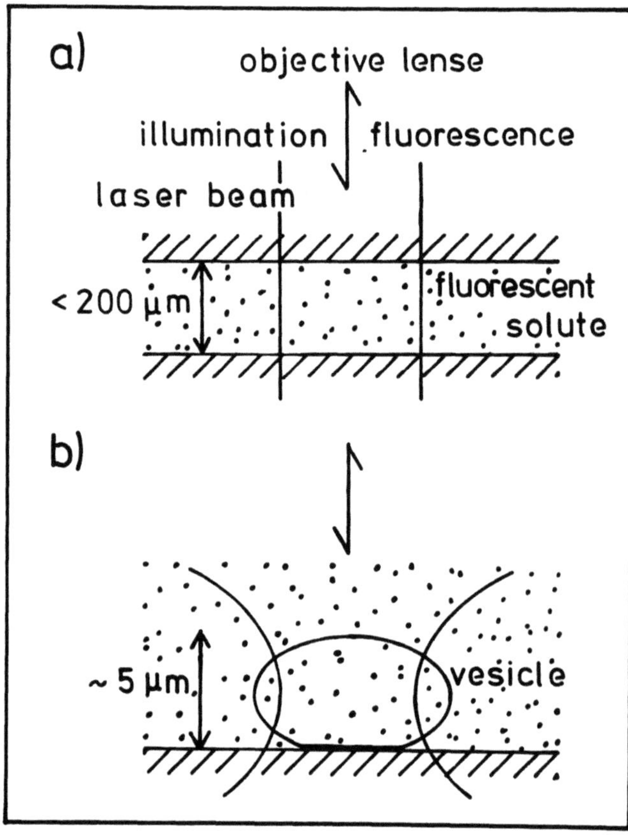

Fig.1 The concept of fluorescence microphotolysis in measurements of a) diffusion and b) membrane transport. For details, see text (from Peters, 1986).

can be scaled down to measure, for instance, diffusion in cytoplasm or nucleus of a single living cell. In that case the fluorescent macromolecule will be usually introduced into the cell by microinjection. The thicknes of the layer is not critical; at the extreme, it can consist of a monomolecular layer. Rotational motion has been measured with linearly polarized beams (Johnson and Garland, 1981) although, at the present stage of development, measurements are restricted to processes µm-s range.

In order to measure transport through membranes (Fig.1b) closed vesicular structures are equilibrated with a fluorescent solute. A single vesicle is illuminated by an attenuated laser beam covering most of the cross-section of the vesicle. Intravesicular fluorescence is measured with sufficient selectivity employing optical sectioning or other means. Then, photolysis and fluorescence measurement follow as in the case of diffusion. Usually, diffusion inside and outside the vesicle is much faster than membrane transport. In that case, fluorescence recovery will be only determined by transport of fluorescently labelled solute across the vesicle membrane and can be conveniently evaluated in terms of an influx rate constant or a permeability coefficient. The procedure can be used, for instance, to measure nucleocytoplasmic flux in living cells. It can be also used to measure membrane transport in isolated nuclei or nuclear envelope vesicles reconstituted in cell-free preparations (Kayne and Peters, unpublished results). Thus, transport properties of the nuclear envelope can be assessed and compared in various states of isolation and reconstitution.

Further methodological aspects including instrumentation, specific procedures for measuring diffusion or membrane tansport, data evaluation and potential artefacts are discussed elsewhere (Peters, 1986).

MOBILITY AND NUCLEOCYTOPLASMIC TRANSPORT OF NUCLEOPLASMIN IN SOMATIC MAMMALIAN CELLS

We isolated nucleoplasmin from Xenopus oocytes essentially as described by Dingwall et al. (1982). In short, 30-40 ml ovar was incubated with collagenase (0.58mg/ml) at room temperature for 2 h. Oocytes were then homogenized and centrifuged once at low speed and twice at high speed to obtain a cytosolic fraction. The cytosolic fraction was twice extracted with 1,1,2-trichlorofluoroethan. From this 'clarified supernatant' nucleoplasmin was purified by chromatography on DEAE-cellulose, phenyl-sepharose and, again, DEAE cellulose. The core pentamer of nucleoplasmin was prepared by digestion of isolated nucleoplasmin with trypsin as described by Dingwall et al. (1982) and Feldherr et al. (1984).

Intact nucleoplasmin and the tryptic nucleoplasmin core were labeled with fluorescein isothiocyanate (FITC) or with tetraethylrhodamin isothiocyanate (TRITC) as follows. Approximately 250 μg of purified nucleoplasmin, contained in 1 ml of a 50 mM carbonate buffer, pH 9.5, was incubated with FITC or TRITC (final concentration 0.4 mg/ml) at 20°C for 30 min. Labelling was stopped by titration to pH

7.0 and repetitive washing on an Amicon filter. In order to determine the degree of labeling small samples of labeled protein were run on 12% SDS gels. Both fluorescence and absorption of nucleoplasmin bands were quantitated by scanning densitometry and scans compared with those of known amounts of FITC-labelled bovine serum albumin. It was estimated that about two moles of FITC or TRITC were bound per mole of nucleoplasmin. In the case of the nucleoplasmin core the degree of labeling was substantially smaller. Other methods such as cell culture, formation of polykaryons by polyethylen-glycol-induced fusion, microinjection and FM measurements were essentially as described (Peters, 1984; Lang et al., 1986). Most frequently, HTC polykaryons were employed. But other mammalian cell lines were also studied (see below). The volume injected into any cell was about 5%-10% of cell volume. The concentration of nucleoplasmin in the injected solution was about 1 g/l yielding an average intracellular concentration of approximately 0.05-0.1 g/l. This corresponds to approximately 50-100 fg of nucleoplasmin per nucleus. The DNA content of an average somatic mammalian cell nucleus, and hence its content of chromosomal proteins, is about 4 pg. Thus, the injected amount of nucleoplasmin was small as compared to DNA and chromosomal proteins.

At 37 $^{\circ}$C fluorescently labelled nucleoplasmin rapidly accumulated in the nuclei of HTC polykaryons after cytoplasmic injection. The typical sequence of events is shown in the micrographs of Fig.2a-d which were obtained with an intesifying video system. After accumulation nuclei showed a perfectly homogeneous fluorescence. This makes an interesting constrast to the nuclear distribution of a viral nuclear protein, the large T antigen of Simian virus 40, visualized in fixed Vero cells by immunofluorescence: Large T was located exclusively in the nucleus; nucleoli, however, were devoid of fluorescence (Lanford and Butel, 1984; Kalderon et al., 1984). The nucleoplasmin core pentamer remained in the cytoplasm after cytoplasmic injection. No permeation into the nuclei was observed even with extended observation times (Fig.2e). Nuclear accumulation of intact nucleoplasmin and exclusion of the core pentamer from the nucleus was also observed with various other cells, namely primary rat hepatocytes, Vero cells, CV-1 cells, BSC cells, LLC-PK1 cells and T24 cells (a lung carcinoma line).

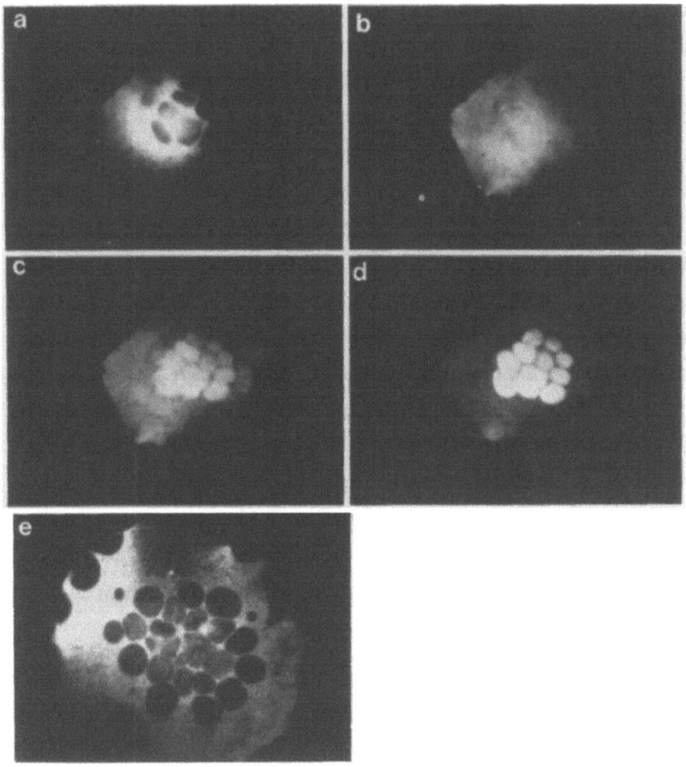

Fig.2 Nucleocytoplasmic transport of nucleoplasmin in polykaryons of HTC cells (37 °C). (a) - (d) Fluorescently labelled nucleoplasmin was injected into the cytoplasm of a HTC polykaryon and micrographs taken at 1min (a), 4 min (b), 10 min (c), and 30 min (d) after injection, respectively. (e) The fluorescently labelled trypsin core pentamer of nucleoplasmin in a HTC polykaryon photographed 30 min after cytoplasmic injection.

Nuclear accumulation was quantitated in HTC polykaryons. In single injected cells fluorescence was measured in a small field of the nuclear region and in a small field of a cytoplasmic region close to the nucleus. The ratio of these measurements constituted the N/C fluorescence ratio ($F_{n/c}$). At 37°C $F_{n/c}$ was about 0.5 immediately after injection. The ratio increased rapidly and, within about 4 min after injection, attained a maximum value of about 2.0 which was stable over an observation time of 30 min. At 10°C $F_{n/c}$ was about 0.5 immediately after injection and remained at about 1.0 for times up to 1 h. Apyrase, a highly active enzyme suited to deplete the cell from ATP pools, also prevented nuclear accumulation. If apyrase (0.3 U/µl) was coinjected with nucleoplasmin $F_{n/c}$ was ~0.4 immediately after injection and 0.8 at 8-45 min after injection (37 ^{o}C).

The concentration ratio $C_{n/c}$ can be derived from the fluorescence ratio $F_{n/c}$. Although a rigorous treatment is possible, we like to keep things simple by introducing a few approximations. We assume that fluorescence efficiency is about the same in cytoplasm and nucleus. We also assume that the height h of the cell, i.e. its vertical diameter, is approximately the same in the nuclear and perinuclear region. This contention is justified by previous experiments with small dextrans which freely permeate into the nucleus; their $F_{n/c}$ values are close to 1.0 (Lang et al., 1986). We recall that, at the employed optical conditions (40-fold, n.a. 0.75 objective lens), the laser beam traverses the cell as cylinder (Peters, 1984). The measured fluorescence is therefore proportional to pathlength times concentration. Then:

$$C_{n/c} = 1 + 1/H\ (F_{n/c} - 1) \qquad (1)$$

where H is the fraction of h taken by the nucleus. Immediately after cytoplasmic injection the concentration of nucleoplasmin in the nucleus is essentially zero and H = $F_{n/c}$ ~ 0.5. This value agrees with previous measurements employing large dextrans (Lang et al., 1986). For nucleoplasmin the maximum $F_{n/c}$ was approximately 2.0 which corresponds to a $C_{n/c}$ value of ~3.0.

The translational mobility of fluorescently labelled nucleoplasmin was measured in cytoplasm and nucleus , before and after nuclear accumulation and at different temperatures. In short, it was found that nucleoplasmin was mobile in both cytoplasm and nucleus and that mobility was not reduced by nuclear accumulation. For instance, at 37°C, 15-30 min after injection, the apparent diffusion coefficient amounted to 2.27+0.86 $\mu m^2/s$ in the cytoplasm and 1.85+0.80 $\mu m^2/s$ in the nucleus; the mobile fraction was 0.90+0.05 in the cytoplasm and 0.89+0.07 in the nucleus. Very similar values apply to the situation immediately after injection. The core pentamer was also mobile in both cytoplasm and nucleus.

In another series of experiments mobility and transport to the nucleus of a macromolecular complex consisting of nucleoplasmin and highly fluorescent phycoerythrin was studied. The pyridyldisulfide derivative of R-phycoerythrin (molecular mass 240 kD) was obtained from Molecular Probes (Junction City, Oregon, USA). Nucleoplasmin was reacted with succinimidyl 4-(p-maleimidylphenyl)butyrate and than coupled to the phycoerythrin derivative. The nucleoplasmin phycoerythrin conjugate was purified on a sephacryl S-300 column.

It was observed that phycoerythrin does not permeate into the nucleus after injection into the cytoplasm of HTC polykaryons. In contrast, the nucleoplasmin-phycoerythrin conjugate permeated and accumulated in the nucleus in a similar manner as fluorescently labelled nucleoplasmin. Both phycoerythrin and its nucleoplasmin conjugate were mobile in cytoplasm and nucleus with no effect of nuclear accumulation.

COMPARISON OF NUCLEOCYTOPLASMIC TRANSPORT IN SOMATIC CELLS AND OOCYTES

Our results establish that nucleoplasmin is recognized as 'karyophilic' protein not only by amphibian oocytes but a variety of somatic mammalian cells. On a qualitative level, nucleocytoplasmic transport of nucleoplasmin is quite similar in Xenopus oocytes and HTC cells. In both systems intact nucleoplasmin but not the core pentamer accumulates in the nucleus. Accumulation is temperature dependent in both systems. Also, large carriers such as colloidal gold particles or phycoerythrin which cannot permeate the nuclear envelope are be targeted to the nucleus by conjugation with nucleoplasmin.

Dingwall et al. (1982) observed by autoradiography that nucleoplasmin was homogeneously distributed in the nucleus of Xenopus oocytes. This argued that nucleoplasmin was probably mobile and not attached to structural elements. We have directly measured intracellular mobility in HTC cells and find that nucleoplasmin is indeed mobile. The mobile fraction of nucleoplasmin was about 90% in both nucleus and cytoplasm and did not depend on nuclear accumulation. The apparent intracellular diffusion coefficient of nucleoplasmin was close to 2 $\mu m^2/s$. This diffusion coefficient is similar to that of a dextran with a Stokes' radius of 91 Å and a molecular mass of 157 kD (Lang et al., 1986). The isolated and purified nucleoplasmin pentamer is smaller; it was visualized as a disc of 75 Å diameter by electron microscopy (Earnshaw et al., 1980). A dextran of that size has an intracellular diffusion coefficient of about 8 $\mu m^2/s$. This comparison suggests that, within the cell, nucleoplasmin engages into interactions with either itself or other cellular constitutents. Complexes of nucleoplasmin with histones have been described for Xenopus oocytes (Kleinschmidt et al., 1985) but not for somatic cells.

Quantitation of nuclear accumulation reveals large differences between oocytes and somatic cells. In the Xenopus oocyte full nuclear accumulation takes about 24 h and proceeds to a $C_{n/c}$ value of 100 and larger (Dingwall et al., 1982). In HTC cells nuclear accumulation is completed in 4 min and the maximum $C_{n/c}$ value is about 3. The time course of accumulation may be simply a matter of nuclear size. The influx rate constant k of a spherical vesicle with radius R is related to the permeability coefficient P by $k = (3/R) P$. Thus, for a given permeability coefficient the rate constant is inversely proportional to vesicle radius. The nuclear radii of Xenopus oocytes and HTC cells differ by a factor of about 100 (e.g. 300 μm/3 μm) which approximates the quoted difference in accumulation times. In the oocyte diffusion in cytoplasm and nucleus may somewhat delay nuclear accumulation. The permeation step proper is also fast in oocytes. Nucleoplasmin-coated gold particles, injected close to the nuclear surface, completely permeated the nuclear envelope within 15 min after injection (Feldherr et al., 1984). The extremely large difference in equilibrium $C_{n/c}$ values of oocytes and HTC cells appears puzzling at first sight. However, it may be recalled that Xenopus oocytes are highly specialized cells. Fertilization triggers a formidable succession of cell divisions which yield, without noticable S-phases, a progeny of about 30,000 cells in 9 h. To this end, the Xenopus oocyte stock-piles many types of nuclear and cytoplasmic components. It has, for instance, a pool of about 135 ng of histone (Woodland and Adamson, 1977) of which only a minor fraction (6 pg) is bound in nucleosomes. Somatic cells undergo cell division at much smaller rates. DNA, RNA and proteins are synthetized in S-phase during the cell cycle. Under these conditions the rate of chromatin assembly is much slower; stock-piling of histones and of other proteins probably involved in chromatin assembly is less important.

POTENTIAL MECHANISMS OF NUCLEOCYTOPLASMIC TRANSPORT

Nucleocytoplasmic transport is a process involving many parameters. Transport through the nuclear envelope is an important but not the only aspect. Other are, for instance, mobility, oligomerization, binding and differential solubility. These partial processes can be combined to yield a variety of potential mechanisms. In this communication we simply describe some plausible models and compare them with the experimental data.

In the simplest model of nucleocytoplasmic transport (Bonner, 1978) the nucleoplasmin pentamer would diffuse from cytoplasm to nucleus and accumulate there by binding to structural elements. Mobility measurements show that binding (if it exsists) must be weak. Otherwise, an immobile fraction would have become apparent. Also, nucleoplasmin (about 150 kD) probably cannot permeate the nuclear envelope by simple diffusion. For exogeneous compounds the nuclear envelope has properties of a molecular sieve characterized by a functional pore radius of about 50 A (Paine et al., 1975; Peters, 1984). The exclusion limit for proteins is close to 50 kD. Bovine serum albumin, for instance , is excluded (Lang et al., 1986).

Notably, the exclusion limit of the nuclear envelope is such that frequently subunits but not complete proteins can pass (Peters, 1986). Since nucleoplasmin is an oligomeric protein composed of relatively small subunits (30 kD) a model based on dynamic oligomerization can be proposed. It is conceivable although purely speculative that nucleoplasmin exsists in the cell in a monomer-oligomer equilibrium. If the nucleoplasmin monomer can freely move around in the cell its concentration should be about equal in cytoplasm and nucleus. The equilibrium distribution of total nucleoplasmin (monomers plus oligomers) would then depend on nuclear and cytoplasmic monomer-oligomer ratios. These might be influenced and actually regulated by subtle physical parameters such as ionic composition and/or chemical factors such as an phosphorylation. Interestingly, several isoelectric variants of nucleoplasmin were observed (Krohne and Franke, 1980a). In order to account for the properties of the nucleoplasmin core one may assume that tryptic cleavage strongly influences the monomer-oligomer equilibrium.

However, most experimental evidence suggests different mechanisms to be effective. Nuclear accumulation of nucleoplasmin depends strongly on temperature and on the availability of ATP. Intact nucleoplasmin but not the nucleoplasmin core can target large carriers such as colloidal gold particles (Feldherr et al., 1984) or phycoerythrin to the nucleus. In a number of viral and yeast proteins nuclear targeting is specified by short sequences of amino acid residues (Lanford and Butel, 1984; Kalderon et al., 1984, Hall et al., 1984; Silver et al., 1984; Davey et al., 1985; Moreland et al., 1985; Richarson et al., 1986; for review, see Smith et al., 1985). The mechanisms by which such nuclear location sequences exert their functions are unknown. A review of the literature suggests (Peters, 1986) that different categories of location sequences exsist. Certain sequences may mediate binding or oligomerisation in the nucleus. Other sequences obviously function as a 'signal' permitting transfer through the pore complex. Sequences which are homologous to the nuclear location sequences of viral and yeast proteins are present in nucleoplasmin (see the article by Dingwall et al. in this volume). Karyophilic proteins which are not composed of subunits but a single large polypeptide chaine accumulate in the nucleus in a manner similar to nucleoplasmin (Bonner, 1975; De Robertis et al., 1978; Dabauvalle and Franke, 1982). In the case of the karyophilic N1-protein from Rana pipiens (single peptide chaine of 160 kD) the kinetics of nuclear uptake were measured and found incompatible with simple diffusion through functional pores (Feldherr et al., 1983). These data suggest that nucleoplasmin, other karyophilic proteins and certain nuclear proteins of viral origin permeate the nuclear envelope by specific mechanisms. These are discussed elsewhere (Peters, 1986).

ACKNOWLEDGEMENT

Support by the Deutsche Forschungsgemeinschaft is gratefully acknowledged.

REFERENCES

Axelrod,D. (1983) J. Membrane Biol. 75, 1-10

Bonner,W.M. (1975) J. Cell Biol. 64, 431-437

Bonner,W.M. (1978) Cell Nucleus 6, 97-148

Cherry,R.J. (1979) Biochim. Biophys. Acta 559, 289-327

Dabauvalle,M.-C. and Franke,W.W. (1982) Proc. Natl. Acad. Sci. USA 79, 5302-5306

Davey,J., Dimmock,N.J. and Colman,A. (1985) Cell 40, 667-675
De Robertis,E., Longthorne,R.F. and Gurdon,J.B. (1978) Nature (Lond.) 272, 254-256
Dingwall,C. (1985) Trends Biol. Sci. 6, 64-66
Dingwall,C., Sharnick,S.V. and Laskey,R.A. (1982) Cell 30, 449-458
Earnshaw,W.C., Honda,B.M. and Laskey,R.A. (1980) Cell 21, 373-383
Feldherr,C.M., Cohen,R.J. and Ogburn,J.A. (1983)J. Cell Biol. 96, 1486-1490
Feldherr,C.M., Kallenbach,E. and Schulz,N. (1984) J. Cell Biol. 99, 2216-2222
Garland,P. and Johnson,P. (1985) in Enzymes of Biological Membranes (Martonosi,A.N., ed.), second edit., vol.1, pp. 421-439, Plenum Press, New York
Hall,M.N., Hereford,L. and Herskowitz,I. (1984) Cell 36, 1057-1065
Kalderon,D., Richardson,W.D., Markham,A.F. and Smith,A.E. (1984) Nature (Lond.) 311, 33-38
Kleinschmidt,J.A., Fortkamp,E., Krohne,G., Zentgraf,H. and Franke,W.W. (1985) J. Biol. Chem. 260, 1166-1176
Krohne,G. and Franke,W.W. (1980a) Proc. Natl. Acad. Sci. USA 77, 1034-1038
Krohne,G. and Franke,W.W. (1980b) Exp. Cell res. 129, 167-189
Lanford,R. and Butel,J.S. (1984) Cell 37, 801-813
Lang,I., Scholz,M. and Peters,R. (1986) J. Cell Biol., in press
Laskey,R.A., Mills,A.D. and Morris,N.R. (1977) Cell 10, 237-243
Laskey,R.A., Honda,B.M., Mills,A.D. and Finch,J.T. (1978) Nature (Lond.) 275, 416-420
Laskey,R.A. and Earnshaw,W.E. (1980) Nature (Lond.) 286, 763-767
Merriam,R.W. and Hill,R.J. (1976) J. Cell Biol. 69, 656-668
Mills,A.D., Laskey,R.A., Black,P. and De Robertis,E.M. (1980) J. Mol. Biol. 139, 561-568
Moreland,R.B., Nam,H.G., Hereford,L.M. and Howard,M.F. (1985) Proc. Natl. Acad. Sci. USA 82, 6561-6565
Paine,P.L., Moore,L.C. and Horowitz,S.B. (1975) Nature (Lond.) 254, 109-114
Peters,R. (1985) Trends Biochem. Sci. 6, 223-227
Peters,R. (1985) EMBO J. 3, 1831-1836
Peters,R. (1986) Biochim. Biophys. Acta, submitted
Peters,R., Peters,J., Tews,K.H. and Bahr,W. (1981) Biochim. Biophys. Acta 367, 282-294
Richarson,W.D., Roberts,B.L. and Smith,A.E. (1986) Cell 44, 77-85
Silver,P.A., Keegan,L.P. and Ptashne,M. (1984) Proc. Natl. Acad. Sci. USA 81, 5951-5955

Smith,A.E., Kalderon,D., Roberts,B.L., Colledge,W.H., Edge,M., Gillett,P. Markham,A. Paucha,E. and Richardson,W.D. (1985) Proc. Roy. Soc. Lond. B 226, 43-58

Ware,B.R. (1984) Am. Labor. 16, 16-28

Woodland,H.R. and Adamson,E.D. (1977) Dev. Biol. 57, 118-135

Amino Acid Sequences Able to Confer Nuclear Location

B. L. Roberts[1], W. D. Richardson[2], D. D. Kalderon[3], S. H. Cheng[1], W. Markland[1], and A. E. Smith[1]

[1]Protein Engineering Laboratory, Integrated Genetics, Framingham, MA 01701, USA

[2]Department of Zoology, University College, Gower Street, London WC1E 6BT, United Kingdom

[3]Department of Biochemistry, University of California, Berkeley, CA 94720, USA

ABSTRACT

The putative nuclear location signals of SV40 VP2, polyoma virus Large-T and adenovirus 72K DNA binding protein have been identified and characterized. The data suggest that nuclear location signals with characteristics similar to the SV40 Large-T prototype may be a general feature of nuclear proteins and several such signals in a given protein can exert co-operative effects. The nuclear location signal of SV40 Large-T has been inserted into various sites within the structure of chicken muscle pyruvate kinase to determine whether the signal can function when introduced into different structural environments within the same protein. We conclude that putative nuclear location signals may be found within any part of the primary structure of a protein, however, they probably need to be exposed at the surface of the protein in order to function.

INTRODUCTION

The nuclear and cytoplasmic compartments of the eukaryotic cell contain largely distinct sets of proteins which are separated by the nuclear envelope (Bonner, 1975; De Robertis et al., 1978). While it is generally assumed that nuclear proteins are synthesized in the cytoplasm from whence they migrate to the nucleus, it is not understood how proteins are partitioned between the two compartments (see Dingwall, 1985 for a review).

Nucleocytoplasmic Transport
Edited by R. Peters and M. Trendelenburg

Microinjection studies have indicated that purified proteins introduced into the cytoplasm of eukaryotic cells can enter the nucleus at a rate governed only by their size (Paine and Feldherr, 1972). It has been proposed that the nuclear envelope behaves as an array of cylindrical pores of internal diameter 7-11 nm (Paine et al., 1975; Peters, 1984). This model agrees well with the dimensions of proteinaceous channels called nuclear pores which have been purified from the nuclear envelope and analyzed by electron microscopy (Unwin and Milligan, 1982).

While the segregation of many nuclear and cytoplasmic proteins can be accounted for on the basis of their size and shape, some means of entry to the nucleus other than diffusion must exist for large proteins. Microinjection studies have shown that purified nuclear proteins have the ability to accumulate rapidly in the nucleus following introduction into the cytoplasm (Bonner, 1975; De Robertis et al., 1978; Dingwall et al., 1982). It has therefore been proposed that nuclear proteins contain inherent information in the form of a signal sequence which specifies nuclear localization (De Robertis, 1983). This is best exemplified by the work of Dingwall et al. (1982) who showed that nuclear localization of the pentameric protein nucleoplasmin (M_r 115,000) is dependent on the presence of at least one copy of a M_r 12,000 carboxy-terminal domain.

The most extensively characterized nuclear location signal described to date is that of SV40 Large-T. The sequence Pro-Lys-Lys(128)-Lys-Arg-Lys-Val is required for the nuclear localization of this protein (Kalderon et al., 1984a) and in addition the sequence is sufficient to bring about the nuclear localization of pyruvate kinase, an otherwise cytoplasmic protein (Kalderon et al., 1984b).

To test whether the sequence Pro-Lys-Lys-Lys-Arg-Lys-Val of SV40 Large-T represents a prototypic nuclear location signal, we have isolated and tested putative nuclear location signals of other known nuclear proteins. In addition, we have inserted the nuclear location signal of SV40 Large-T into various sites within the structure of pyruvate kinase to determine whether the signal can function when introduced into different structural environments within the same protein. The data suggest that nuclear location signals with characteristics similar to the SV40 prototype may be a general feature of nuclear proteins and that several such signals in a given protein can exert co-operative effects. We conclude that putative nuclear location signals may be found within any part of the primary structure of a protein, however, they almost certainly may need to be exposed at the surface of the protein to function.

NUCLEAR LOCATION SIGNALS IN OTHER PROTEINS

(a) SV40 Capsid Proteins

Previously we showed, using deletion and point mutants, that the minimal sequence of the nuclear location signal of SV40 Large-T is Pro-Lys-Lys-Lys-Arg-Lys-Val (Kalderon et al., 1984a; Kalderon et al., 1984b). A computer search revealed that while only the closely related Large-T antigen of BK virus contains exactly the same sequence, many other known nuclear proteins contain similar sequences (Smith et al., 1985 and Table 1). Here we describe experiments designed to identify nuclear location signals of other proteins.

The late region of the SV40 genome codes for the 3 capsid proteins VP1, VP2, and VP3. In virus-infected cells, these proteins rapidly accumulate in the nucleus following their synthesis in the cytoplasm (Kasamatsu and Nehorayan, 1979). We sought to identify the putative nuclear location signals of VP1 and VP2.

SV40 VP2 contains the sequence Pro-Asn-Lys-Lys(319)-Lys-Arg-Lys-Leu proximal to its carboxy terminus (Tooze, 1981). A DNA fragment encoding this sequence was isolated and, following addition of suitable linkers, it was fused to the 5'-end of the chicken muscle pyruvate kinase gene (PK) (Kalderon et al., 1984b). The resultant plasmid pVP2-PK encoded amino acids 312 to 325 of SV40 VP2 fused to the amino terminus of pyruvate kinase. Plasmid DNA was injected into the nucleus of Vero (African green monkey) kidney cells and approximately 16 hours later the fusion protein was visualized by indirect immunofluorescence as described previously (Kalderon et al., 1984b). The fusion protein was located exclusively in the nucleus of microinjected cells as judged by immunofluorescence. Thus, the sequence of amino acids between residues 312 and 325 of SV40 VP2 can suffice to bring about the nuclear localization of pyruvate kinase.

The coding sequences of SV40 VP1 and VP2 overlap in this region of the genome. Thus the putative nuclear location signal of VP2 can be translated in a different reading frame to code for the basic amino terminal sequence of VP1 (Smith et al, 1985). Hence, the same DNA fragment used to construct pVP2-PK was fused to the 5'-end of the pyruvate kinase gene in an alternate reading frame. The resultant plasmid pVP1-PK encoded amino acids 1 to 11 of SV40 VP1 fused to the amino terminus of pyruvate kinase. By the microinjection assay, the fusion protein encoded by pVP1-PK was found to be localized exclusively to the cytoplasm. A second tract of basic amino acids appears at residues 16 to 19 of SV40 VP1 (Smith et al., 1985). We considered the possibility that the putative nuclear location signal of VP1 might encompass a larger proportion of the entire protein. Thus, we generated a plasmid pVP1-PKL which encoded the amino terminal 42 residues of VP1 fused to the amino terminus of pyruvate kinase. In this instance, the fusion protein was located in the cytoplasm of the majority of microinjected cells, however, bright nuclear fluorescence was observed in a small percentage of cells. We are presently investigating whether this amino terminal portion of VP1 contains an inefficient nuclear location signal or whether it is able to target pyruvate kinase to the nucleus by some other means.

(b) Adenovirus 72K DNA binding protein.

In an analogous study, we isolated a small DNA fragment encoding amino acids 81 to 95 of the 72K DNA binding protein of adenovirus type 5 which corresponds to a basic amino acid tract similar to the SV40 nuclear location signal (Table 1). Following the addition of appropriate linkers, the fragment was fused to the 5'-end of the pyruvate kinase gene. The fusion protein encoded by the resultant plasmid p72K-PK was localized exclusively to the nucleus as judged by microinjection assays.

Thus, we have identified two additional nuclear location signals from two different nuclear proteins. These signals are similar in sequence to the SV40 Large-T prototype. Current investigations are underway to ascertain whether these sequences fulfill a similar role in the proteins from which they were derived.

(c) Polyoma virus Large-T antigen

The nuclear Large-T antigen of polyoma virus bears considerable sequence homology to SV40 Large-T (Tooze, 1981). To identify the nuclear location signal of this protein, DNA fragments encoding amino terminal portions of polyoma virus Large-T were fused to the 5'-end of the pyruvate kinase gene (Richardson et al., submitted for publication). The subcellular distribution of these fusion proteins was determined by the microinjection assay. A fusion protein containing the first 295 amino acids of polyoma Large-T appended to pyruvate kinase was localized to the nucleus. Surprisingly, a fusion protein containing only the first 243

SV40 LT	128	P	P	K	K	K	R	K	V
BK LT	130	P	P	K	K	K	R	K	V
BPV E1	106	P	V	K	R	R	K	S	G
Adeno 72K DBP	87	P	K	K	K	K	K	R	P
Murine PY LT	282	P	P	K	K	A	R	E	D
	192	V	S	R	K	R	P	R	P
Hamster PY LT		T	P	K	K	P	P	P	T
		S	S	R	K	R	K	R	V
SV40 VP2	319	P	N	K	K	K	R	K	L
Murine PY VP2	315	P	Q	K	K	K	R	R	L
Hamster PY VP2		P	K	K	K	K	R	R	F
SV40 VP1	1	M	A	P	T	K	R	K	G
Murine PY VP1	1	M	A	P	K	R	K	S	G

TABLE 1 - Selected amino acid sequences which were identified on the basis of a shared homology to the SV40 Large-T prototype nuclear location signal (Smith et al., 1985) are listed here. In addition, homologous sequences present in the putative Large-T and VP2 proteins of a recently characterized hamster papovavirus (Delmas et al., 1985) are shown here for comparison. Since the open reading frames of hamster papovavirus Large-T and VP2 have not been definitively established, amino acid residue numbers have been omitted. Note that murine polyoma virus Large-T and hamster papovavirus Large-T contain 2 putative nuclear location signals each. Adeno 72K DBP refers to the 72K DNA binding protein of adenovirus type 5.

amino acids of polyoma Large-T, and consequently lacking the region of the antigen which bears direct homology to the SV40 Large-T nuclear location signal (Table 1) (Smith et al., 1985), was located both in the nucleus and cytoplasm of microinjected cells. The amino terminal 243 residues of polyoma Large-T were examined for the presence of an additional basic amino acid sequence which could act as a nuclear location signal. A second basic sequence, centered on amino acid 192, was identified (Table 1), and a protein consisting of only the first 185 residues of polyoma Large-T fused to pyruvate kinase was located solely in the cytoplasm.

A DNA fragment encoding amino acids 183 to 195 of polyoma Large-T was isolated and fused to the 5'-end of the pyruvate kinase gene in order to determine whether this sequence would target pyruvate kinase to the nucleus. This fusion protein was found to be located entirely in the nucleus of micro-injected cells. However, a DNA fragment encoding amino acids 279 to 290 of polyoma Large-T was not able to impart a nuclear location on pyruvate kinase.

Deletion mutants of polyoma Large-T were constructed such that either one or both of the basic amino acid tracts were removed. Mutant Large-T proteins which lacked either one of these putative signal sequences were located in both the nuclear and cytoplasmic compartments of microinjected cells. However, a Large-T protein which lacked both basic tracts was confined to the cytoplasm. The nuclear localization of mutant polyoma Large-T proteins was not perceptably altered by the introduction of deletions which flanked but did not impinge upon the basic amino acid tracts. This data suggests that two mutually independent sequence elements contribute to the nuclear localization of polyoma Large-T.

(d) Positional Effect of Nuclear Location Signals

The observations described above suggest that putative nuclear location signals may be found within any portion of the primary structure of a known nuclear protein. In the cases of SV40 and polyoma Large-T antigens, we have two examples of instances where nuclear location signals can function when present at internal sites. However, it is not clear whether nuclear location signals need to be integrated into a specific region of a protein in order to ensure that they can function.

To test this relationship directly, the SV40 Large-T nuclear location signal was inserted into various sites within the primary structure of

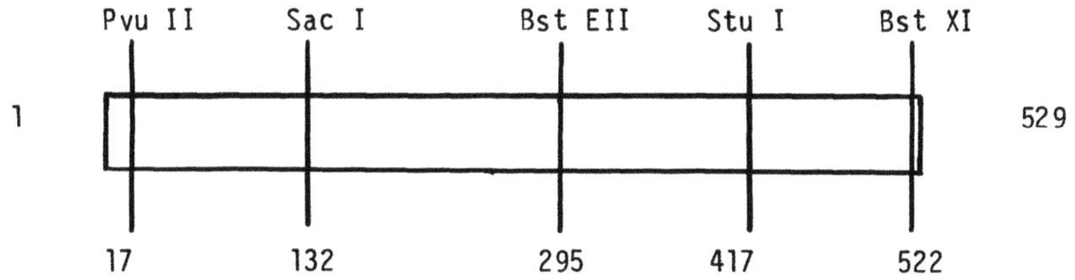

	SUBCELLULAR LOCATION	
Site	WT	d10
Pvu II	N	C
Sac I	C	C
Bst EII	N	C
Stu I	N	C
Bst XI	N	C

FIGURE 1 - A schematic diagram of the chicken muscle pyruvate kinase cDNA is shown here indicating the unique restriction enzyme sites that were selected for insertion of the nuclear location signal. The approximate amino acid residues which correspond to these restriction enzyme sites are shown. SV40 Large-T nuclear location signals containing either lysine (WT) or threonine (d10) at the 128 equivalent position of the sequence were inserted into the various sites. The fusion proteins were judged to be either nuclear (N) or cytoplasmic (C) by indirect immunofluorescence.

chicken muscle pyruvate kinase (Lonberg and Gilbert, 1983) and the subcellular distribution of the fusion proteins was determined by microinjection.

Plasmid RL142PK8F (Richardson et al., submitted for publication) consists of a chicken muscle pyruvate kinase cDNA placed under the control of the SV40 early promoter and early transcription termination signals.

Either Eco RI or Xho I linkers of three different lengths were inserted into unique restriction sites within the pyruvate kinase gene. Hence by the combination of appropriate DNA fragments, nuclear location signals can be inserted within the pyruvate kinase gene at defined locations.

A Pvu II and a Bst XI site were selected for linker insertion in order to permit the addition of nuclear location signals to the amino or carboxy terminii of pyruvate kinase respectively (Figure 1). Three internal sites were also selected: the Sac I site centered on codon 132, the Bst EII site centered on condon 295 and, lastly, the Stu I site centered on codon 417. A DNA fragment encoding the minimal nuclear location signal sequence of SV40 Large-T bounded on the 5'-end by an Xho I linker and on the 3'-end by an Eco RI linker was inserted into each of these restriction enzyme sites. As a control, a defective nuclear location signal containing Thr at the Lys-128 equivalent position of the signal (Kalderon et al., 1984b) was also inserted into the various sites within the pyruvate kinase gene.

When the wild type nuclear location signal was added to either the amino or carboxy terminus of pyruvate kinase, the resultant fusion proteins were localized in the nucleus (Figure 1). Addition of the defective nuclear location signal to either end of the pyruvate kinase gene did not discernably alter the normal cytoplasmic distribution of the protein. Similarly, when the nuclear location signal was inserted into internal sites of the protein corresponding to either the Stu I or Bst EII sites of the pyruvate kinase gene, the fusion proteins were located in the nucleus of microinjected cells. The substitution of a defective nuclear location signal for the wild type counterpart abolished the observed nuclear accumulation and fusion proteins were found to reside only in the cytoplasmic compartment as judged by immunofluorescence.

In contrast to these observations, a wild type nuclear location signal inserted into the Sac I site of the pyruvate kinase gene was unable to direct the encoded fusion protein to the nucleus of microinjected cells and, as expected, the defective nuclear location signal did not alter the normal cytoplasmic location of pyruvate kinase.

DISCUSSION

Identification of Putative Nuclear Location Signals

The data presented here suggests that nuclear location signals similar to that of SV40 Large-T may be a general feature of at least some nuclear proteins. The 39K capsid protein VP2 of SV40 virus contains a putative nuclear location signal near its carboxy terminus while the 72K DNA binding protein of adenovirus type 5 contains a similar signal proximal to its amino terminus. In spite of the fact that these putative signals differ in sequence (Table 1) and are derived from different proteins, both function to target pyruvate kinase to the nucleus. Current investigations using the techniques of deletion and site-directed mutagenesis are underway to ascertain whether these sequences fulfill a similar role in the proteins from which they were derived.

Our results show that the amino terminal 42 residues, but not the amino terminal 11 residues, of SV40 VP1 are sufficient to target pyruvate kinase to the nucleus at least in some cells. This is in contrast to the results obtained with the putative nuclear location signals of SV40 VP2 and adenovirus 72K DNA binding protein which can target pyruvate kinase to the nucleus in all microinjected cells when fused to the 5'-end of the pyruvate kinase gene. This suggests that the fusion protein encoded by the plasmid pVP1-PKL accumulates in the nucleus of microinjected cells via a mechanism which is less efficient than that involved in the nuclear accumulation of the fusion proteins encoded by plasmids pVP2-PK and p72K-PK. Possibly, the amino terminal 42 amino acids of VP1 enable the fusion protein encoded by pVP1-PKL to be retained in the nucleus once it has gained access to the nuclear space following transient dissolution of the nuclear envelope during mitosis.

The putative nuclear location signal of SV40 VP2 is highly conserved in the VP2 capsid proteins of both murine polyoma virus and hamster papovavirus (Table 1). The basic amino acid sequence contained within the amino-terminal 11 residues of SV40 VP1 is conserved in murine polyoma VP1, however, it is not present in hamster papovavirus VP1. This suggests that this sequence is not absolutely required for the nuclear localization of VP1 capsid proteins. SV40 VP1 may contain other, as yet, unidentified signal sequences which are responsible for its accumulation in nuclei. Alternatively, SV40 VP1 may gain access to the nucleus principally by passive diffusion and its accumulation in the nucleus may be facilitated by its ability to bind to fixed nuclear components. Present studies are underway to ascertain whether the amino terminal 11 residues of SV40 VP1 contribute towards the nuclear accumulation of the native capsid protein.

Multiple Nuclear Location Signals

The Large-T antigen of polyoma virus appears to contain two signal sequences which co-operate together to confer a nuclear location on this protein (Richardson et al., submitted). The deletion of either one of these sequences appears to impair the ability of the mutant protein to accumulate in nuclei while the deletion of both sequences results in the generation of a mutant which is totally unable to accumulate in nuclei. Of these two sequences, only the basic sequence centered on Lys-192 (Table 1) was able to target pyruvate kinase to the nucleus.

Our experiences with polyoma Large-T serve to illustrate a few features which may be characteristic of many nuclear location sequences. Firstly, nuclear proteins may contain more than one region which is involved in nuclear localization. These sequences may act independently as appears to be the case with polyoma Large-T or, alternatively, they may be interdependent constituting part of a larger structure or domain specifying a nuclear location. Why some nuclear proteins should require two co-operating nuclear location sequences, while only a single sequence of eight amino acids can suffice to bring about the nuclear localization of SV40 Large-T, is unclear. In the case of polyoma Large-T, the nuclear location signals are encoded in a region of the Large-T gene which overlaps with middle-T coding sequence. Hence, it may be difficult to reconcile the coding requirements of a single effective nuclear location signal in the Large-T reading frame with the requirements for a functional middle-T protein in an alternative frame. Polyoma Large-T may have circumvented this problem by evolving two less effective sequences which together fulfill the same task.

Secondly, sequences which confer nuclear location in the context of a given nuclear protein may not exhibit similar characteristics when isolated and fused to the coding sequence of a cytoplasmic protein such as pyruvate kinase. This appears to be the case with the sequence centered on Lys-282 of polyoma Large-T (Table 1). It is possible that the sequences which flank this basic tract in Large-T enhance its properties by forcing it to adopt a specific configuration which is lacking in the pyruvate kinase fusion protein. As a corollary to this observation, it may be possible to identify putative nuclear location signals on the basis of their ability to target pyruvate kinase to the nucleus, however, such sequences may fail to fulfill a similar role in the context of the proteins from which they were derived. Hence, basic amino acid sequences such as those found in the cytoplasmic core proteins of cauliflower mosiac virus and Southern bean mosaic virus and which bear homology to the SV40 nuclear location signal (Smith et al., 1985) may act as nuclear location signals only when transposed onto other proteins.

Sequence Requirements for Nuclear Localization

The studies described here have shown that, while sequences which appear to be homologous to the SV40 prototype signal do not necessarily target pyruvate kinase to the nucleus, other basic sequences which bear little apparent homology to the SV40 prototype can mimic its action. This is best exemplified by the sequence contained within the amino terminal 11 amino acids of SV40 VP1 and the sequence centered on Lys-192 of Polyoma Large-T, respectively (Table 1). These observations are in keeping with results obtained from a mutational analysis of the SV40 Large-T signal which illustrate that while Lys-128 is of crucial importance, numerous single amino acid substitutions can be accomodated by this sequence without significantly altering its ability to function (Smith et al., 1985). Perhaps the sequences described here which act as nuclear location signals share an underlying structural trait and may participate in the same mechanism of nuclear accumulation. As more nuclear location signals are identified, a consensus sequence may emerge and may provide insight as to the nature of this underlying feature.

Signal Sequences Inserted into Pyruvate Kinase

The sequence Pro-Lys-Lys-Lys-Arg-Lys-Val-Glu-Asp-Pro of SV40 Large-T can act as a nuclear location signal when inserted into various sites within the primary structure of pyruvate kinase.

In contrast, proteins destined to enter the lumen of the endoplasmic reticulum or mitochondria contain signal sequences which are generally appended to their amino terminii and translocation is accompanied by a proteolytic processing event (Blobel, 1980). No such proteolytic processing event coincident with nuclear envelope translocation has been described for any nuclear protein.

One might expect that in order for a signal sequence to be recognized as such, it must be exposed on the surface of a protein. Since the amino and carboxy terminal ends of the pyruvate kinase polypeptide are exposed (Lonberg and Gilbert, 1985), it is not surprising that the SV40 Large-T nuclear location signal can direct pyruvate kinase to the nucleus when added at these positions. However, the signal sequence could only function when inserted into 2 of the 3 selected internal sites of pyruvate kinase (Figure 1). It is interesting to note that a hydrophilicity plot of the pyruvate kinase structure predicts that both the Bst EII and Stu I sites of the pyruvate kinase gene correspond to strongly hydrophilic regions of the primary structure of the protein. In addition, if one can extrapolate from the crystallographic data obtained for cat muscle pyruvate kinase, the Bst EII and Stu I sites of the gene are predicted to correspond to exposed portions of the polypeptide chain (Lonberg and Gilbert, 1985). In contrast, the hydrophilicity plot predicts that the Sac I site of the pyruvate kinase gene corresponds to a hydrophobic region within the primary structure and by comparison to the three dimensional structure of cat muscle pyruvate kinase, the Sac I site of the gene is predicted to correspond to a portion of the polypeptide chain which is buried within the protein (Lonberg and Gilbert, 1985). These observations support the contention (Smith et al., 1985) that nuclear location signals may need to be situated within exposed regions of proteins in order to be recognized.

Mechanisms of Nuclear Accumulation

In the present study, we have not attempted to discern the mechanism(s) involved in the nuclear accumulation of proteins containing sequences analogous to that of the SV40 Large-T nuclear location signal. Possible mechanisms include unidirectional transport across the nuclear envelope, passive entry into nuclei followed by binding to fixed nuclear components or protection against degradation in the nucleus. In the case of SV40 Large-T, the failure of cytoplasmic variants of this protein to accumulate in nuclei cannot be attributed to a failure to bind to SV40 origin DNA (Paucha et al., 1985a) while conversely, mutants which are unable to bind to SV40 origin DNA can nevertheless accumulate in nuclei (Paucha et al., 1985b).

On the other hand, it is known that nucleoplasmin enters nuclei via nuclear pores (Feldherr et al., 1984) and this translocation event is absolutely dependent on the presence of a M_r 12,000 carboxy terminal domain (Dingwall et al., 1982). It is not known whether SV40 Large-T also gains access to the nucleus via nuclear pores nor is it proven that this translocation event is mediated by the nuclear location signal.

Many questions remain to be answered concerning the mechanisms of action of nuclear location signals. The development of an *in vitro* system would greatly facilitate these studies. The nuclear and cytoplasmic variants of SV40 Large-T and pyruvate kinase described here and elsewhere (Kalderon et al., 1984b; Smith et al., 1985) should greatly assist in the characterization of such *in vitro* systems.

References

1. Blobel, G. (1980). Intracellular protein topogenesis. Proc. Nat. Acad. Sci. USA 77, 1496-1500.

2. Bonner, W.M. (1975). Protein migration into nuclei. II. Frog oocyte nuclei accumulate a class of microinjected oocyte nuclear proteins and exclude a class of microinjected oocyte cytoplasmic proteins. J. Cell Biol 64, 431-437.

3. Delmas, V., Bastein, C., Sheneck, S. and Feuteun, J. (1985). A new member of the polyomavirus family:the hamster papovavirus. Complete nucleotide sequence and transformation properties. EMBO J. 4, 1279-1286.

4. De Robertis, E.M., Longthorne, R.F. and Gurdon, J.B. (1978). Intracellular migration of nuclear proteins in Xenopus oocytes. Nature 272, 254-256.

5. De Robertis, E.M. (1983). Nucleocytoplasmic segregation of proteins and RNAs. Cell 32, 1021-1025.

6. Dingwall, C. (1985). The accumulation of proteins in the nucleus. TIBS 10, 64-66.

7. Dingwall, C., Sharnick, S.V. and Laskey, R.A. (1982). A polypeptide domain that specifies migration of nucleoplasmin into the nucleus. Cell 30, 449-458.

8. Feldheer, C.M., Kallenback, E. and Schultz, N. (1984). Movement of a karyophilic protein through the nuclear pores of oocytes. J. Cell Biol. 99, 2216-2222.

9. Kalderon, D., Richardson, W.D., Markham, A.F. and Smith, A.E. (1984a). Sequence requirements for nuclear localisation of SV40 Large-T antigen. Nature 311, 33-38.

10. Kalderon, D., Roberts, B.L., Richardson, W.D. and Smith, A.E. (1984b). A short amino acid sequence able to specify nuclear location. Cell 39, 499-509.

11. Kasamatsu, H. and Nehorayan, A. (1979). Intracellular localization of viral polypeptides during Simian Virus 40 infection. Journal of Virology 32, 648-660.

12. Lonberg, N. and Gilbert, W. (1983). Primary structure of chicken muscle pyruvate kinase mRNA. Proc. Natl. Acad. Sci. USA 80, 3661-3665.

13. Lonberg, N. and Gilbert, W. (1985). Intron/Exon Structure of the chicken pyruvate kinase gene. Cell 40, 81-90.

14. Paine, P.L. and Feldherr, C.M. (1972). Nucleocytoplasmic exchange of macromolecules. Exp. Cell Res. 74, 81-98.

15. Paine, P.L., Moore, L.C. and Horowitz, S.B. (1975). Nuclear envelope permeability. Nature 254, 109-114.

16. Paucha, E., Kalderon, D., Richardson, W.D., Harvey, R.W. and Smith, A.E. (1985a). The abnormal location of cytoplasmic SV40 Large-T is not caused by failure to bind to DNA or to p53. EMBO J. (in press).

17. Paucha, E., Kalderon, D. and Smith, A.E. (1985b). The SV40 origin DNA binding domain on Large-T antigen. J. Virology (in press).

18. Peters, R. (1984) Nucleocytoplasmic flux and intracellular mobility in single hepatocytes measured by fluorescence microphotolysis. EMBO J. 3, 1831-1836.

19. Richardson, W.D., Roberts, B.L. and Smith, A.E. Two independent nuclear location signals in polyoma virus Large-T (submitted for publication).

20. Smith, A.E., Kalderon, D., Roberts, B.L., Colledge, W.D., Edge, M., Gillett, P., Markham, A.F., Paucha, E. and Richardson, W.D. (1985). The nuclear location signal. Proc. Royal Soc. Lond. B 226, 43-58.

21. Tooze, J. (1981). Molecular biology of tumor viruses. Part 2. DNA tumor viruses, revised edn. 2. (Cold Spring Harbor New York:Cold Spring Harbor Laboratory.

22. Unwin, P.N.T. and Milligan, R.A. (1982). A large particle associated with the perimeter of the nuclear pore complex. J. Cell Biol. 93, 63-75.

Distinct Domains of Simian Virus 40 Large T Antigen Mediate Its Association with Different Cellular Targets

W. DEPPERT and M. STAUFENBIEL[1]
Department of Biochemistry, University of Ulm, D-7900 Ulm, FRG
[1]Present Address: Max-Planck-Institut für Zellbiologie, D-6802 Ladenburg, FRG

Introduction

The large tumor antigen encoded by simian virus 40 (SV40 large T) is a multifunctional protein. During SV40 lytic infections it is required for several steps of viral replication. Furthermore, large T is able to fully transform primary cells, i.e. it can mediate both transforming functions defined so far, immortalization of cells and expression of the transformed phenotype, (reviewed in (1)). The multitude of functions performed by large T requires a complex regulation of its activities in the cell. This may, in part, be achieved by the various posttranslational modifications described for large T, like phosphorylation (2), ADP-ribosylation (3), glycosylation (4), and fatty acid acylation (5). In addition, oligomerization of large T and complex formation with the cellular protein p53 might be involved in modulating its activities (reviewed in (6)).

Beyond these biochemical modifications the intracellular topology of large T could be an important means to regulate its multiple functions. Large interacts with various structural systems of the cell and is present in at least four discrete subcellular locations: A small percentage of large T is associated with the plasma membrane of SV40 infected and transformed cells (5, 7-18) and shows some properties characteristic for a transmembrane protein (11, 19, 20). This subclass can be distinguished from the bulk of nuclear large T by a posttranslational modification, the fatty acid acylation (5, 11). Nuclear large T, which accounts for more than 95 percent of total large T is further subcompartmented as it is present in the nucleoplasm, associated with chromatin, and tightly bound to the

Nucleocytoplasmic Transport
Edited by R. Peters and M. Trendelenburg

nuclear matrix (21). Biochemically, the nuclear subclasses of large T differ in metabolic stability and in their degree of complex formation with the cellular protein p53 (21); functionally, these subclasses are drastically different in their binding to regulatory sequences on the SV40 genome (M. Hinzpeter, W. Deppert, manuscript in preparation).

Another aspect seems to be very important for our understanding of how large T performs its multiple functions in viral replication and in cellular transformation: Genetic analyses have clearly established that different functions of large T can be allocated to different segments of the molecule and that at least some of these segments are functional when isolated (22-28). Thus, large T appears to be comprised of partially independent functional domains. Given this and assuming that indeed the performance of large T functions requires its interaction with different cellular targets we hypothesize that a domain on large T responsible for a certain function(s) will also mediate its association with the corresponding cellular system(s). To test this hypothesis we have analyzed the subcellular location(s) of individual functional domains of large T generated either naturally (AD2^{+}SV40-hybrid viruses), or by molecular cloning techniques.

Subcellular locations of carboxy-terminal large T fragments.

The Ad2^{+}SV40 hybrid viruses Ad2^{+}ND1, Ad2^{+}ND2 and Ad2^{+}ND4 are nondefective type 2 adenoviruses (Ad2), which contain overlapping fragments of the SV40 genome inserted into the Ad2 genome (Fig. 1). The integrated SV40 DNA has a common endpoint at 0.11 SV40 m.u. in the late region of the SV40 genome, and extends for various length towards the early region of the SV40 genome. At the integration sites, 4 to 6% of the Ad2 genome to the left of Ad2 map position 85.5 is deleted. During productive infection, the SV40 information in the hybrid viruses is expressed and codes for overlapping fragments of SV40 large T. All fragments contain the carboxyterminus of large T and extend for various length towards its aminoterminus (reviewed in (1, 29)).

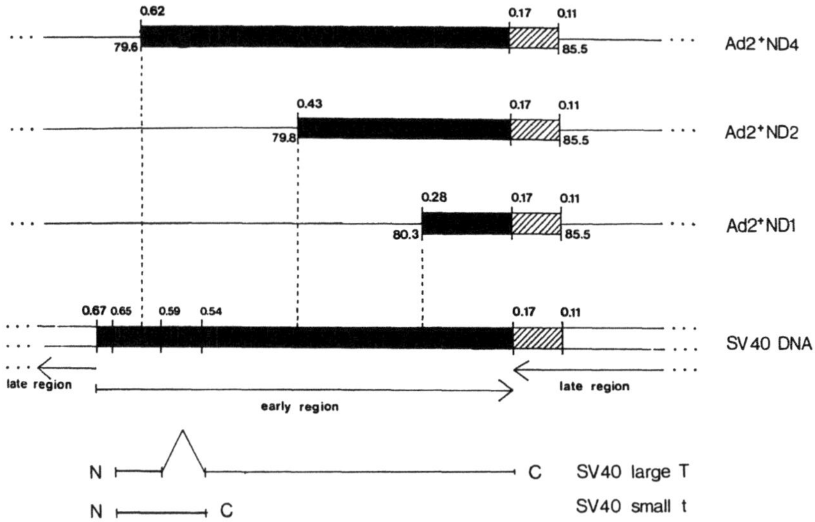

Fig. 1: SV40 DNA segments in the nondefective Ad2$^+$SV40 hybrid viruses Ad2$^+$ND1, Ad2$^+$ND2 and Ad2$^+$ND4. Map coordinates of SV40 DNA fragments and their integration sites in the Ad2 genome were derived from data compiled by Patch et al. (29) and in Tooze (1). Arrows indicate the direction of transcription.The Ad2$^+$SV40 hybrid viruses code for the following SV40 specific proteins: 28K (Ad2$^+$ND1); 42K, 56K (Ad2$^+$ND2); 64K, 94K (Ad2$^+$ND4).

The largest SV40 specific protein encoded by the Ad2$^+$SV40 hybrid viruses, the 90K large T fragment of Ad2$^+$ND4 (see Fig. 7), lacks only about 20 amino acids from the very amino-terminal end of large T. This fragment exhibits the properties and functions of authentic large T and is present in all those subcellular locations which are occupied by authentic large T (W. Deppert, unpublished).

The smallest of the SV40 specific proteins, the 28K carboxy-terminal large T fragment of Ad2$^+$ND1, performs only two of the described large T functions: (i) the SV40 helper function necessary for growth of this virus on monkey cell. This function involves transcriptional and posttranslational control mechanisms for generating functional "late" adenovirus mRNA in monkey cells (30) and was postulated to be nuclear event (31); (ii) expression of SV40 TSTA, which by definition is a cell surface function.

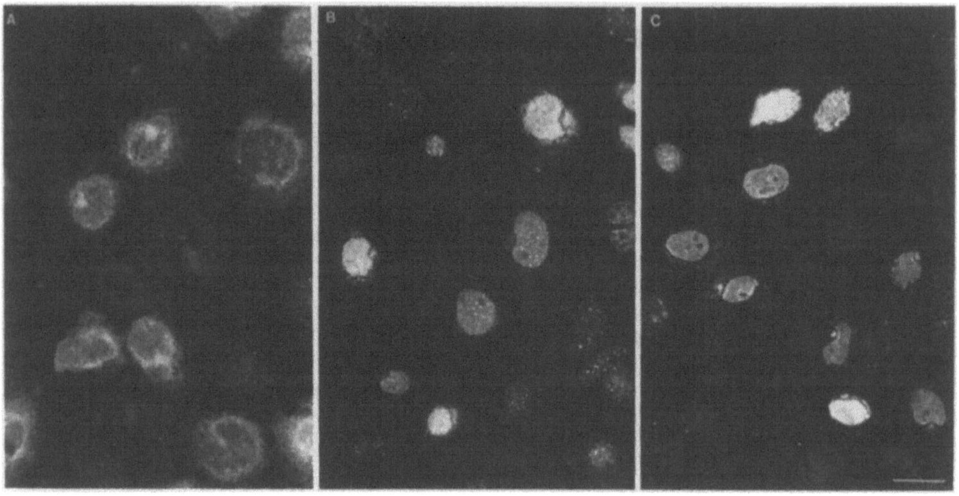

Fig. 2: Intracellular locations of the SV40 specific 28K protein in $Ad2^{+}ND1$ infected monkey cells.
TC7 cells infected with $Ad2^{+}ND1$ were subfractionated in situ 24 h postinfection and analyzed for the 28K protein by immunofluorescence microscopy. A: cells; B: NP40 nuclei; C: nuclear matrices.

By biochemical cell fractionation (19, 32) and by immunofluorescence analyses (9, 33) the 28K protein of $Ad2^{+}ND1$ has been located on the plasma membrane of $Ad2^{+}ND1$ infected cells. A second subcellular location defined for this protein by biochemical cell fractionation was the nuclear matrix (34). However, it was not found in the nucleoplasm and in the chromatin fraction of $Ad2^{+}ND1$ infected cells. Analysis of the intracellular location of the 28K protein by immunofluorescence microscopy (Fig. 2A), in addition to a weak nuclear fluorescence, shows a clearly structured, predominantly cytoplasmic fluorescence, extending from a perinuclear area into the cytoplasm (Fig. 2A). We were able to demonstrate that the cytoplasmic structural system stained in $Ad2^{+}ND1$ infected cells with antibodies against SV40 large T (Fig. 2A) corresponds to the endoplasmic reticulum (19). We also provided evidence that the association of the 28K protein with intracellular membranes not simply reflects newly synthesized 28K molecules at membrane-bound polysomes of the rough endoplasmic reticulum (35). This indicates that SV40 large T in its carboxy-terminal region contains a domain mediating its membrane association. Of particular importance in this respect might be a stretch of 15 hydrophobic amino acids extending from Ile_{568} through Tyr_{582} (19).

The ability of the 28K protein to associate with the nuclear matrix is visualized in Fig. 2B, C, where $Ad2^{+}ND1$ infected TC7 cells were subfractionated using an in situ fractionation protocol (21, 36). Analysis by immunofluorescence microscopy of the structures prepared shows that extraction of the cells with NP40 (Fig. 2B) solubilized the 28K protein associated with internal membrane systems, including the nuclear membranes. This treatment, however, did not abolish the nuclear fluorescence in these cells, demonstrating an additional association of the 28K protein with a structural system of the nucleus. After removal of the chromatin (DNase-digestion/2 M NaCl-extraction) the 28K protein was still present on the isolated nuclear matrices (Fig. 2C). This indicates the presence of an additional domain of this protein, and thus of large T, mediating its nuclear matrix association. We thus tentatively conclude that domains in the carboxy-terminal region of large T mediate its membrane location as well as its association with the nuclear matrix. Interestingly, the domain responsible for nuclear matrix association appears to act independently from a domain(s) responsible for chromatin association or signals for an accumulation of large T in the nucleoplasm, since the 28K protein is not present in these nuclear subcompartments.

Association of Lys_{128} mutant large T with the nuclear matrix

To test the above conclusion we used the Lys_{128}-mutants of large T which were developed independently by Drs. R.E. Lanford and A.E. Smith (37-39). These mutants have retained the important biological function of large T to fully transform established cell lines but fail to transport large T into the nucleus (37-39). Therefore, the mutant large T accumulates in the cytoplasm and on the cell surface of mutant infected or transformed cells (37-39). Work in Dr. A.E. Smith laboratory (see preceeding article) has shown that the sequence of SV40 large T around Lys_{128} corresponds to a nuclear translocation signal. Therefore, it was of interest to analyze whether Lys_{128}-mutant large T could still associate with the nuclear matrix. This seemed possible despite the fact that these mutants were described as being negative for nuclear location, since the bright cytoplasmic staining observed in these cells might have obscured a weak nuclear matrix staining. Therefore, we have analyzed several cell lines transformed with large T Lys_{128}-mutants (kindly provided by Drs. A.E. Smith and R. Lanford) for the presence of nuclear matrix associated large T, using our in situ cell fractionation protocol.

The results of these experiments are shown in Fig. 3: In all Lys_{128}-mutant large T transformed cells nuclear fluorescence was not, or only barely visible if whole cells were analyzed (Fig. 3 D, G, K). A faint, but in many cell already clearly discernable nuclear

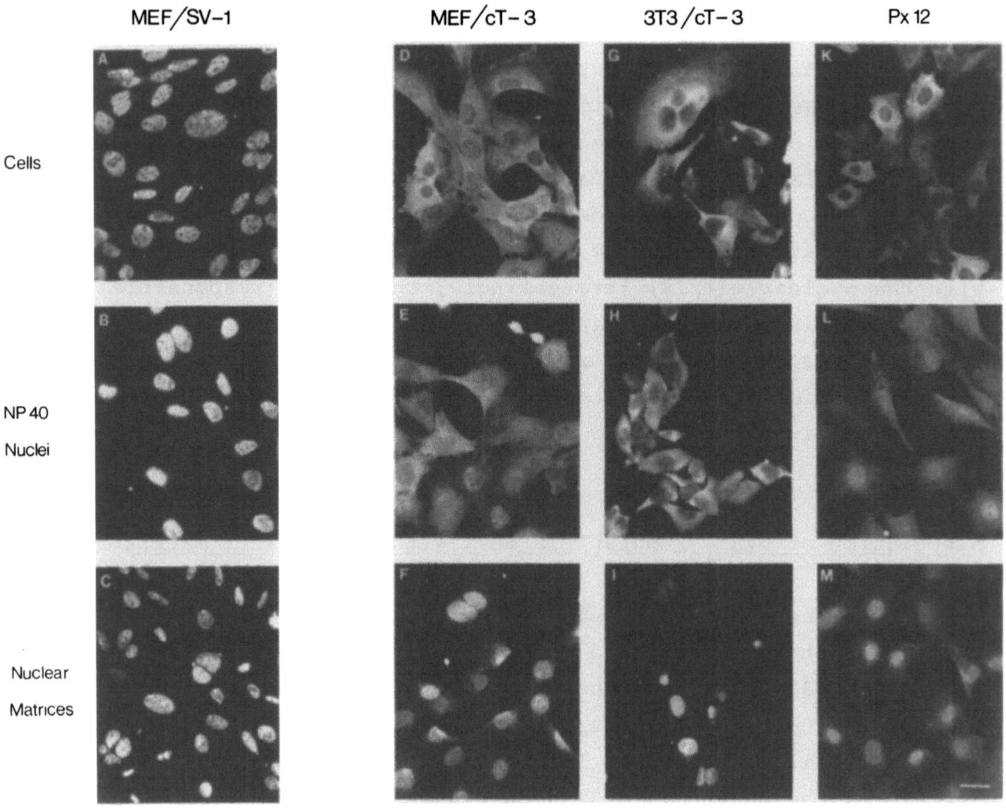

Fig. 3: Intracellular locations of Lys_{128}-mutant large T in transformed cells.
Lys_{128}-mutant transformed cells (MEF/cT-3 cells, 3T3/cT-3 cells (39), and Px12 cells (38)) were subfractionated in situ and analyzed for the presence of large T by immunofluorescence microscopy. Wild type SV40 transformed cells MEF/SV1 (39) were analyzed in parallel as a control. A, D, G, K: cells; B, E, H, L: NP40 nuclei; C, F, I, M: nuclear matrices.

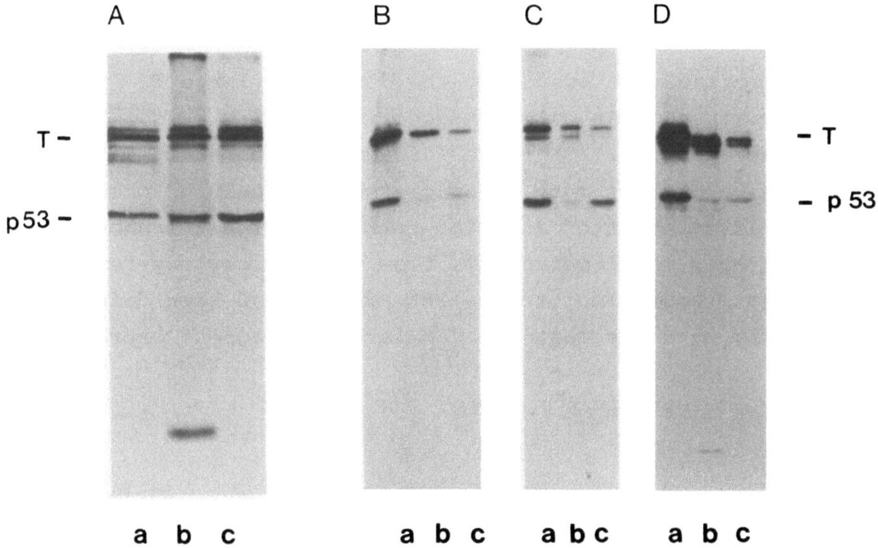

Fig. 4: Distribution of wild type large T and Lys_{128}-mutant large T in subcellular fractions of transformed cells. MEF/SV-1 cells(A), MEF/cT-3 cells (B), 3T3/cT-3 cells (G) and Px 12 cells (D) were labeled for 2 h with ^{35}S-methionine and subfractionated in situ. Extracts were immunoprecipitated with anti-SDS T serum (33) and immunoprecipitates analyzed by SDS-PAGE followed by fluorography. The origin of the cells is as described in the legend to Fig. 2. a: NP40 extract; b: DNase1 / 2M NaCl extract; c: Empigen BB extract.

staining became visible after extraction of the bulk cytoplasm by hypotonic detergent treatment of the cells (Fig. 3E, H, L); analysis of the isolated nuclear matrices (Fig. 3F, I, M) then clearly demonstrated association of some mutant large T at this structure. The association of Lys_{128}-mutant large T with the nuclear matrix of mutant transformed cells was also confirmed by biochemical analysis of nuclear subfractions of ^{35}S-methionine labeled cells. Fig. 4 shows that the nuclear matrix fraction of all mutant transformed cells (Fig. 4B-D, c) contained large T, although considerably less than SV40 wild type transformed cells (Fig. 4A, c).

In conclusion, these data demonstrate that large T is able to associate with the nuclear matrix even if its major nuclear transport signal is impaired.

A putative nuclear matrix association domain on SV40 large T

We tried to define more precisely the domain of large T responsible for its nuclear matrix association.
Since the carboxy-terminal 28K large T fragment encoded by $Ad2^{+}ND1$ already is nuclear matrix associated, we analyzed the nuclear matrix association of an amino-terminal large T fragment (p113; Assum et al., in preparation), from which a carboxy-terminal portion of large T equivalent to the 28K protein had been deleted (see Fig. 7). Fig. 5 shows that the 72K large T fragment expressed

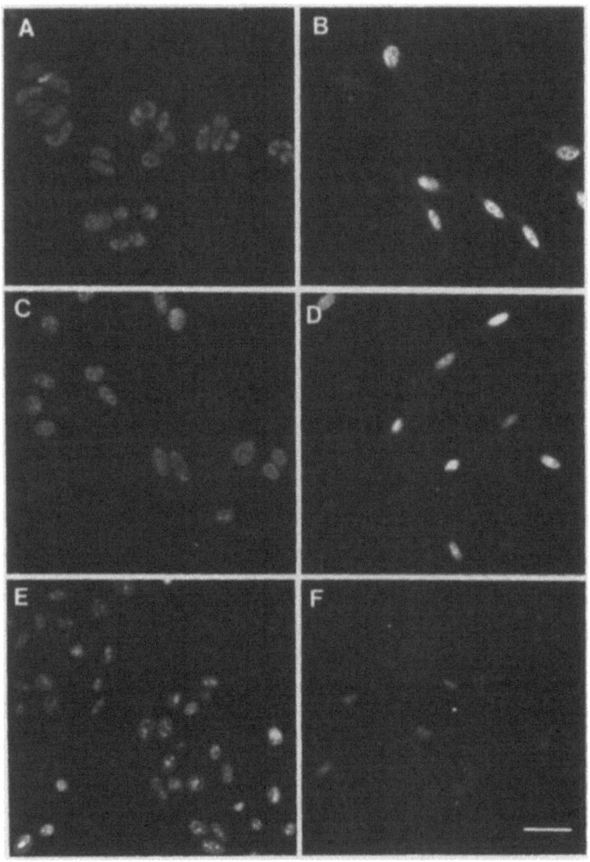

Fig. 5: Subnuclear locations of the amino-terminal 72K large T fragment in p113 transfected cells.
p113 plasmid DNA was transfected into mouse LTK^{-}cells together with the cloned HSV TK gene. Stable transformants were selected, subfractionated in situ and analyzed for the 72K protein by immunofluorescence microscopy (B; D, F). SV40 wild type transformed monkey cells (mkSA; A, C, E) were analyzed in parallel as a control. A, B: cells; C, D: NP40 nuclei; E, F: nuclear matrices.

in p113 transfected cells still was nuclear matrix associated, although at a somewhat reduced level compared to wild type large T. Therefore, the domain of large T mediating nuclear matrix association must extend further into the amino-terminal region of the protein than the coding sequences of the 28K protein of $Ad2^{+}ND1$. This result is in agreement with the fact that only a small fraction of the 28K protein of $Ad2^{+}ND1$ is associated with the nuclear matrix and that the association of the protein with the nuclear matrix of human cells (where its helper function is not required for Ad2 replication) depends on the cell line used for

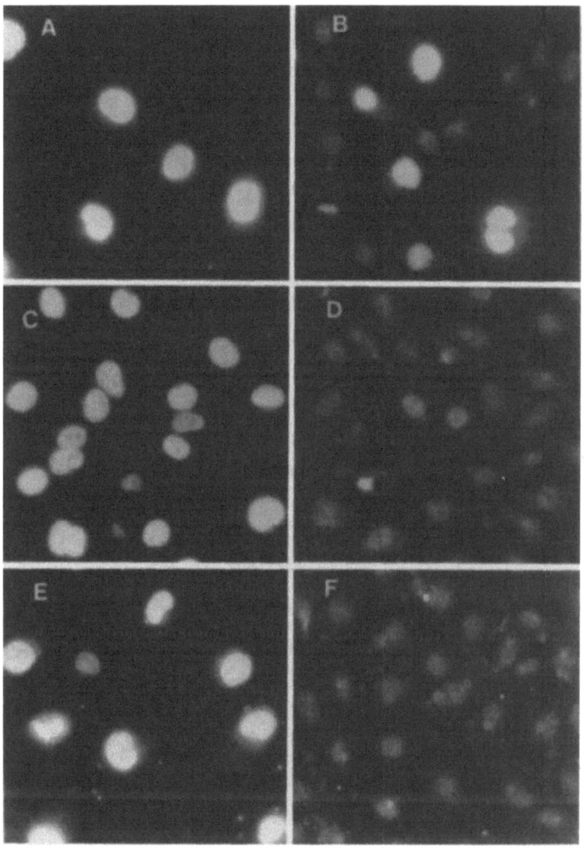

Fig. 6: Subnuclear locations of the amino-terminal 33K large T fragment in dl 1001 infected monkey cells.
SV40 dl 1001 virus was prepared in COS1 cells. TC7 cells were infected with either wild type SV40 virus (A, C, E) or dl 1001 virus (B, D, F). Cells were subfractionated in situ 40 h post infection and analyzed for SV40 large T(A, C, E) or the dl 1001 33K protein (B, D, F) by immunofluorescence microscopy. A, B: cells; C, D: NP40 nuclei; E, F: nuclear matrices.

infection (W. Deppert, unpublished results). In contrast, nuclear matrix associaton of the 42K and 56K large T fragments encoded by $Ad2^{+}ND2$, which extend further into the amino-terminal region of large T, is much more prominent (34; and W. Deppert, unpublished data).

Therefore, we analyzed the nuclear matrix association of the amino-terminal large T fragment encoded by the SV40 deletion mutant dl 1001 (22, 27, 40). dl 1001 codes for a truncated 33K large T fragment which extends from the aminoterminus of large T to nucleotide 4002 (22, 27, 40), i.e. roughly to the aminoterminus of the 42K large T fragment encoded by $Ad2^{+}ND2$ ((1), see Fig. 7). Fig. 6 shows that the 33K large T fragment encoded by dl 1001 still was found in the nuclei of TC7 cells infected with dl 1001 (Fig. 6B). Most of it seemed to accumulate in the nucleoplasm, since NP40 extraction of the cells removed most of the protein (compare Fig. 6B, D), but a minor fraction was associated with the chromatin (Fig.

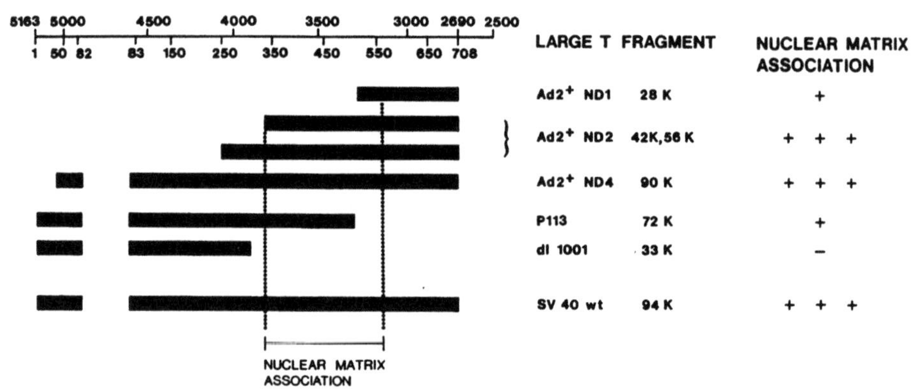

Fig. 7: Identification of a nuclear matrix association domain of SV40 large T.
The data on the nuclear matrix association of large T fragments presented in this paper are summarized. Heavy bars represent the coding sequences of the various large T fragments. A domain on large T, roughly bordered by amino acids 320 and 550, mediates nuclear matrix association.

6D). However, no dl 1001 33K protein could be detected at isolated nuclear matrices (Fig. 6F).

The association with the nuclear matrix of the various large T fragments analyzed is summarized in Fig. 7. These analyses roughly indicate the boundaries of a putative nuclear matrix association domain of large T. This domain is located approximately between amino acids 320 and 550, assuming that only the very amino-terminal region of the $Ad2^{+}$ND1 28K protein is involved in nuclear matrix association.

Discussion

From the experiments presented here and in previous studies (19, 33-35) we have developed a model for the subcellular localization of SV40 large T. We postulate that the association of large T with its cellular targets is mediated by distinct domains of the large T molecule. These domains are located in different parts of large T and determine different subcellular locations. Domains in its carboxy-terminal region mediate the membrane location of large T as well as its association with the nuclear matrix. Domains in its amino-terminal half, on the other hand, cause the accumulation of large T in the nucleoplasm and its association with chromatin. We speculate that domains mediating a certain subcellular location of large T might correspond to biologically functional domains of large T, as is the case with the carboxy-terminal large T fragments encoded by the $Ad2^{+}$SV40 hybrid viruses (reviewed in (1, 29)).

An important feature of our model is that these domains are able to act more or less independently from one another. Thus, impairment of the major nuclear translocation site on large T still allowed for an association of Lys_{128}-mutant large T with the nuclear matrix. The question of how the mutant large T enters the nucleus is not clear, but several mechanisms can be envisioned:

(i) Low levels of the Lys_{128}-mutant large T may diffuse into the nucleus, despite this protein lacking the nuclear translocation signal. Once in the nucleus, it associates with the nuclear matrix.

(ii) The Lys_{128}-mutant large T still has the large T membrane association domain. It is possible that it may associate with cytoplasmic membranes, is transferred to the outer nuclear membrane and from there into the nucleus and to the nuclear matrix.

(iii) Finally, it is conceivable that the Lys_{128}-mutant large T does not enter the nucleus during interphase, but associates with nuclear matrix components during mitosis, when the nuclear envelope is dissolved.

Experiments to distinguish among these or additional alternatives are in progress.
Analysis of the nuclear matrix association of large T fragments allowed us to identify a putative nuclear matrix association domain of large T between amino acids 320 and 550. Since, so far, this domain is bordered only roughly, more precise analyses may allow to narrow down the segment of large T in question. However, we assume that the nuclear matrix association domain is much larger than the nuclear translocation signal around Lys_{128} (see preceeding paper). This is based on the fact that the latter is a signal sequence important for the translocation of large T into the nucleus. Most probably this signal seguence does not function to bind large T to a subnuclear structural system but may be sufficient for its accumulation in the nucleoplasm. On the other hand, the association of large T with the nuclear matrix appears to involve a larger segment (or multiple points) of the molecule as is indicated by the poor nuclear matrix association of both the carboxy-terminal 28K and the 72K amino-terminal large T fragments. Binding of the entire domain to components of the nuclear matrix then leads to a tight association of large T with this subnuclear structure (21).

Since the nuclear matrix is involved in several important basic biological functions, like DNA replication, transcription and RNA processing (reviewed in (41)), the association of large T and its carboxy-terminal fragments with the nuclear matrix is of considerable interest. The identification of a putative nuclear matrix association domain on large T should allow to generate large T mutants which no longer associate with the nuclear matrix.

Analysis of the biological properties of such mutants could help towards an understanding of the biological role of the nuclear matrix association of large T.

Acknowledgements

We thank Drs. R.E. Lanford and A.E. Smith for providing Lys_{128}-mutant large T transformed cells and for stimulating discussions. A detailled account of the nuclear matrix association of Lys_{128}-mutant large T will be given elsewhere (W. Deppert, R.E. Lanford, and A.E. Smith, manuscript in preparation).

This study was supported by grants from the Deutsche Forschungsgemeinschaft.

Experimental Procedures

Cells and viruses. The following cell lines were used in this study: monkey TC7 cells; the SV40 transformed Balb/c mouse tumor line mkSA; SV40 wild type transformed mouse embryo fibroblasts (MEF/SV-1; 39); mouse embryo fibroblasts transformed by SV40 Lys_{128}-mutant SVcT-3 (MEF/cT-3; 39); 3T3 cells transformed by SVcT-3 (3T3/cT-3; 39); Rat-1 cells transformed by SV40 Lys_{128}-mutant px12 (Px12 cells; 28). MEF/SV-1 cells, MEF/cT-3 cells and 3T3/cT-3 cells were kindly provided by Dr. R.E. Landford, Southwest Foundation for Biomedical Research, San Antonio, Texas, U.S.A.; Px12 cells were kindly provided by Dr. A.E. Smith, Integrated Genetics, Framingham, Mass., U.S.A. $Ad2^+$SV40 hybrid viruses were obtained from Dr. A. Lewis, National Institutes of Health, Bethesda, Maryland, U.S.A. Experiments with these viruses were conducted in containment laboratories according to NIH guidelines for handling these viruses.

Plasmid p113 containing an SV40 fragment that spans from nucleotide 1782 counter clockwise to nucleotide 3204 of the SV40 genome codes for an aminoterminal 72K large T fragment (Assum et al., manuscript in preparation). p113 was transfected into mouse LTK^- cells together with the cloned HSV TK gene as a selectable marker and stable transformants expressing nuclear fluorescence were selected. The construction of p113, the selection of stable p113 transformants and the biological properties of the 72K amino-terminal large T fragment encoded by p113 will be described elsewhere (Assum et al., manuscript in preparation).

SV40 mutant dl 1001, originally constructed by Pipas et al. (40) and further characterized by Pipas et al. (27) was obtained as plasmid in pBR 322 from Dr. D. Müller, Dept. of Biochemistry, University of Ulm. dl 1001 virus stocks were prepared in COS1 cells after transfection of plasmid DNA. SV40 wild type virus was grown in TC7 cells.

Cell fractionation. A detailed description of the cell fractionation procedure as well as a characterization of extracts and structures has been given elsewhere (21, 36). Briefly, cells grown to confluency were washed with KM buffer (10 mM morpholinopropane sulfonic acid (MOPS) pH 6.8; 10 mM NaCl; 1.5 mM $MgCl_2$; 1 mM EGTA; 5 mM DTT; 10% glycerol) and lysed in KM buffer containing 1% NP40 ("nucleoplasmic extract"). NP40 nuclei still attached to the substratum were extracted with KM buffer without DTT and EGTA, containing 100 ug/ml DNaseI (Sigma, no. D-5010) for 15 min at 37 ^{o}C. Then an equal volume of 4 M NaCl was added, and the incubation continued for further 30 min at 4 ^{o}C ("chromatin extract"). Nuclear matrices were solubilized in TK-buffer (40 mM Tris-HCl, pH 9.0, 25 mM KCl, 5 mM DTT; 10% glycerol) containing 1% Empigen BB (Albright and Wilson) for 60 min at 4 ^{o}C. All buffers contained 30 ug aprotinin (200 KIU; Trasylol, Bayer) and immediately after fractionation phenylmethylsulfonyl fluoride was added to 1 mM.

Immunofluorescence microscopy. Cells grown on coverslips (0 12 mm) were subfractionated as described above. Cells and nuclear structures were fixed in methanol-acetone and immunofluorescence was performed as described previously (21). Fluorescence was viewed with a Zeiss photomicroscope.

Labeling of cells and immunoprecipitation of extracts. Cells were labeled with ^{35}S-methionine (100 uCi/plate/ml) prior to fractionation as described previously (21). All extracts were cleared by centrifugation at 130.000 x g for 30 min at 4 ^{o}C. Extracts were then precipitated with 10 ul of anti-SDS-T-serum and 200 ul of settled protein A-sepharose as described previously (21).

Samples were processed for SDS-PAGE and analyzed on SDS-polyacrylamide gels as described (21).

References

1. Tooze, J. Molecular Biology of Tumor Viruses: DNA tumor viruses. Cold Spring Harbor Laboratory. Cold Spring Harbor, N.Y., 2nd revised edition (1981).

2. Scheidtmann, K.H., B. Echle, and G. Walter. J. Virol. 44: 116-133 (1982).

3. Goldman, N., M. Brown, and G. Khoury. Cell 24: 567-572 (1981).

4. Jarvis, D.L., and J.S. Butel. Virology 141: 173-189 (1985).

5. Klockmann, U., and W. Deppert. FEBS Lett. 151: 257-259 (1983).

6. Rigby, P.W.J., and D.P. Lane. In "Advances in Virol. Oncology" (G. Klein, ed.) Vol. 3: 31-57, Raven Press, New York (1983).

7. Deppert, W., K. Hanke, and R. Henning. J. Virol. 35: 505-518 (1980).

8. Deppert, W., and R. Henning. Cold Spring Harbor Symp. Quant. Biol. 44: 225-234 (1979).

9. Deppert, W., and G. Walter. Virology 122: 56-70 (1982).

10. Henning, R., J. Lange-Mutschler, and W. Deppert. Virology 108: 325-337 (1981).

11. Klockmann, U., and W. Deppert. EMBO J. 2: 1151-1157 (1983).

12. Lanford, R.E., and J.S. Butel. Virology 97: 295-306 (1979).

13. Santos, M., and J.S. Butel. Virology 120: 1-17 (1982).

14. Santos, M., and J.S. Butel. J. Cell. Biochem. 19: 127-144 (1982).

15. Santos, M., and J.S. Butel. J. Virol. 41: 50-56 (1984).

16. Soule, H.R., and J.S. Butel. J. Virol. 30: 523-532 (1979).

17. Soule, H.R., R.E. Lanford, and J.S. Butel. J. Virol. 33: 887-901 (1980).

18. Whittaker, L., A. Fuks, and R. Hand. J. Virol. 53: 366-373 (1985).

19. Klockmann, U., M. Staufenbiel, and W. Deppert. Mol. Cell. Biol. 4: 1542-1550 (1984).

20. Klockmann U., and W. Deppert. J. Virol. 56: in press (1985).

21. Staufenbiel, M. and W. Deppert. Cell 33: 173-181 (1983).

22. Clark, R., K.W.C. Peden, J.M. Pipas, D. Nathans, R. Tijan. Mol. Cell. Biol. 3: 220-228 (1983).

23. Cosman, D.J., and M.J. Tevethia. Virology 112: 605-624 (1981).

24. Gluzman, Y., and B. Ahrens. Virology 123: 78-92 (1982).

25. Kalderon, D., and A.E. Smith. Virology 139: 109-137 (1981).

26. Peden, K.W.C., and J.M. Pipas. J. Virol. 55: 1-9 (1985).

27. Pipas, J.M., K.W.C. Peden, and D. Nathans. Mol. Cell. Biol. 3: 203-213 (1983).

28. Stringer, J.R. J. Virol. 42: 854-864 (1982).

29. Patch, C.T., A.S. Levine, and A.M. Lewis, Jr. In "Compressive Virology" (H. Fraenkel-Conrat and R. Wagner, eds.) Vol. 13: 459-542. Plenum Press, New York (1979).

30. Anderson, K.P., and D.F. Klessig. J. Mol. Appl. Genet. 2: 31-43 (1983).

31. Zorn, G.A., and C.W. Anderson. J. Virol. 37: 759-769 (1981).

32. Jay, G., F.T. Jay, C. Chang, R.M. Friedman, and A.S. Levine. Proc. Natl. Acad. Sci. U.s.A. 75: 3055-3059 (1978).

33. Deppert, W., and R. Pates. J. Virol. 31: 522-526 (1979).

34. Deppert, W. J. Viorl. 26: 165-178 (1978).

35. Deppert, W., and M. Staufenbiel. In "Biochemical and Biological Markers of Neoplastic Transformation" (P. Chandra, ed.) pp. 537-554. Plenum Press, New York (1983).

36. Staufenbiel, M., and W. Deppert. J. Cell. Biol. 98: 1886-1894 (1984).

37. Lanford, R.E., and J.S. Butel. Cell 27: 801-813 (1984).

38. Kalderon, D., W.D. Richardson, A.F. Markham, and A.E. Smith. Nature (London) 311: 33-38 (1984).

39. Lanford, R.E., C. Wang, and J.S. Butel. Mol. Cell. Biol. 5: 1043-1050 (1985).

40. Pipas, J.M., S.P. Adler, K.W.C. Peden, and D. Nathans. Cold Spring Harbor Symp. Quant. Biol. 44: 285-291 (1979).

41. Berezney, R. In "Chromosomal Nonhistone Proteins" (Hnilica, ed.) Vol. 10: 120-180. CRC Press, Boca Raton (1984).

Identification of a Sequence in Influenza Virus Nucleoprotein Which Specifies Nuclear Accumulation in *Xenopus* Oocytes

A. COLMAN and J. DAVEY[1]
Dept. of Biological Sciences, University of Warwick, Coventry CV4 7AL, United Kingdom
[1]Present Address: University of Dundee, Dept. Biochemistry, Dundee DD1 4HN, Scotland, United Kingdom

It has been known for a long time that a subclass of cellular proteins accumulate in the nucleus of the interphase eukaryotic cell. Experimental interest has focussed on two aspects of the accumulation process. First, studies have been directed at understanding the cellular mechanisms which underlie the observed nuclear accumulation. The selectivity of the uptake process (Bonner, 1975a,b; De Robertis et al., 1978; De Robertis and Black, 1981; Dabauvalle and Franke, 1982) led to the proposal that the accumulation is controlled by a signal contained within the mature molecular structure of the protein and two types of approach have been used to identify and localize such a karyophilic signal. The first is exemplified by the work of Dingwall et al. (1982), who used partial proteolysis of nucleoplasmin, the most abundant nuclear protein in Xenopus oocytes (Mills et al., 1980; Krohne and Franke, 1980), to produce two fragments, only one of which retained the ability to migrate into and accumulate in the oocyte nucleus. The second approach involves the use of DNA recombinant methodology to construct genes in which various regions have been deleted or altered. In this way Kalderon et al. (1984) have identified a region of the SV40 large T antigen necessary for its accumulation in the nuclei of transformed Rat-1 cells. A further modification of this approach involves the construction and expression of chimeric genes containing various regions of a nuclear protein fused to a cytosolic protein. In this way Hall et al. (1984) have identified an amino-terminal region of the yeast protein α2, which can transport bacterial β-galactosidase into the yeast cell nucleus.

Nucleocytoplasmic Transport
Edited by R. Peters and M. Trendelenburg

In this paper we have examined the accumulation of the influenza nucleoprotein (NP) in Xenopus oocyte nuclei following microinjection of NP protein, mRNA and cloned cDNA. Using DNA recombinant technology we have extended these studies to produce mutant NPs that retain or lack the ability to accumulate in nuclei and have thereby defined a region of the molecule necessary for this accumulation. When a portion of NP cDNA including this region was fused to α1-globin cDNA the resulting fusion protein accumulated in the oocyte nucleus, whereas fusion proteins lacking this region of the NP entered but did not accumulate in the nucleus. We discuss these findings in relation to nuclear signals identified in other proteins and in terms of existing models which attempt to explain nuclear accumulation.

METHODS

Exact details of all methods used can be found in Davey et al. (1985a,b). The general procedures were as follows: 25-50 ul aliquots of protein (1 mg/ml), mRNA (1mg/ml) or NP plasmid DNAs (0.2 mg/ml) were microinjected into either the cytoplasm (protein and mRNA) or nucleus (DNA) of intact Xenopus oocytes. Oocytes were then cultured for varying times in a frog saline with or without ^{35}S-methionine, as appropriate. At the end of the culture period, oocytes were either homogenized directly or manually enucleated and the separated nuclei and cytoplasms then homogenized. Homogenates were then prepared immediatly for SDS polyacrylamide electrophoresis or immunoprecipitated first; often homogenates of nuclei were sonicated and the sonicate clarified by centrifugation before analysis. After fluorography gel bands corresponding to the NP proteins were excised and the radioactive content quantitated by scintillation counting. In order to compute the relative concentrations of NP protein between the nuclear and cytoplasmic compartments, we have taken the figure of 12% as the percentage of the total oocyte volume occupied by the nucleus, which is readily accessible to most macromolecules (Bonner, 1975a). Thus, if equal quantities of NP protein are found when the contents of equal numbers of nuclei (N) and cytoplasms (C) are compared this will reflect a N/C concentration ratio of 7.33 (i.e. 88%-12%).

RESULTS

A) Injection of NP protein

Figure 1a shows the gradual nuclear accumulation of NP protein observed after the injection of oocytes with a radiolabelled lysate prepared from chick embryo fibroblast (CEF) cells infected 3h earlier with influenza virus [A/FP/Rostock/34 (FP/R)(H7N1)]. Quantitation of these data are shown in Fig. 1b. A four-fold accumulation was established within 45 h. Significantly (see discussion), equilibration of NP between nuclear and cytoplasmic compartments (i.e. N/C ratio=1) was achieved within 2h. Two other influenza proteins present in the lysate, the matrix protein (M) and a non-structural protein (NS1), hardly entered the nucleus during this period (data not shown but see Fig. 2 and Davey et al., 1985b).

B) Injection of influenza mRNAs

Similar results to the above were obtained when poly A^+ RNAs prepared from CEF cells infected as above were injected into oocytes (Fig. 2). However, the nuclear accumulation of NP was somewhat higher (i.e. N/C ratio = 7-8), a result also found after NP DNA injections (see below). The difference is possibly attributable to denaturation of NP protein during preparation of the lysate from CEF cells. The experiment in Fig. 2 also shows that an apparent accumulation of matrix protein in the nucleus is due to its interaction with a sedimentable fraction of the nucleus. Further work, which involved the manual removal of the nuclear envelope, identified this structure as the element responsible (data not shown). Its interaction with matrix protein is probably non-specific and stems from the membrane-sticky nature of the matrix protein (Gregoriades, 1980).

C) Injection of full length cloned NP cDNA

Recently we have modified existing eukaryotic expression vectors to optimize the expression in oocytes of cloned cDNAs. The vector pTK2, (Krieg et al., 1984) contains the herpes simplex thymidine kinase

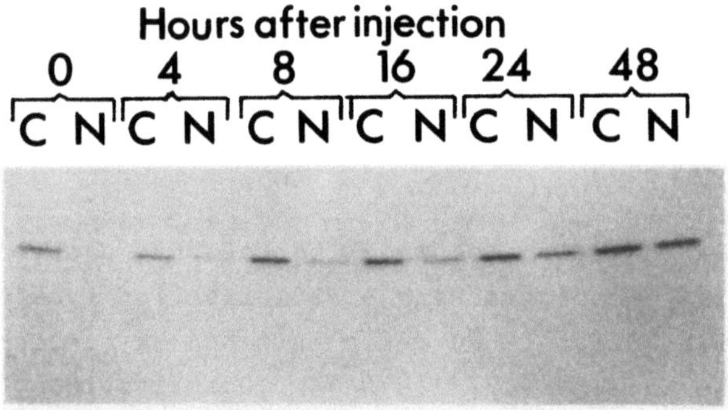

Figure 1 SDS-PAGE analysis of oocytes injected with radiolabelled NP
a) Radiolabelled lysate from CEF cells infected with FP/R virus was injected into the cytoplasm of *Xenopus* oocytes. The oocytes were enucleated at various times after injection and the separated nuclear (N) and cytoplasmic (C) fractions analysed by SDS-PAGE. The region of the gel showing NP protein is shown. A minimum of ten nuclei and ten cytoplasms were pooled for each time point and the equivalent of one nucleus and one cytoplasm loaded per track.

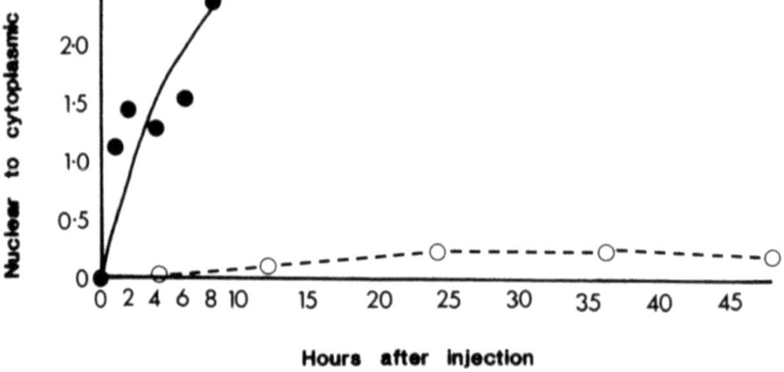

b) The bands in (a) were excised and the radioactive contects assessed in a scintillation counter. o, incubation temp. 4°C; •, incubation temp. 20°C. From Davey et al., (1985b) with permission.

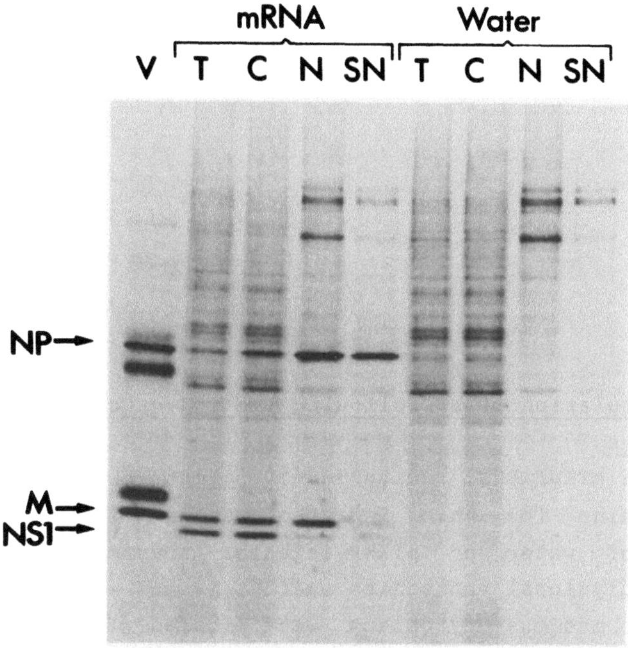

Figure 2 The nuclear association of influenza virus proteins following the injection of viral mRNA into oocytes

Total (T), cytoplasmic (C), nuclear (N) fractions and the supernatant from the centrifugation of the nuclear fraction (SN) of oocytes injected with water or mRNA were analysed by SDS-PAGE. Approximately equal amounts of radioactivity in each track were achieved by loading the equivalent of one total oocyte (T), 1.5 cytoplasms (C) and 10 nuclei (N and SN). Track V contains purified ^{35}S-methionine FP/R virus as marker. From Davey et al., (1985b) with permission.

promotor, SV40 poly adenylation and termination signals and both eukaryotic and prokaryotic origins of replication. In Fig. 3 we show the accumulation of NP protein in oocyte nuclei 24h after the injection of such a vector containing a full length cloned NP cDNA (pTKNP); the cDNA encodes the NP from influenza strain A/NT/60/68 (Huddleston and Brownlee, 1982). The accumulation seen (N/C=8) did not change when the oocytes were chased for 24h, indicating that equilibrium had been achieved. In addition, in this and all the experiments in this paper, no degradation of NP (or its variants, see below) was detected. Thus the results seen reflect a genuine accumulation of NP and not the effect of differential degradation of the protein between the cytoplasm and nucleus as recently demonstrated for calf thymus testis H1 (Dingwall and Allan, 1984).

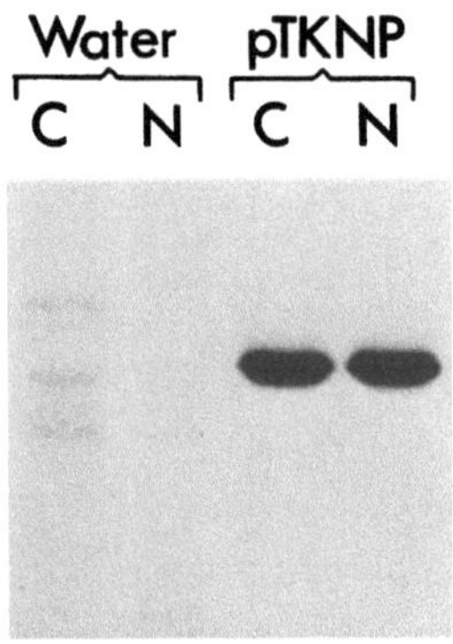

Figure 3 The nuclear accumulation of NP following the injection of pTKNP into oocytes

Oocytes were injected with pTKNP DNA, cultured overnight and then labelled in ^{35}S-methionine for 6h. Isolated nuclear (N) and cytolasmic (C) fractions of water or pTKNP-injected oocytes were immunorecipitated using polyclonal antibodies against NP and analysed by SDS-PAGE. One nucleaus and one cytoplasm were loaded in each track. From Davey et al., (1985b).

D) Injection of mutant NP DNAs

We have demonstrated above that the full length cloned NP cDNA could be used in oocytes as a template for the eventual production of NP protein. In fact, as much as 7% of the newly synthesized protein of the oocytes can be represented by NP (Davey et al., 1985b), indicating the efficiency of the coupled transcription-translation assay using this particular DNA. This finding has facilitated the following studies, fully reported in Davey et al., (1985a), where we have investigated the effect of various deletions on the nuclear accumulation of NP. Two classes of deletion were constructed; C-terminal and internal deletions. The constructs were injected into oocyte nuclei and the fate of the encoded protein monitored for up to 48h. Figure 4 shows the type of data obtained with representatives of each class whilst Figure 5 tabulates the results obtained with all the deletions tested: full length NP protein contains 598 amino acids. A C-terminal deletion which encoded a protein of 386 amino acids (pTKNP HAE 1474) had no effect on NP accumulation, whilst a mutant NP containing the N-terminal 345 amino acids (pTK HIND 1348) showed a reduced though nevertheless significant nuclear

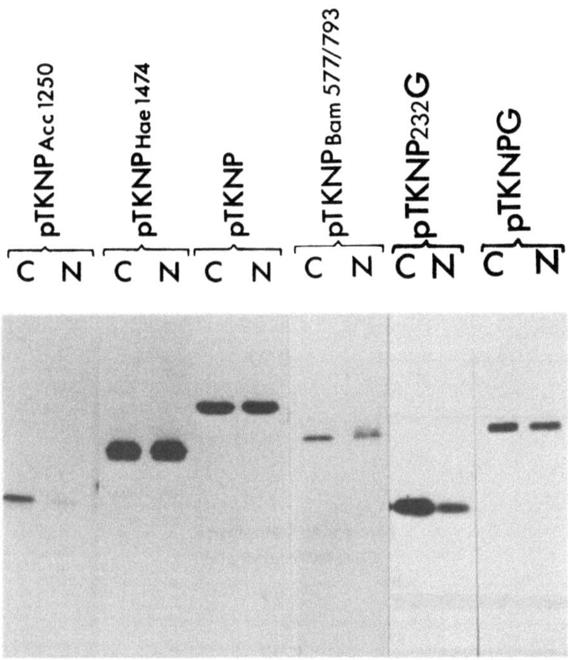

Figure 4 The nuclear association of mutant NP proteins
Oocytes were injected with various mutant NP DNAs and labelled and processed exactly as in Fig. 3. The exact mutations are shown in Figs. 5 and 6. Modified from Davey et al., (1985a) with permission.

accumulation. However, proteins in C-terminal deletions containing less than the first 327 amino acids failed to accumulate in the nucleus and became equally partitioned between the nucleus and the cytoplasm (i.e. N/C =1).

From these data we conclude that a region responsible for NP accumulation may be located between amino acids 327-345. This conclusion is further substantiated by the demonstration that an internal deletion encoded by pTK BGL 1078/1333, which spans this region, greatly affects accumulation whilst the accumulation of proteins lacking other regions was unaffected (Fig. 5).

E) Injection of NP-globin fusion DNA.

The fact that mutant proteins lacking the region from amino acids 327-340 had a reduced ability to accumulate in the nucleus suggested

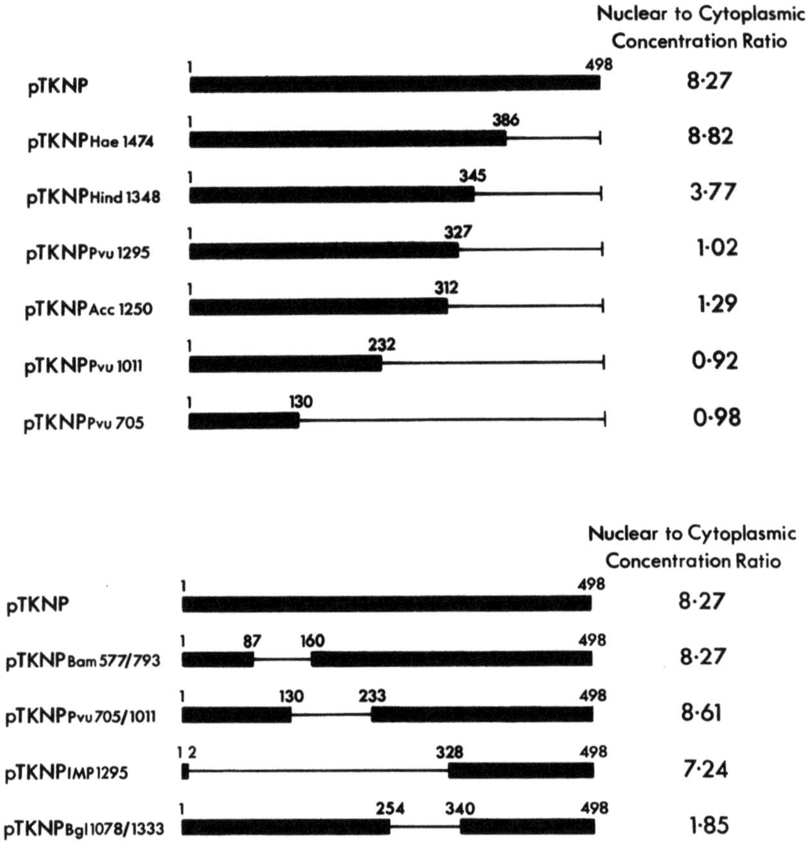

Figure 5 <u>Nuclear to cytoplasmic concentration ratios of mutant NP proteins</u>

The immunospecific bands shown in Figure 4 together with others not shown were excised and their radioactive content was assessed by scintillation counting. The thin line represents the region missing from the wild-type NP and the numbers refer to the positions of the terminal amino acids. For convenience, the amino acids encoded by the translational stop region in the vector [see Davey et al. (1985a)] are omitted. From Davey et al., (1985a) with permission.

that the information necessary for the nuclear accumulation of NP resides wholly or partially within this region. This strategy however suffers from the criticism that the removal of large regions of the NP may cause gross changes in its conformation which could impair nuclear accumulation. The presence of a karyophilic signal in a region of the NP can be tested positively by constructing chimeric proteins containing various regions of the NP fused to a protein that ordinarily does not accumulate in nuclei. The accumulation of the fusion product in the nucleus would then indicate that the NP region present in the fusion protein contained the information necessary for this process. The same strategy has been used to study signal sequences in prokaryotes (Silhavy et al., 1977; Bedoulle et al., 1980; Emr et al., 1980), in in vitro systems (Lingappa et al., 1984) and in eukaryotes (Tabe et al.,1984).

An important consideration in such experiments is the choice of cytosolic protein. Globins are ideal candidates since they exhibit no affinity for membranes either in vitro (Meek et al., 1982) or in vivo in Xenopus oocytes (Zehavi-Willner and Lane, 1977) and have recently been used successfully in fusions involving secretory proteins (Lingappa et al., 1984; Tabe et al., 1984).

Four plasmids were constructed in which the globin cDNA was fused to various amino-terminal regions of the NP cDNA. Proteins of expected size and composition were produced after injection of the DNAs into oocytes. The concentration of the proteins in each cellular compartment was determined and used to calculate the nuclear to cytoplasmic concentration ratios, which are summarized in Figure 6. Although all the fusion proteins entered the oocyte nucleus only the protein produced from pTKNPG accumulated there.

These data indicate that an autonomous nuclear transport signal is located beyond amino acid 312. When combined with the previous data, the location of this signal appears to reside between amino acids 328 and 340.

F) Does the nuclear signal have a role in cytotoxic T cell recognition?

During infection of mice by influenza and other viruses, cytotoxic T lymphocytes (CTL) are induced which can recognize infected cells by virtue of the surface expression of certain viral antigens (see Zinkernagel and Rosenthal, 1981). Recent work using murine CTL has shown that during influenza infection subpopulations of CTL can specifically recognize infected cells via the surface presentation of NP (Townsend et al., 1984). These results have raised the question

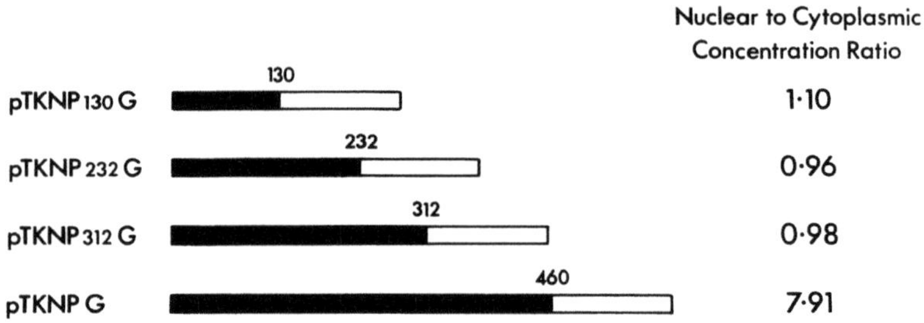

Figure 6 <u>Nuclear to cytoplasmic concentration ratios of they NP-globin fusion proteins</u>

Immunospecific bands from Fig. 4, together with others not shown, were processed as in Fig. 5. The solid box represents the NP region of the fusion protein and the open box the globin region (full length chimpanzee α1-globin has 145 amino acids). The numbers refer to the positions of the terminal amino acids of the NP region. Bracketted ratios are from a separate experiment.

of how NP is transported to the surface. Recently, Townsend et al. (1985) have used some of the pTKNP plasmids displayed in Fig. 5 to see which region of NP is required for the CTL recognition and also to ask whether or not any regions are needed for the transport to the plasma membrane. A composite of the nuclear accumulation data and the results of Townsend et al., (1985) are shown in Fig. 6. Townsend and co-workers studied the recognition process using influenza-infected L cells challenged with CTL populations derived from two different strains of mice. The results show that the NP recognition signal must be different for the two CTL populations and that, in the case of the CTL from CBA mice, the NP accumulation signal is not required for CTL recognition. More important however is the fact that since non-overlapping NP proteins can be "seen" by at least one of the CTL populations, it follows that the plasma membrane transport signal is either present in both halves of the protein or that there is no specific signal. Whatever the answer, it is clear that transport to the cell surface can occur in the absence of the nuclear accumulation signal.

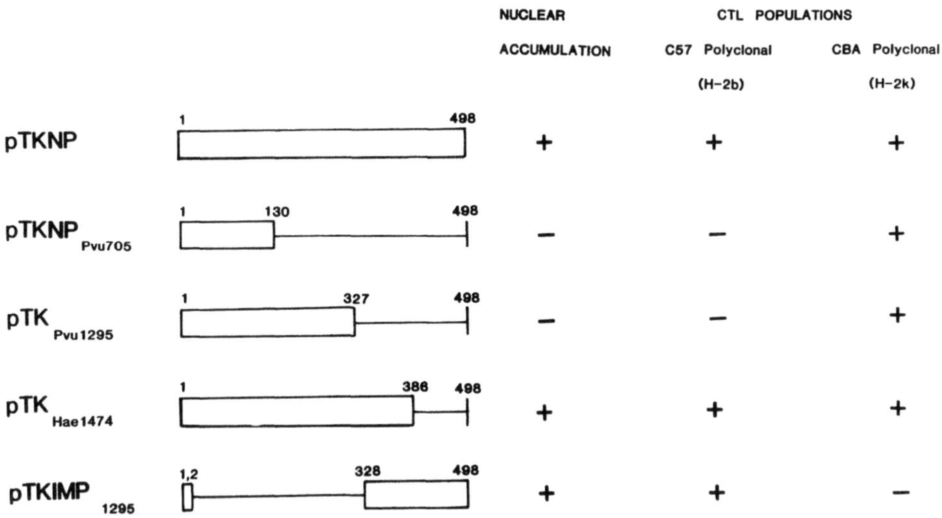

Figure 7 Recognition of NP-transfected L cells by cytotoxic lymophocyte populations from two strains of mice

Mouse L cells transfected with various NP plasmids were tested for their sensivity to virus-primed polyclonal CTL populations raised in C57 or CBA mice. A + sign indicates sensivity. The nuclear accumulation data from Fig. 5 is included for comparison. the histocompatibility specificity of the CTLs are shown in brackets. Adapted from Townsend et al., (1985).

DISCUSSION

A) Current models for nuclear accumulation

Much of the current speculation on the molecular mechanisms responsible for nuclear accumulation are based on experiments using *Xenopus* oocytes. There are two reasons for the popularity of this system. First, the large size of the oocyte and its nucleus have allowed perturbations of the system by either the microinjection of various dextrans (Paine et al., 1975), colloidal particles (Feldherr, 1969), proteins (Gurdon, 1970; Bonner 1975a,b), etc., or by microneedle-induced disruption of the nuclear envelope (Feldherr and Pomerantz, 1978). Today, microinjection techniques are no longer confined to large cells and transformation and transfection techniques in yeast and cultured cells can substitute for microinjection in many types of experiments. However, the large size

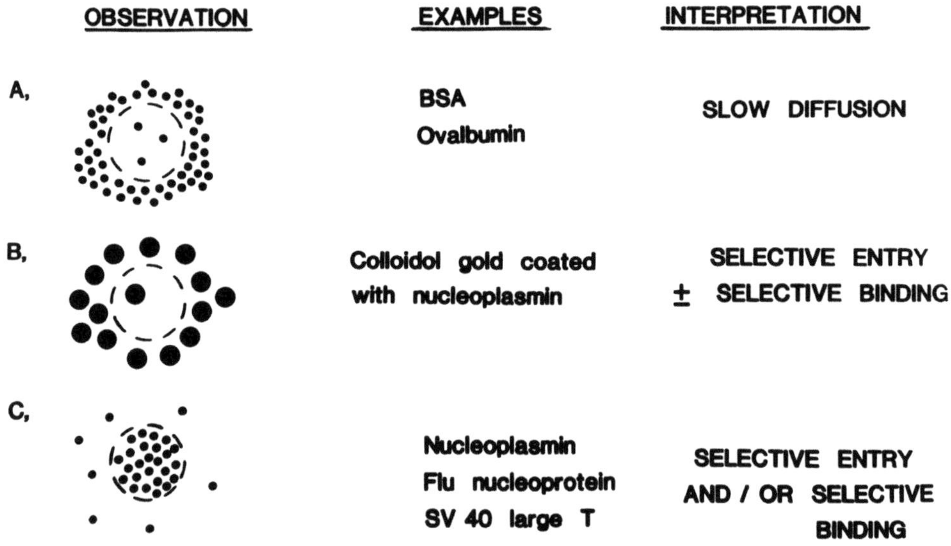

Figure 8 Models for protein movement into the oocyte nucleus
The various "phenotypes" exhibited by proteins (solid circles) which enter nuclei are shown above with possible mechanistic interpretations shown alongside.

of the nucleus has allowed the spatial distribution of the injected material to be followed temporally using ultra-low temperature autoradiography (Paine et al., 1975). This has enabled events occuring at the nuclear envelope to be studied and to be dissociated from the subsequent events occuring after nuclear entry.

Pioneering work on establishing ground rules for nuclear entry was performed by Paine et al. (1975). Using microinjection of Xenopus oocytes with radioactive dextrans of different diameters, they established that the oocyte nuclear envelope acted as a sieve in restricting molecular movement into the nucleus. The permanent radius of the pores of this sieve was determined to be about 45 A. Similar experiments (Gurdon, 1970; Bonner, 1975a,b; Dabauvalle and Franke, 1984) found, that after injection into the cytoplasm, most non-nuclear proteins (i.e. proteins which do not accumulate in the nucleus) would enter the oocyte nucleus at rates predicted by the calculations of Paine et al., (1975). These predictions were correct as long as the longest molecular dimension of the protein did not exceed the diameter of the pore. So, for example, ferritin (diameter

>95A) did not enter the nucleus (Gurdon, 1970; Paine and Feldherr, 1972) whilst bovine serum albumin (150 x 38 A, Young, 1963) entered more slowly than predicted (Bonner, 1975a,b). When the experiments were repeated using nuclear proteins it was found that proteins with molecular dimensions <90A entered for more rapidly than predicted. In addition, colloidal gold particles with diameters greater than 160A but coated with nucleoplasmin can still enter the nucleus (Feldherr et al., 1984). A consideration of a subclass of small, cytoplasmic, soluble proteins which do not enter the nucleus (Dabauvalle and Franke, 1984) is beyond the scope of this review.

Two alternative models are commonly considered to explain the nuclear accumulation of certain proteins (Dingwall et al., 1982; De Robertis, 1983; Hall et al., 1984). The first envisages the free diffusion of all proteins into the nucleus with the subsequent retention of nuclear proteins by their binding to nondiffusible nuclear elements. The second model suggests that the accumulation of nuclear proteins is due to their selective transport across the nuclear envelope. Since nuclear accumulation is selective and occurs posttranslationally with no apparent modification of the proteins (Gurdon, 1970; Bonner, 1975a,b; De Robertis et al., 1978; De Robertis and Black, 1981; Dabauvalle and Franke, 1982; Dingwall et al., 1982), it has been proposed that nuclear proteins contain within their mature molecular structure a signal sequence (or other property such as conformation) that controls their accumulation (De Robertis et al., 1978). In the first model this karyophilic signal would be responsible for the nuclear retention of the protein while, in the second model, its interaction with the nuclear envelope would allow selective transport of the protein into the nucleus. The two models are not mutually exclusive (Fig. 8). In our opinion it is still difficult, even in the best studied cases such as nucleoplasmin accumulation, to be certain which model applies. Xenopus nucleoplasmin (a pentameric protein with a molecular weight of 168,000) rapidly moves into the oocyte nucleus after cytoplasmic injection (Dingwall et al., 1982). Selective proteolysis cleaves nucleoplasmin into a "core" region (M.W.115,000) and monomeric "tails" (M.wt, 12,000). Whilst the tails rapidly migrated into, and accumulated in the nucleus, the cores remained in whichever compartment they were injected (Dingwall et al., 1982). These results are cited as evidence for the selective entry of nucleoplasmin (Dingwall, 1985) However if this were so, one might have expected core fragments (<45A in diameter) injected into the nucleus to slowly diffuse out. Alternatively, selective entry could be ratchet-like and irreversible. Nucleoplasmin clearly does contain a selective entry signal since colloidal gold particles larger than 160A in diameter can enter the nucleus when coated with nucleoplasmin (Feldherr, 1984). One possibility is that an additional contribution by selective binding cannot a priori be excluded. Two additional experiments are needed to resolve this issue. First, if the injection of core nucleoplasmin into the nucleus was followed by puncturing of the nuclear envelope (Feldherr and Pomerantz, 1978) then the diffusion of the nucleoplasmin into the cytoplasm to achieve an equal concentration would exclude a selective binding contribution. Secondly, if nucleoplasmin injected into oocytes containing a demembranated oocyte nucleus (cited in De Robertis, 1983) was to accumulate in the nucleus, this would be evidence for a contribution from selective binding. Interestingly, in the cases of two endogenous oocyte proteins, B3 and B4 (Feldherr and Ogburn, 1980), which rapidly enter but do no accumulate in the nucleus, selective entry alone must explain the kinetics of entry.

B) Which model applies to influenza NP?

Although all of the mutant proteins were able to enter the oocyte nuclei and at least become equally distributed between this compartment and the cytoplasm, only some mutants accumulated there. Since the loss of nuclear accumulation was associated with the loss of amino acids 327-345 of the wild-type NP, we conclude that the information necessary for the nuclear accumulation of this protein is contained wholly or partially within this region. By definition, then, this region contains the karyophilic signal (De Robertis et al., 1978) itself or is responsible for maintaining the integrity of a signal elsewhere in the protein. The amino-terminal limit is set by the protein encoded by pTKNP IMP 1295, which contains amino acids 327-498 and accumulated in the nucleus to a similar extent to wild-type NP. The carboxy-terminal limit is less well defined and based principally on the nuclear accumulation of the protein containing the first 345 amino acids of NP. However, the fact that this mutant accumulated to a lesser extent than the wild-type NP suggests it may not contain all of the karyophilic signal. These observations of the accumulation of different NP mutant proteins whose only common feature is the segment of amino acids 327-345 argue against the indirect signalling role of this region described above. The results obtained with the NP-globin fusion proteins are consistent with this proposed location of the karyophilic signal, since the fusion protein containing this region acumulated in the nucleus, while those lacking this region did not accumulate but became evenly distributed between this compartment and the cytoplasm. However a definitive proof of the function of this region will require the addition of just this region to a non-nuclear protein with the subsequent demonstration of nuclear accumulation of the fusion protein. Such experiments have been performed using SV40 T antigen (see below). All the mutant NP proteins which do not accumulate in the nucleus nevertheless enter and become evenly distributed between the nucleus and the cytoplasm within 6h. Even assuming each of the mutant NP's is spherical in shape the rate of entry is too rapid to be accounted for by unmediated diffusion. Since there is no accumulation only selective entry can be responsible. We therefore speculate that a region of the NP polypeptide that is distinct from that involved in nuclear accumulation is involved in the rapid entry. In addition, since mutants containing non overlapping portions of NP show similar rapid entry kinetics (cf NPimp 1295 and NPpvu1295) this recognition signal must be duplicated. In contrast to the non-accumulating mutants, all mutants which contain amino acids 328-340 accumulate in the oocyte nucleus. The simplest explanation for this behavior is that amino acids 328-340 specify a binding site for some non-diffusible nuclear element. The alternative possibility that this NP region markers entry irreversible is ruled out by the fact that NP injected into the nucleus diffuses into the cytoplasm (Davey et al., 1985b).

C) Sequence homologies between nuclear proteins

The amino acid sequence of wild-type NP from residues 325-350 is shown in Fig. 9.

Although there is no tract of basic residues similar to that which may be important in the nuclear accumulation of the SV40 large T antigen (Kalderon et al., 1984), there is a sequence similar to one within the proposed karyophilic signal-containing region of the yeast α_2 protein (Hall et al., 1984). The α_2 sequence has two basic amino

NP SEQUENCE HOMOLOGY

ACTUAL SEQUENCES

STRAIN	331														345
A/	MET	ALA	CYS	HIS	SER	ALA	ALA	PHE	GLU	ASP	LEU	ARG	VAL	LEU	SER
B/	◆	CYS	◆	PHE	GLY	◆	◆	TYR	◆	◆	◆	◆	◆	◆	◆
C/	◆	◆	◆	PHE	GLY	LEU	◆	TYR	◆	◆	PHE	SER	LEU	VAL	◆

CONSENSUS SEQUENCES

A/B	MET	aa	CYS	aa	aa	ALA	ALA	aa	GLU	ASP	LEU	ARG	VAL	LEU	SER
A/B/C	MET	aa	CYS	aa	aa	aa	ALA	aa	GLU	ASP	aa				

↓

336				340
ALA	h	GLU	ASP	h

Figure 9 Sequence homologies between different NP proteins
The published amino acid sequences form A/NT/60/68 and A/PR/8/34 from anino acids 331-345 (strain A) are shown against the most homologous regions from B/Sing/222/79 (strain B) and the protein encoded on RNA segment 5 from C/Col/78 (strain C). Abbreviations, aa, amino acid, L, hydrophobic residue.

acids flanking three hydrophobic residues, while the NP sequence between amino acids 341-348 has two basic amino acids flanking five residues, four of which are hydrophobic, although this NP sequence has no proline residue in significant contrast to the α_2 sequence.

Whether these sequences have any importance in nuclear accumulation will only become evident after the analysis of mutants altered in these residues. The α_2 sequence is also found in other yeast nuclear proteins (see Hall et al., 1984), but its proposed involvement in the nuclear accumulation of at least one of these proteins, the yeast histone H2B, is difficult to reconcile with deletion analysis apparently showing this region to be unimportant to its *in vivo* function (Wallis et al., 1983).

When the protein sequences of the NP proteins of the influenza virus used in this study (A/NT/60/68; Huddleston and Brownlee, 1982),

A/PR/8/34 (Winter and Fields, 1981) and B/Singapore/222/79 (Londo et al., 1983) are aligned to satisfy the criteria of maximum amino acid homology, a highly conserved sequence (amino acids 336-345) within the proposed karyophilic signal-containing region is revealed (Fig. 9). Although identical for the two type A NPs, it may not be significant since they are highly conserved throughout their entire length. However it represents the second most highly conserved stretch of ten residues between type A and type B NPs, differing only at residue 338 (phe to tyr), but retaining hydrohobicity. This is remarkable because type A and type B NPs differ not only in Mr (56,000 and 61,500, respectively) but are antigenically distinct. Recently (Nakada et al., 1984) the putative NP sequence from a strain C virus has been published. The major area of homology with strains A and B is again in the same region except that the best homology is slightly N-terminally displaced (see Fig. 9). When all three sequences are considered together the best consensus is the pentapeptide ALA-h- GLU- ASP-h where h is a hydrophobic residue. These homologies, and particularly this last one, could therefore reflect the need to retain a sequence necessary for a fundamental feature of the protein, such as its accumulation in nuclei.

CONCLUSION

We suggest that the nuclear accumulation of proteins may require two signals, the first allowing the protein to enter nuclei at a rate faster than that predicted form its size, and the second allowing its nuclear accumulation. It is this second signal that we appear to have identified in the influenza virus NP. Existence of the first signal could be tested by demonstrating that rapid nuclear entry can be conferred on a large nonnuclear protein by its fusion with a region of NP that allows entry but not accumulation in the nucleus.

REFERENCES

Bedoulle, H., Bassford, P., Fowler, A., Zabin, I., Beckwith, J., and Hofnung, M. (1980). Nature 285, 78-821.

Bonner, W.M. (1975a). J. Cell Biol. 64, 421-430.

Bonner, W.M. (1975b). J. Cell Biol. 64, 431-437.

Dabauvalle, M.C. and Franke, W.W. (1982). Proc. Natl. Acad. Sci. USA 79, 5302-5306.

Dabauvalle, M.C. and Franke, W.W. (1984). Exp. Cell Res. 153, 308-326.

Davey, J. Colman, A. and Dimmock, N.J. (1985b). J. Gen. Virol. In press.

Davey, J., Dimmock, N.J. and Colman, A. (1985a). Cell 40, 667-675.

De Robertis, E.M. (1983). Cell 32, 1021-1025.

De Robertis, E.M., and Black, P. (1981). Fortschr. Zool. 26, 49-58.

De Robertis, E.M., Longthorne, R.F. and Gurdon, J.B. (1978). Nature 272, 254-256.

Dingwall, C. (1985). Trends in Biochem. Sci. 64-66.

Dingwall, C. and Allan, K. (1984). EMBO J., 3, 1933-1937.

Dingwall, C., Sharnick, S.V., and Laskey, R.A. (1982). Cell 30, 449-458.

Emr, S.D., Hedgpeth, J., Clement, J.M. Silhavy, T.J. and Hofnung, M. (1980). Nature 285, 82-85.

Feldherr, C. (1969). J. Cell Biol. 42, 841-845.

Feldher, C.M. and Ogburn, J.A. (1980). J. Cell Biol. 87, 589-593.

Feldherr, C.M. and Pomerantz, J. (1978). J. Cell Biol. 78, 168-175.

Feldherr, C., Kallenbach, E. and Schultz, N. (1984). J. Cell Biol

Gurdon, J.B. (1970). Proc. Roy. Soc. Lond. B. 176, 303-314.

Gregoriades, A. (1980). J. Virol. 36, 470-479.

Hall, M.N., Hereford, L., and Herskowitz, I. (1984). Cell 36, 1057-1065.

Huddleston, J.A., and Brownlee, G.G. (1982). Nucl. Acids Res. 10, 1029-1038.

Kalderon, D., Richardson, W., Markham, A., and Smith, A. (1984). Nature 311, 33-38.

Krieg, P., Strachan, R., Wallis, E., Tabe, L., and Colman, A. (1984). J. Mol. Biol. 180, 615-643.

Krohne, G., and Franke, W.W. (1980). Exp. Cell Res. 129, 167-189.

Lingappa, V.R., Chaidez, J., Yost, C.S. and Hedgpeth, J. (1984). Proc. Natl. Acad. Sci. USA 81, 456-460.

Londo, D.R., Davis, A.R., and Nayak, D.P. (1983). J. Virol. 47, 642-648.

Meek, R.L., Walsh, K.A., and Palmiter, R.D. (1982). J. Biol. Chem. 257, 2245-2251.

Mills, A.D., Laskey, R.A., Black, P, and De Robertis, E.M (1980). J. Mol. Biol. 139, 561-568.

Nakade, S., Creager, R., Krystal, M. and Palese, P. (1984). Virus Res. 1, 433-441.

Paine, P. and Feldherr, C. (1972). Exp. Cell Res. 74, 81-98.

Paine, P.L., Morre, L.C., and Horowitz, S.B. (1975). Nature 254, 109-114.

Silhavy, T.J., Benson, S.A. and Emr, S.D. (1983). Microbiol. Rev. 47, 313-344.

Tabe, L., Krieg, P., Jackson, D., Strachan, R. Waliis, E., and Colman, A. (1984). J. Mol. Biol. 180, 645-666.

Townsend, A., Gotch, F. and Davey, J. (1985). Cell, in press.

Townsend, A., McMichael, A.J., Carter, N.P., Huddleston J.A., and Brownlee, G.G. (1984). Cell 39, 13-25.

Wallis, J.W., Rykowski, M., and Grunstein, M. (1983). Cell 35, 711-719.

Winter, G., and Fields, S. (1981). Virology 114, 423-428.

Zehavi-Willner, T., and Lane, C.D. (1977). Cell 11, 683-693.

Zinkernagel, R. and Rosenthal, K. (1981). Immunol. Rev. 58, 131-155.

Adenovirus E1A: Implications of a Post-Translational Modification for Nuclear Localization and Stimulation of Transcription

J. D. Richter[1] and N. C. Jones[2]
[1] Worcester Foundation for Experimental Biology, Shrewsbury, MA, 01545 USA
[2] Imperial Cancer Research Fund, Lincolns Inn Fields, London WC2 3PX, United Kingdom

The E1A gene is one of the first viral genes to be expressed following adenovirus infection of permissive cells. Several functions have been ascribed to the E1A gene product which include the *trans* activation of other viral genes (1-3) and certain cellular genes (4,5), the immortalization and partial transformation of cells (6), and, when acting in concert with other oncogenes, the complete transformation of cells (7). Two mRNAs which are encoded by the E1A gene are synthesized during the early phase of a productive infection. These mRNAs, 12S and 13S, are derived from the same primary transcript by in-frame alternative splicing and encode proteins which contain 243 and 289 amino acid residues, respectively (8). The proteins have identical termini and differ solely by the number of internal amino acid residues. Analysis of the E1A proteins by SDS polyacrylamide gel electrophoresis reveals four prominant species (9-11). It has been suggested that post-translational modifications of each of the primary proteins lead to these molecular weight variants (10). The biological significance of these modifications is not known.

The Xenopus Oocyte as an Assay System for trans Activation by E1A

We have attempted to elucidate the role the modification might play in the function of the E1A 41 kd (243 residues) and 43 kd (289 residues) proteins. In one approach, we have used the *Xenopus laevis* oocyte as a system in which a wide variety of macromolecules may be injected and subsequent biochemical events may be examined. The initial experiments with E1A have been directed toward the determination of whether this *trans* acting transcription factor retains its biological activity in injected oocytes.

Several adenovirus genes are activated by E1A. The promoter region of one of these genes, E3, was linked to a marker gene encoding the bacterial enzyme chloramphenicol acetyltransferase (CAT). This chimeric gene was injected into the nuclei (germinal vesicles or GVs) of oocytes together with either pBR322 or a plasmid containing the E1A gene. Following incubation, the level of CAT activity

Nucleocytoplasmic Transport
Edited by R. Peters and M. Trendelenburg

in each group of injected oocytes was assessed. Relative to E3CAT-pBR322 co-injected controls, there was a greater than eight fold stimulation of CAT activity in oocytes co-injected with E3CAT and the E1A gene (12). Further experiments demonstrated that a protein extract from adenovirus type-5 infected HeLa cells also was capable of stimulating CAT activity in injected oocytes. These studies (12) therefore demonstrated that the *Xenopus* oocyte could serve as an efficacious system for examining *trans* activation by E1A.

In the experiments described above, it was not determined conclusively whether the E1A 41 or 43 kd proteins, or both, was responsible for enhanced CAT activity in injected oocytes. As an initial step in the examination of the activity of each of the proteins, DNAs encoding the 41 and 43 kd proteins were linked to the bacteriophage lambda P_L promoter for overexpression in transformed *E. coli* (13,14). Following their isolation from bacteria, the purified proteins were injected into *Xenopus* oocytes previously injected with E3CAT. Both proteins were shown to stimulate CAT activity (15). An example of such an assay in injected oocytes is demonstrated in Figure 1. Oocyte nuclei were injected with E3CAT and the cytoplasms injected with either control protein (AS1) or the E1A 41 or 43 kd proteins. Following 24 hours of incubation, oocyte homogenates were prepared for CAT assays in which the acetylation of ^{14}C-chloramphenicol is monitored. The autoradiogram in the left portion of the figure shows that the E1A 41 (12 S mRNA) and 43 (13 S mRNA) kd proteins stimulated the acetylation of chloramphenicol by 5-6 fold relative to control protein (AS1). The right portion of the figure shows, more quantitatively, the stimulation of CAT activity by the E1A proteins. In this case, the actual rates of chloramphenicol acetylation were measured. The same approximate increases in CAT activity by the E1A proteins are evident.

E1A Undergoes a Post-translational Modification in the Oocyte Cytoplasm

In adenovirus infected cells, the E1A proteins are located predominantly in nuclei. We have attempted to determine whether the *E. coli*- expressed E1A 43 kd protein localizes to the nucleus in injected oocytes. Figure 2 shows an immunoblot in which the E1A 43 kd protein was detected in nuclei and cytoplasms following cytoplasmic injection. Immediately after injection, E1A migrates as a single 43 kd protein. After one hour, two species of E1A are evident in the cytoplasm, the original 43 kd form and a new 45 kd form. With continued incubation, progressively more of the 43 kd form is modified to the 45 kd form. The examination of nuclei (GV) reveals that one hour following cytoplasmic injection, only the modified 45 kd form of E1A is detected. The same steady state level of the modified E1A is maintained throughout the 12 hour duration of the experiment. The two molecular weight variants of E1A described above are identical in molecular weight to the two

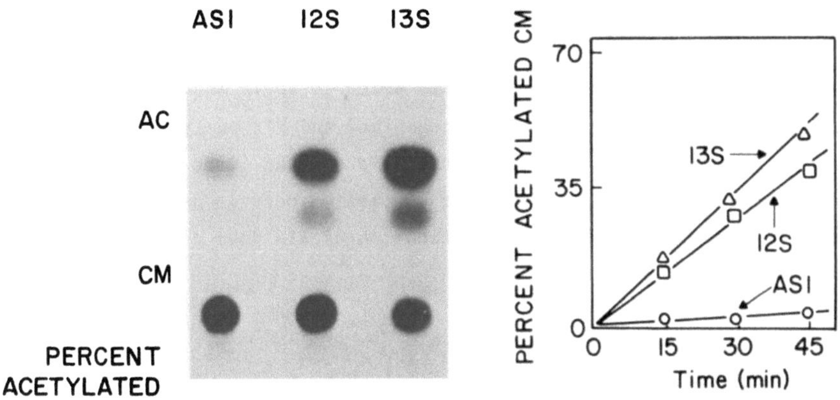

Figure 1 Stimulation of CAT activity of by E1A. Oocytes were injected with E3CAT and either contol protein (AS1) or the E1A 41 (12 S) or 43 (13 S) kd proteins. Following a 24 hour incubation, the oocytes were homogenized and the homogenates were incubated with acetyl coenzyme A and ^{14}C-chloramphenicol. The chloramphenicol then was extracted and the acetylated (AC) and nonacetylated (CM) forms were separated by thin layer chromatography and visualized by autoradiography. The left part of the figure shows an autoradiogram and the right part of the figure shows a time course of incubation during the assay. The percent of acetylated ^{14}C-chloramphenicol was determined by liquid scintillation spectrometry.

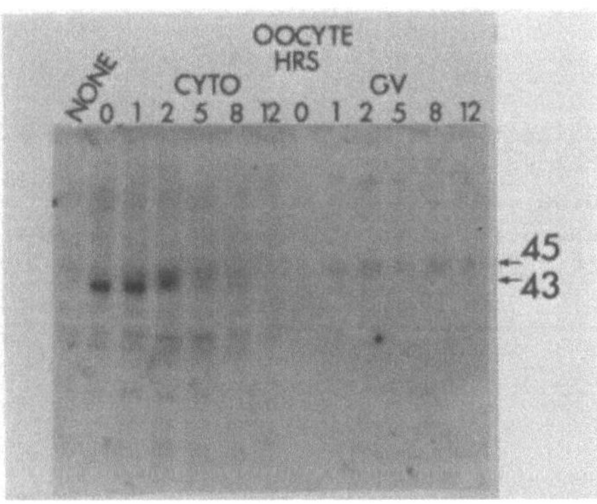

Figure 2 Immunoblot of E1A following oocyte injection. The cytoplasm of oocytes was injected with the E1A 43 kd protein and the cytoplasms and nuclei (GV) were separated at several subsequent time periods. The protein in each fraction then was resolved on a SDS 10% polyacrylamide gel, transferred to nitrocellulose paper, and probed with E1A specific antisera and ^{125}I-rabbit anti-mouse IgG.

E1A forms in HeLa cells transfected with DNA specific for the E1A 43 kd protein (15).

The data in Figure 2 indicate that E1A is modified in the cytoplasm prior to nuclear localization. To determine whether the nuclear envelope is important for the modification of E1A, oocytes were enucleated prior to their injection.

Following incubation, the oocytes were homogenized and E1A analyzed by an immunoblot (Figure 3). Both the E1A 43 kd (lane 1) and 41 kd (lane 3) proteins were modified (lanes 2 and 4, respectively) in enucleated oocytes as assessed by their apparent 2 kd increase. Thus, the modification of E1A does not require any nuclear component.

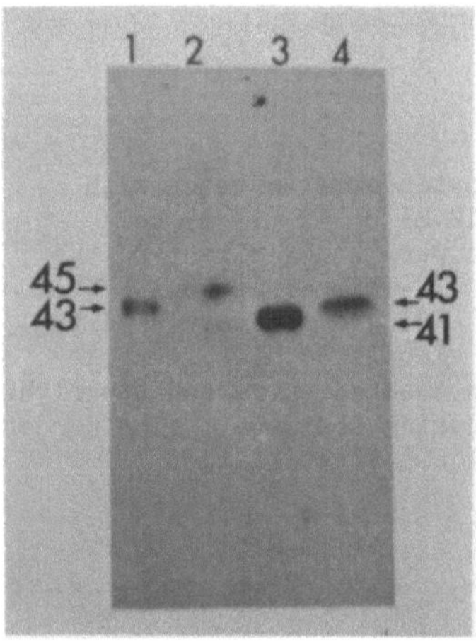

Figure 3 Immunoblot of E1A following injection into enucleated oocytes. Oocytes were enucleated and then injected with the E1A 43 kd or 41 kd proteins. Oocytes then were homogenized, total protein was extracted, and E1A was resolved by an immunoblot as described in Figure 2. Lanes 1 and 3 refer to noninjected E1A 43 S and 41 kd proteins, respectively. Lanes 2 and 4 refer to injected E1A 43 kd and 41 kd proteins, respectively.

E1A is Modified in a First Exon Encoded Domain

In order to examine what portion of the E1A protein is required for the modification, nuclear localization, and stimulation of E3CAT transcription, a series of E1A DNA deletions were constructed. Following transformation of *E. coli* with the mutant DNAs, truncated E1A proteins were isolated and used for oocyte injection experiments. The characteristics of the mutants are listed in Table 1.

The next group of experiments was designed to determine whether the truncated E1A proteins would be modified and accumulate in the nucleus. Figure 4 shows an immunoblot in which oocyte cytoplasms were injected with the truncated E1As and the resulting nuclear forms of the proteins were compared to the noninjected proteins.

Table 1 Salient Features of E1A Mutants

Mutant	Product	Features
AS1	control	no E1A
410	13 S product	288 amino acid residues, lacks the amino terminal residue
610	13 S product	222 amino acid residues, lacks 66 carboxy terminal residues
420	13 S product	267 amino acid residues, lacks 21 amino terminal residues
620	13 S product	207 amino acid residues, lacks 66 carboxy and 21 amino terminal residues
412	12 S product	242 amino acid residues, identical to 410 except lacks 46 internal residues
410X	13 S product	167 amino acid residues, lacks 139 carboxy terminal residues, but has an added 18 residues at the carboxy terminus
412C	12 S product	167 amino acid residues, lacks 169 carboxy terminal residues, but has an added 48 residues at the carboxy terminus

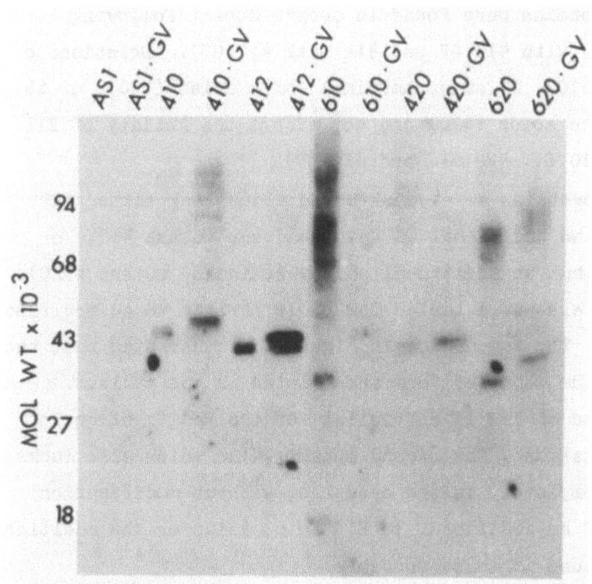

Figure 4 Immunoblot of truncated E1A proteins in injected oocytes. Oocyte cytoplasms were injected with the mutant E1A proteins described in Table 1. Following incubation, the nuclei (GV) of the oocytes were isolated and the presence of E1A was determined as described in Figure 2. AS1, 410, 412, 610, 420, and 620 refer to noninjected E1A proteins. The suffix GV refers to the nuclei from oocytes whose cytoplasms had been injected with the E1A proteins.

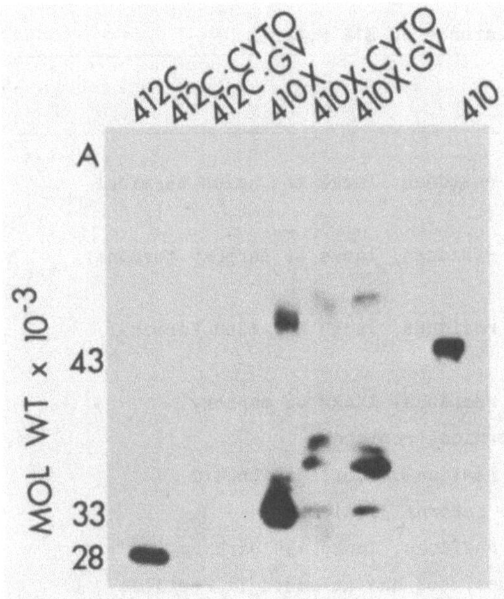

Figure 5 Immunoblot of truncated E1A proteins which contain only first exon-encoded amino acids. Oocytes were injected with proteins 412C and 410X (Table 1) and were analyzed 24 hours after injection as described in Figure 4.

The modified forms of both E1A proteins were found in oocyte nuclei following cytoplasmic injection (compare 410 with 410.GV and 412 with 412.GV). Deletions of 66 carboxy terminal amino acids (610), 21 amino terminal amino acids (420), or 66 carboxy and 21 amino terminal amino acids (620) did not affect the ability of E1A to be modified or enter nuclei (610.GV, 420.GV, and 620.GV).

Two additional truncated E1A proteins were constructed which lack either 139 carboxy terminal amino acids (but an additional 18 new residues, mutant 410X) or 169 carboxy terminal amino acids (but an additional 48 new residues, mutant 412C) (Table 1). These mutant proteins also were tested for their ability to be modified and to localize in oocyte nuclei. The immunoblot in Figure 5 demonstrates that the 410X mutant is modified and that the modified form accumulates in the nucleus. Thus, the first exon-encoded region of E1A is sufficient for the modification to take place and for nuclear localization. The second mutant, 412C which also lacks first exon material, turns over completely in the cytoplasm without modification. This may be due to the deletion of an additional 30 E1A amino acids or the addition of 48 "nonsense" residues which cause it to be unstable.

A First Exon Encoded Region of E1A is Sufficient for trans Activation of E3CAT

In addition to nuclear localization studies, the mutant E1A proteins have been used to map the domain which is responsible for stimulating transcription of E3CAT. Oocytes injected with E3CAT and the truncated E1A proteins were used for a series of CAT assays. Figure 6 demonstrates that both the E1A 43 kd (410) and 41 kd (412) proteins stimulate CAT expression by the same amount (5-6 fold). The mutant proteins which lack the carboxy terminus (610), the amino terminus (420), or both termini (620) retain their ability to stimulate CAT expression. Another mutant which contains only the first exon-encoded amino acids (410X) also stimulates CAT expression by the same amount as E1A 43 kd protein (Figure 7). Mutant 412C, which turns over in the cytoplasm does not stimulate E3CAT (Figure 7). These data demonstrate that the first exon of E1A contains sufficient information for the induction of adenovirus promoter expression.

Discussion

The results reported in this study demonstrate that the *E. coli* expressed E1A 41 and 43 kd proteins both undergo a modification in the cytoplasm of injected *Xenopus* oocytes. When analyzed by SDS polyacrylamide gel electrophoresis, the modified forms of the proteins appear to have increased in mass by 2 kd. Fractionation of injected oocytes into nuclear and cytoplasmic compartments reveals that only the modified forms of the proteins are localized in the nucleus. Deletion mapping with truncated E1A proteins demonstrates that the modification takes place in a first exon-encoded region. This same region also contains sufficient information for the induction of adenovirus promoter expression.

In addition to *Xenopus* oocytes, the same apparent modification of E1A occurs in adenovirus infected HeLa cells, injected Vero cells, and in a soluble rabbit reticulocyte lysate (15). Thus, the activity which catalyzes the modification is general in nature, and implies that many different proteins undergo the same modification. Although there appears to be no widespread occurance of proteins which increase in molecular weight when they enter nuclei, it is possible that many proteins are modified in a manner similar to E1A except that there is no resulting obvious change in molecular weight. For example, many proteins are known to migrate abberantly in SDS gels. The E1A 43 kd protein, for one, actually is a 32 kd protein as deduced by DNA sequence analysis. Modified E1A might appear as an obvious molecular weight variant because the unmodified form behaves anomalously in SDS polyacrylamide gels. Other proteins which might be modified need not necessarily appear as molecular weight variants if the unmodified forms migrate normally in SDS polacrylamide gels. It is noteworthy that another nuclear protein,

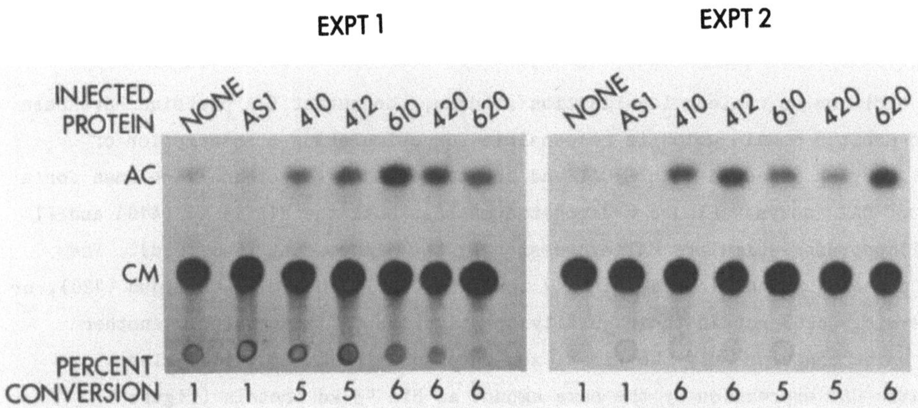

Figure 6 CAT assays with truncated E1A proteins. Oocytes were injected with E3CAT and the truncated E1A proteins (Table 1). After incubation, CAT assays were performed as described in Figure 1.

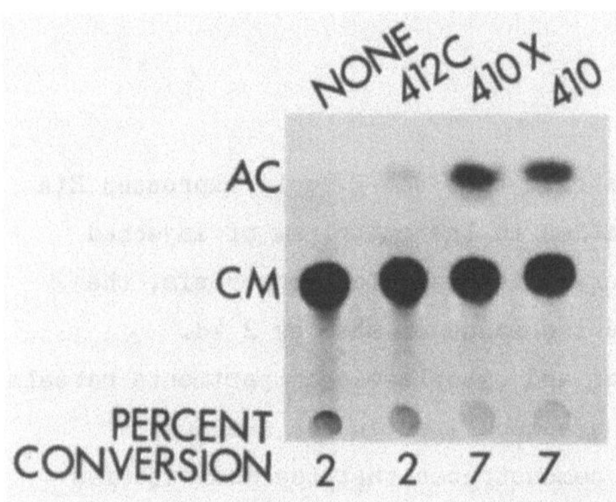

Figure 7 CAT assays with truncated E1A proteins which contain only first exon-encoded amino acids. Oocytes were injected with the truncated E1A proteins 412C and 410X, as well as the E1A 43 kd protein (410) (Table 1). After incubation, CAT assays were performed as described in Figure 1.

that encoded by the c-myc oncogene, migrates anomolously in SDS polyacrylamide gels (13,16) and also appears to be modified (17,18). Further experiments are needed to determine the nature of the modification of E1A as well as other related proteins such as that encoded by the c-myc gene.

Three general hypotheses regarding the localization of proteins in the nucleus have been put forth. The first holds that proteins diffuse freely across the nuclear membrane, but that only certain proteins are retained in the nucleus by an interaction with some nondiffusable substrate. The second also states that all proteins diffuse across the nuclear membrane, but that selective degradation in the cytoplasm increases the apparent nuclear concentration of a protein. In the third hypothesis, only certain proteins have an intrinsic quality, a "signal", which directs them to the nucleus. The data in Figure 2 show clearly that within one hour following cytoplasmic injection, modified E1A is distributed asymmetrically between nucleus and cytoplasm. Taking into account the differences in volume between the oocyte nucleus and cytoplasm (19), we estimate that at one hour following injection, modified E1A is concentrated about 50 fold in the nucleus. The speed and magnitude of this difference would argue against the free diffusion of E1A but does not speak directly to the question of whether it is selectively retained in the nucleus once there. The hypothesis that modified E1A is selectively degraded in the cytoplasm is inconsistent with the data in Figure 3 since modified E1A is stable in the cytoplasm if it does not have a nucleus to enter. We believe the data presented in this report are consistent with the hypothesis that E1A contains a nuclear signal, but with the added pronouncement that the modification influences the signal. The signal need not actually be the modification and the amino acid residue(s) it modifies, but may be at some residues distal to the site of modification which only become exposed after the protein is modified. This theory does not exclude the possiblity that the modification also is important for retaining the protein once inside the nucleus (20). Indeed, the observation that the modification may be important for <u>trans</u> activation of genes might indicate that it is multifunctional.

The results obtained with the E1A deletion mutants indicate that the modification occurs in the first exon-encoded region (residues 22-167). This would suggest that all signals which are required for nuclear localization are located in this region. However, in injected monkey kidney cells, removal of the carboxy terminus of E1A slows its rate of nuclear accumulation (21). This difference may be due to cell-specific factors since no obvious change in the rate of nuclear accumulation of the truncated E1As is observed in injected oocytes.

In addition to nuclear localization, this report presents evidence that the first exon-encoded region of E1A also is important for <u>trans</u> activation of the adenovirus E3 promoter. These data are consistent with other reports which indicate that a first exon-encoded region of E1A is adequate for the expression of other adenovirus promoters (22,23). In most studies dealing with adenovirus infected cells, however, both exon-encoded domains must be present for the efficient expression of other adenovirus genes (24). These differences may reflect cell specificity with respect to E1A function. For example, the E1A mutants described in this report also have been injected into monkey kidney (Vero) cells and tested for their ability to complement the E1A deficient adenovirus-5 mutant <u>dl</u>312. The assay in this case is stimulation of expression of the adenovirus major late transcription unit. Mutant 410X stimulated dl312 expression 10-20 fold less than wild type E1A (43 kd) (21). In injected oocytes, however, both proteins stimulated E3CAT expression by the same amount. Whether the modification of E1A affects directly its function as a *trans* acting transcription factor in *Xenopus* oocytes or mammalian cells remains to be demonstrated.

In summary, the post-translational modification of E1A may be important for its nuclear localization and stimulation of transcription. Further elucidation of the role of this modification will require its identity and its exact location(s) in the E1A protein.

Acknowledgements

We thank Drs. B. Ferguson and M. Rosenberg for providing the E1A protein samples. This work was supported in part by grants from the NIH (GM 34554 and CA 40189) to JDR.

References

1. Berk, A.F., Lee, F., Harrison, T., Williams, J., and Sharp, P.A. (1979). Cell 17, 935-944.
2. Jones, N.C., and Shenk, T. (1979). Proc. Natl. Acad. Sci., USA 76, 3665-3669.
3. Nevins, J.R. (1981). Cell 26, 213-220.
4. Kao, H.-T., and Nevins, J.R. (1983). Mol. Cell. Biol. 3, 2058-2065.
5. Stein, R., and Ziff, E.B. (1984). Mol. Cell. Biol. 4, 2792-2801.
6. Houweling, A., van den Elsen, P.J., and van der Eb, A.J. (1980). Virology 105, 537-550.
7. Ruley, H. E. (1983). Nature (London) 304, 602-606.
8. Perricaudet, M., Akusjarvi, G., Virtanen, A., and Pettersson, U. (1979). Nature (London) 281, 694-696.
9. Smart, J.E., Lewis, J.B., Mathews, M.B., Harter, M.L., and Anderson, C.W. (1981). Virology 112, 703-713.
10. Yee, S., Rowe, D.T., Tremblay, M.L., McDermott, M.L., and Branton P.E. (1983). J. Virol. 46, 1003-1013.
11. Green, M., Wold, W.S.M., Brachmann, K.H., and Cartas, M.A. (1979). Virology 97, 75-286.
12. Jones, N. C., Richter, J.D., Weeks, D.L., and Smith, L.D. (1983). Mol. Cell. Biol. 3, 2131-2142.
13. Ferguson, B., Krippl, B., Jones, N., Richter, J., Westphal, H., and Rosenberg, M. (1985). In Cancer Cells 3: Growth Factors and Transformation (eds. Feramisco, J., Ozanne, B., and Stiles, C.) Cold Spring Harbor Laboratory, Cold Spring Harbor, N.Y.
14. Ferguson, B., Jones, N., Richter, J., and Rosenberg, M. (1984). Science 224, 1343-1346.
15. Richter, J.D., Young, P., Jones, N.C., Krippl, B., Rosenberg, M., and Ferguson, B. (1985). Proc. Natl. Acad. Sci. USA (in press).
16. Ferguson, B., Krippl, B., Andrisani, O., Jones, N., Westphal, H., and Rosenberg, M., (1985). Mol. Cell. Biol. 5, 2653-2666.
17. Hann, S.R., Abrams, H.D., Rohrschneider, L.R., and Eisenman, R.N. (1983). Cell 34, 789-798.

18. Persson, H., Hennighausen, L., Taub, R., DeGrado, W., and Leder, P. (1984). Science 225, 687-693.
19. Davey, J., Dimmock, N.J., and Colman, A. (1985). Cell 40, 667-675.
20. Feldman, L.T., and Nevins, J.R. (1983). Mol. Cell. Biol. 3, 829-838.
21. Krippl, B., Ferguson, B., Jones, N., Rosenberg, M., and Westphal, H. (1985). Proc. Natl. Acad. Sci. USA (in press).
22 Solnick, D., and Anderson, M.A. (198). J. Virol. 42, 106-113.
23. Bos, J.L., Jochemsen, A.G., Bernards, R., Scrier, P.I., van Ormondt, H., and van der Eb, A.J. (1983). Virology 129, 393-400.
24. Velcich, A., and Ziff, E. (1985). Cell 40, 705-716.

Considerations on the Mechanism of Nuclear Protein Localization in Yeast

M. N. Hall
Department of Biochemistry and Biophysics, University of California, San Francisco, CA 94143, USA

The past ten years have witnessed tremendous advances in an understanding of how proteins are selectively localized to organelles or the cell exterior (1-5). However, the mechanism by which specific proteins accumulate in the nucleus is relatively poorly understood (6). Until recently, the commonly accepted model for nuclear protein localization has been that all proteins freely diffuse into the nucleus with subsequent retention of nuclear proteins by binding to a non-diffusible substrate (e.g., DNA) (7). The general acceptance of this model is perhaps responsible for the tardiness of recent observations which indicate that nuclear localization may involve more than simple diffusion. A second model, for which evidence is now emerging, is that proteins are selectively translocated across the nuclear envelope. This second model is based on observations that nuclear proteins contain a determinant which identifies them as proteins destined to be taken into the nucleus (8-12). Here I briefly review one such observation.

A Localization Signal

To identify a possible nuclear localization determinant within a nuclear protein, we have taken a genetic approach with the yeast *Saccharomyces cerevisiae* (9). We constructed a set of gene fusions that code for hybrid proteins containing varying amounts of the yeast nuclear protein α2 at the amino terminus and a constant, enzymatically active portion of *Escherichia coli* β-galactosidase at the carboxy terminus. The α2 protein is a sequence-specific DNA-binding, regulatory protein (13). The rationale for identifying a localization determinant made the assumption that should α2 contain such a signal, then fusion of β-galactosidase to the appropriate amount of α2 would convert β-galactosidase from a cytoplasmic protein to a nuclear protein. The portion of the α2 amino acid sequence required for this conversion would give an indication of the site of the localization determinant within α2.

Nucleocytoplasmic Transport
Edited by R. Peters and M. Trendelenburg

	+	HYDROPHOBIC	+		
MET ASN	LYS 3	ILE PRO ILE	LYS 7	ASP LEU LEU ASN PRO GLN	α2 (210A.A.)
MET PHE THR SER	LYS 5	PRO ALA PHE	LYS 9	ILE LYS ASN LYS ALA SER	α1 (175A.A.)
PRO ALA GLU LYS	LYS 18	PRO ALA ALA	LYS 22	LYS THR SER THR SER THR	H2B-1 (131A.A.)
GLN GLY ILE THR	LYS 32	PRO ALA ILE	ARG 36	ARG LEU ALA ARG ARG GLY	H4 (103A.A.)
ASN LYS LYS THR	ARG 79	ILE ILE PRO	ARG 83	HIS LEU GLN LEU ALA ILE	H2A (132A.A.)
SER VAL TYR GLU	LYS 411	PHE ALA PRO	LYS 415	GLY LYS GLN LEU SER MET	RAD52 (504A.A.)
HIS ALA PRO LEU	LYS 112	PRO VAL VAL	ARG 116	LYS LYS PRO GLU PRO ILE	SPT2 (333A.A.)

Figure 1. A sequence present within the amino-terminal thirteen residues of the α2 protein and in other known or presumed nuclear proteins. The number underneath an amino acid indicates the position of the amino acid within the respective protein. Numbers in parenthesis indicate the total number of amino acids in the designated proteins.

The hybrid proteins contain 3, 13, 25, 67, or all 210 amino acids of α2, and are referred to as 3α2-LacZ, 13α2-LacZ, 25α2-LacZ, 67α2-LacZ, and 210α2-LacZ, respectively.

Indirect immunofluorescence studies with cells containing the hybrid proteins indicate that the amino-terminal thirteen residues of α2 are sufficient for targeting β-galactosidase to the nucleus (see Table 1). Comparison of amino acid sequences of other nuclear proteins with these thirteen amino acids of α2 reveals a sequence that may be important for nuclear targeting, lys_3-ile-pro-ile-lys_7 (9). A similar sequence of two positively charged amino acids flanking three hydrophobic residues, one of which is proline, is present in several other yeast nuclear proteins (see Figure 1). The sequence is not present in any yeast cytoplasmic protein currently known.

Role of the Signal

According to a passive diffusion and selective retention model of nuclear localization, the role of the signal within the amino-terminal thirteen residues of α2 would be to bind DNA, the intra-

HYBRID	NUCLEAR LOCALIZATION	DNA-BINDING
3α2-LacZ	-	-
13α2-LacZ	+	n.d.
25α2-LacZ	+	-
67α2-LacZ	+	-
210α2-LacZ	+	+

Table 1. Summary of Properties of Hybrid Proteins

nuclear substrate of α2. This prediction was tested by posing the following question. Is there a correlation between nuclear localization of the hybrid proteins and an ability to bind DNA? The different hybrid proteins were partially purified from yeast and assayed for non-specific DNA-binding by DNA-cellulose chromatography. Determination of non-specific binding should be adequate to test the retention model since 1) there is no precedent for a DNA-binding protein that binds specifically but not non-specifically, and 2) specific binding alone would presumably not be sufficient to retain α2 in the nucleus due to the relative scarcity of specific binding sites. The results are presented in Table 1. In summary, the 210α2-LacZ hybrid binds DNA; the 67α2-LacZ, 25α2-LacZ, and 3α2-LacZ hybrids do not bind DNA. These results indicate that the role of the localization signal in α2 is not to bind intra-nuclear substrate; nor is such binding necessary for nuclear localization, since the 67α2-LacZ and 25α2-LacZ hybrids do not bind DNA but are localized to the nucleus. The above results taken together argue against passive diffusion and retention as the mechanism of nuclear localization. Of course, these results are not necessarily significant nor is the above conclusion correct should there be a second intra-nuclear substrate (other than DNA) which could be a retainer component in the absence of DNA-binding; however, no existing data directly justify or require invoking a second intra-nuclear substrate for α2. Additional experiments using the α2-LacZ hybrids as tools may indicate that the role of the localization determinant within α2 is to interact with the nuclear envelope.

A Second Signal?

Is the signal within the amino-terminal thirteen residues of α2 necessary for nuclear localization? A deletion of amino acids three through twenty of α2 does not affect, by the indirect immunofluorescence assay, nuclear localization of hybrid 210α2-LacZ (M.N.H., G. Mullenbach, and C. Craik, unpublished). Thus, the amino-terminal thirteen residues of α2 are sufficient but not necessary for targeting β-galactosidase to the nucleus. This raises the unexpected possibility that α2 has a second localization determinant which, like the first, is sufficient but not necessary.

Should α2 have two nuclear localization determinants, they are not necessarily functionally equivalent (the sequence at the amino terminus is not reiterated). For reasons described below, the signal at the amino terminus is likely a specialized signal directly involved in mediating nuclear localization, perhaps by binding a receptor on the nuclear envelope. The presumed second signal could be 1) a DNA-binding domain, 2) a protein-protein interaction or multimerization domain, or 3) a second specialized localization signal. A commonly invoked mechanism of nuclear localization is that all proteins freely diffuse into the nucleus with selective retention by binding to a non-diffusible intra-nuclear substrate. Accordingly, the second signal could be a DNA-binding domain mediating retention of α2 in the nucleus. This is consistent with the observation that a deletion of the amino terminus of α2 does not affect DNA binding (M.N.H. and A. Johnson, unpublished). The work of Dingwall et al. (8) on the pentameric protein nucleoplasmin suggests that a defective subunit can "piggyback" into the nucleus by multimerization with a functional subunit. In the case of α2, the notion of multimerization and subsequent ride to the nucleus is supported by the observation that the 210α2-LacZ hybrid containing a deletion of the amino-terminal localization signal confers a dominant $\alpha 2^-$ phenotype i.e., the mutated protein interacts with wild-type α2. Experiments to test directly whether the mutated hybrid protein requires wild-type α2 for nuclear localization have not been possible since the mutated hybrid is, for unknown reasons, toxic in the absence of wild-type α2. Thus, the second localization signal is not necessarily a specialized signal directly involved in mediating entry into the nucleus. The following evidence suggests that the signal within the amino-terminal thirteen residues is a specialized localization signal. First, this signal can mediate nuclear localization in the absence of wild-type α2. Second, the amino terminus of α2 is neither required nor sufficient for DNA binding, as determined by the non-specific DNA-binding assay with the different hybrid proteins and with the 210α2-LacZ hybrid containing a small amino-terminal deletion.

Conclusions

The results reviewed above indicate the following. 1) The amino terminus of the yeast nuclear protein α2 is sufficient but not necessary for nuclear localization. 2) The role of the amino terminus is not to bind DNA, the intra-nuclear substrate of α2. DNA-binding is not necessary for nuclear localization. 3) The α2 protein contains a second localization determinant that is also sufficient but not necessary for targeting to the nucleus.

The results further suggest that nuclear proteins are localized by a selective uptake; however, the results do not rule out a selective retention mechanism. Additional experiments are necessary to clarify the existence, roles, and sites of the possibly multiple nuclear localization signals in α2. Clarification will require not only further analysis of α2 itself, but also identification of the cellular components with which the localization signal(s) interact.

ACKNOWLEDGEMENTS: This work was supported by a Research Grant (GM35284) from the National Institute of General Medical Sciences awarded to M.N.H.

REFERENCES

1. Sabatini, DD, Kreibich, G, Morimoto, T, and Adesnik, M (1982). Mechanisms for the incorporation of proteins in membranes and organelles. J Cell Biol 92:1-22.

2. Schekman, R, and Novick, P (1982). The secretory process and yeast cell-surface assembly. In Strathern, JN, Jones, EW, and Broach, JR (eds): "The Molecular Biology of the Yeast Saccharomyces: Metabolism and Gene Expression" Cold Spring Harbor, New York: Cold Spring Harbor Laboratory, p 361-393.

3. Walter, P, Gilmore, R, and Blobel, G (1984). Translocation across the endoplamsic reticulum. Cell 38:5-8.

4. Benson, SA, Hall, MN, Silhavy, TJ (1985). Genetic analysis of protein export in *Escherichia coli* K12. Ann Rev Biochem 54:101-134.

5. Dunphy, WG, and Rothman JE (1985). Compartmental organization of the golgi stack. Cell 42:13-21.

6. De Robertis, EM (1983). Nucleocytoplasmic segregation of proteins and RNAs. Cell 32:1021-1025.

7. Bonner, WM (1978). Protein migration and accumulation in nuclei. In Busch, H (ed): "The Cell Nucleus" Vol 6 New York: Academic Press, p 97-148.

8. Dingwall, C, Sharnick, SV, and Laskey, RA (1982). A polypeptide domain that specifies migration of nucleoplasmin into the nucleus. Cell 30:449-458.

9. Hall, MN, Hereford, L, and Herskowitz, I (1984). Targeting of E. coli β-galactosidase to the nucleus in yeast. Cell 36:1057-1065.

10. Lanford, RE, and Butel, JS (1984). Construction and characterization of an SV40 mutant defective in nuclear transport of T antigen. Cell 37:801-813.

11. Kalderon, D, Richardson, WD, Markham, AF, and Smith, AE (1984). Sequence requirements for nuclear location of simian virus 40 large-T antigen. Nature 311:33- 38.

12. Kalderon, D Roberts, BL, Richardson, WD, and Smith, AE (1984). A short amino acid sequence is able to specify nuclear location. Cell 39:499-509.

13. Johnson, AD, and Herskowitz, I (1985). A repressor (MATα2 product) and its operator control expression of a set of cell type specific genes in yeast. Cell 42:237-247.

Regulation of the Chicken Ovalbumin Gene Expression by Steroid Hormones in a Transient Assay

A. Dierich, M.P. Gaub, D. Astinotti, I. Touitou, J.P. Le Pennec, and P. Chambon
Laboratoire de Génétique Moléculaire des Eucaryotes du CNRS, Unité 184 de Biologie Moléculaire et de Génie Génétique de l'INSERM, Institut de Chimie Biologique, Faculté de Médecine, 67085 Strasbourg, France

INTRODUCTION

The chick oviduct is a particularly suitable system to study the molecular mechanisms regulating gene expression, because the synthesis of the major egg-white proteins (e.g. ovalbumin and conalbumin) can be modulated in the tubular gland cells by the administration and withdrawal of estrogens, progestins, glucocorticoids and androgens, each acting through a distinct hormone receptor (see Shepherd et al. 1980 and refs. therein). Following hormonal stimulation, there is an accumulation of the egg-white protein mRNAs, which results at least in part from an increased rate of transcription of the corresponding genes (see McKnight and Palmiter 1979 and refs. therein). Arrest of stimulation (withdrawal) leads to a cessation of synthesis of these mRNAs, which can be restimulated by administrations of anyone of the four steroid hormones.

The molecular mechanisms by which steroid hormones induce transcription are poorly understood, but it is assumed that it is in some way mediated by the binding of nuclear steroid hormone-receptor complexes to regulatory sequences in the vicinity of the induced genes (see for instance Chandler et al. 1983). However, it is not known whether the interaction between the hormone-receptor complex and the regulatory sequences leads to the activation of promoter elements otherwise intrinsically inactive (positive regulation) or to a relief of repression on promoter elements "constitutively" active unless repressed (negative regulation).

We have previously cloned the ovalbumin gene which is specifically transcribed in the oviduct and determined its structure (see Heilig et al. 1982 and refs. therein). We have also shown that it is neither accurately nor efficiently expressed when transferred into a variety of non-oviduct cells (Breathnach et al. 1980; Chambon et al. 1984). Since there is no permanent chicken cultured cell

Nucleocytoplasmic Transport
Edited by R. Peters and M. Trendelenburg

line derived from oviduct or other chicken tissues which contains estrogen and progestin receptors, and in which the cloned ovalbumin gene could be transferred to study its function and the mechanisms of its hormonal induction, we have microinjected recombinants containing the ovalbumin promoter region into the nucleus of primary cultured oviduct tubular gland cells.

RESULTS

A) A microinjection assay for ovalbumin promoter activity

5'-flanking fragments of increasing lengths of the ovalbumin promoter (from position -1 to positions -56, -134, -295, -425 or -1348 upstream to the capsite - Fig. 1, the pTOT series) were coupled to the SV40 T-antigen coding sequence in such a manner that T-antigen expression is under the control of the ovalbumin promoter. pTOT series DNAs were microinjected into the nuclei of primary cultured oviduct tubular gland cells. 24 hrs later the function of the ovalbumin promoter region was monitored by indirect immunofluorescence against T-antigen. The validity of this promoter assay has been previously discussed (Moreau et al. 1981; Wasylyk et al. 1983). Since variations of fluorescence intensity which cannot be validly measured are not taken into account, the assay which scores the number of fluorescent nuclei rapidly reaches a saturation level as the amount of synthesized

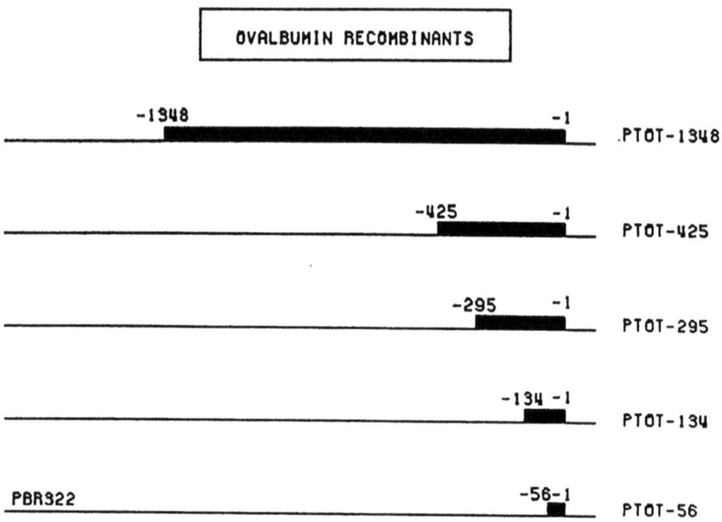

Fig. 1

RNA increases. Large decreases in the efficiency of a promoter are accompanied by very significant decreases of the number of immunofluorescent nuclei, whereas increases of promoter efficiency above a certain level are reflected by only small increases in this number. The wild-type SV40 early transcription unit recombinant pSV1 (Benoist and Chambon, 1981), which functions in all cell types and is not sensitive to steroid hormones, was microinjected in parallel as reference gene.

B) Effect of steroid hormone antagonists on expression of the ovalbumin promoter-recombinants.

All five pTOT recombinants (pTOT-56 to pTOT-1348) were expressed to approximately the same level (50% compared to PSV1 taken as 100%) after microinjection in nuclei of tubular gland cells maintained in a medium containing fetal calf serum. In contrast, none of the pTOT recombinants was expressed when microinjected into the fibroblasts which were also present in primary cultures of chicken oviduct cells, whereas the SV40 recombinant pSV1 was similarly expressed in both oviduct cell types. The possible effect of the steroid hormones present in fetal calf serum on expression of the ovalbumin promoter recombinant was investigated using the anti-estrogen, Tamoxifen, and a new compound, RU486, which is a specific anti-glucocorticoid in chicken (E. Baulieu, and M. Govindan, personal communications). Tamoxifen and RU486 did not significantly affect the expresion of pTOT-56, pTOT-134 and pTOT-295, but the expression of pTOT-425 and of pTOT-1348, was drastically inhibited (10-fold inhibition) by these antagonists. The residual level of activity of pTOT-425 (8%) may correspond to the presence of variable amounts of progestins in fetal calf serum, since the expression of pTOT-425 was "reactivated" by addition of progesterone. Microinjection of the "TATA-box"-mutated recombinants pTOT-134M and pTOT-425M, instead of their wild-type counterparts, resulted in an approximately 5-fold decrease in expression, indicating that RNA synthesis was indeed initiated under the control of the ovalbumin promoter.

C) Expression of pTOT-425, but not of pTOT-295, pTOT-134 and pTOT-56, is blocked in the absence of steroid hormones.

To demonstrate directly that expression of pTOT-425, but not of pTOT-295, pTOT-134 and pTOT-56 is dependent on steroid hormones, primary cultured oviduct cells were maintained in a stripped medium (SM) from which these hormones were removed by treatment with dextran-charcoal, which removes most, but not all, of the steroid hormones present in serum. Expression of pTOT-56, pTOT-134 and pTOT-295 was unsensitive to steroid hormone or antagonist additions. On the contrary, microinjection of pTOT-425 resulted in a lower level of expression (26% in SM medium instead of 53% in FCS medium). Addition of both Tamoxifen and RU486 caused a further decrease of expression down to 4% on average, which most likely reflects a residual amount of progestins in the stripped medium.

The repression exerted on expression of pTOT-425 in the absence of steroid hormones was relieved by the addition of estradiol or progesterone. Both of them increased pTOT-425 expression to a level similar to that observed for pTOT-134 and pTOT-295. A similar level was obtained by adding progesterone to the stripped medium in the presence of RU486. The results obtained after addition of the specific glucocorticoid antagonist RU486 suggest that glucocorticoids can also relieve the repression exerted in the absence of steroid hormones. pTOT-1348 gave results identical to those obtained with pTOT-425.

D) Orientation specificity for the negative regulation.

When the ovalbumin sequence between -134 to -425 was inserted in pTOT-425 in inverse orientation to the transcription start site, the expression after micro-injection of the resulting recombinant pOVB in oviduct cells in MFCS became insensitive to the addition of both hormone antagonists Tamoxifen and RU486 indicating that the action of the negative regulatory element is dependent of its orientation.

E) Ability of the ovalbumin sequence (-134 to -425) to confer negative regulation and steroid hormone inducibility on heterologous promoters.

Insertion of the ovalbumin sequence (-134 to -425) upstream of the chicken conalbumin and adult β-globin promoter sequences (see Fig. 2) resulted in the

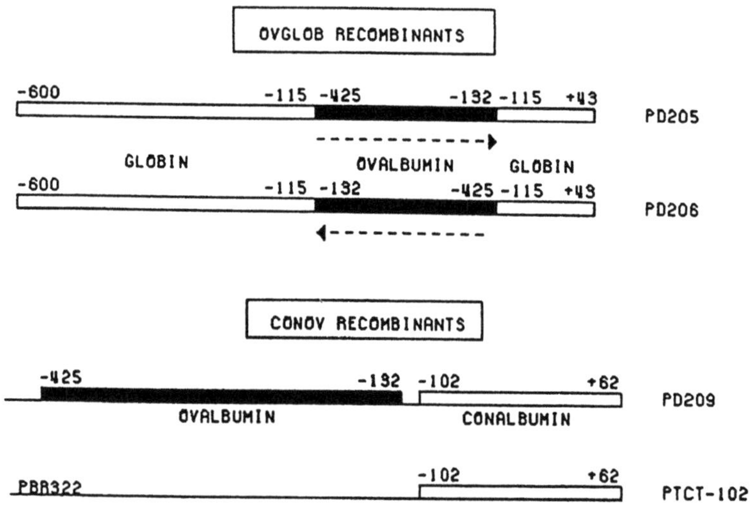

Fig.2

a repression of expression of the respective recombinants pD209 and pD205 after microinjection in oviduct tubular gland cells in presence of Tamoxifen and RU486. Addition of progesterone in the presence of these inhibitions resulted in an increase of expression, whereas the expression of the conalbumin recombinant pTCT-102 and the wild type β-globin recombinant was insensitive to the hormone antagonists Tamoxifen and RU486 and to progesterone. The repressive effect was lost in the ovglob recombinant pD206 (Fig. 2) when the ovalbumin sequence was cloned in opposite orientation.

CONCLUSIONS

We have previously shown that none of the pTOT recombinants functions when microinjected into a variety of heterologous non-chicken or chicken cells, including the fibroblasts present in primary cultured oviduct or liver cells (Chambon et al. 1984 and unpublished results). However, anyone of these recombinants was efficiently expressed in primary cultures of chicken embryo hepatocytes, irrespective of the presence of steroid hormones, although the ovalbumin gene is never expressed in chicken liver in spite of the presence of estradiol receptor. These results together with those presented here, lead to the following conclusions :

1. Chicken hepatocytes contain cell-specific factor(s) which permit the constitutive activity of the microinjected ovalbumin promoter in the absence of steroid hormones. The corresponding control element(s) is contained within 134 bp upstream from the capsite.
2. Oviduct tubular gland cells contain similar cell-specific factor(s), which permit the "constitutive" expression of the ovalbumin promoter region up to 295 bp upstream from the capsite, in the absence of steroid hormones. This cell-specific factor(s) is apparently not present in fibroblasts from primary cultures of liver or oviduct cells.
3. A negative control is exerted on the activity of the ovalbumin promoter region in oviduct tubular gland cells and involves sequences located between -295 and -425.
4. Hepatocytes lack the factor responsible for this negative control.
5. The repression effect was abolished when the ovalbumin sequence -134 to -425 was cloned in opposite orientation to the sense of transcription suggesting an orientation specific action.
6. Insertion of the ovalbumin sequence -134 to -425 upstream from two heterologous promoters chicken conalbumin and β-globin confer to both genes the negative control mechanism.

7. The presence of estradiol and/or progesterone (and most likely of glucocorticoids) relieves the "repression" in oviduct tubular gland cells.

Therefore the inactivity of the ovalbumin promoter in chicken tubular gland cells in the absence of steroid hormones appears to be due to the existence of a negative regulatory element (a blocker), involving sequences located between -295 and -425. Since there is no repression in hepatocytes, it is most likely that the blocker element is functional only in the presence of a "trans-acting" repressor. Thus the non-expression of the ovalbumin gene in the liver of the whole animal is not due to the absence of cell-specific transcription factor(s) which acts positively on promoter elements contained within 134 bp upstream from the capsite or to the presence of the repressor, but must be due to a cis-acting negative mechanism which results from the "liver developmental history" of the gene and causes its permanent "closing". This "closing" mechanism can be bypassed by microinjection of plasmid recombinants. In other chicken cells, like fibroblasts, the cell-specific transcription factor(s) appears to be absent, since pTOT-134 is not expressed.

Our findings raise a number of interesting questions. First, what is the nature of the repressor and how steroid hormones relieve its negative effect? Recent studies indicate that steroid hormone receptor molecules may be permanently located in the nucleus (see Schrader 1984 for refs), which suggests that the unfilled receptor molecule may be involved in the repression. Second, do the various steroid hormone-receptor complexes interact with the blocker element or with a positive dominant regulatory element(s) when repression is relieved ? Is such a positive regulatory element(s) an enhancer(s), as it has been shown to be the case for regulation of MMTV promoter function by glucocorticoids (see Chambon et al. 1984 for refs.) ? In this respect, it is interesting that the insertion of the -134 to -425 ovalbumin promoter region upstream from the +62 to -102 conalbumin promoter element (recombinant pOVCON in Fig. 2) which functions "constitutively" when microinjected alone (pTCT-102) into oviduct tubular gland cells (Chambon et al. 1984), results in "repression" of the latter in these cells in the absence of steroid hormones, and in its induction in their presence. Thus it is clear that at least some of the sequences involved in regulation of the ovalbumin promoter by steroid hormones are located between positions -295 and -425; this is upstream from the -95 to -222 region, which has been identified as responsible for a five-fold activation of the ovalbumin promoter by progesterone or estrogens (Dean et al. 1984). It is possible that the effect of this latter region has been missed in our present study, because of the relative insensitivity of the immunofluorescence assay (see above). On the other hand, it is likely that the "blocker" effect of the -295 to -425 sequence has not been observed by Dean et al. (1984) because their ovalbumin promoter chimeric recombinant contains the SV40 enhancer. We have indeed observed that the effect of the "blocker" sequence is not seen when the SV40 enhancer is introduced into pTOT-425 or pTOT-1348 (recombinants pTOT-425S and pTOT-1348S).

To our best knowledge, the present results constitute the first clear demonstration of the existence of a far upstream negative regulatory element (a blocker) participating in the regulation of gene expression in a higher eukaryote. Together with enhancer (for review see Chambon et al. 1984), blockers may provide the large number of combinatorial possibilities which are required for multifactorial regulation of gene expression in complex organisms.

ACKNOWLEDGEMENTS

We thank Roussel-Uclaf (Paris) for providing RU486.

REFERENCES

Benoist, C. and Chambon, P. (1981). In vivo sequence requirements of the SV40 early promoter region. Nature 290: 304-310.

Breathnach, R., N. Mantei and P. Chambon (1980). Correct splicing of a chicken ovalbumin gene transcript in mouse L cells. Proc. Natl. Acad. Sci. USA 77: 740-744.

Chambon, P., A. Dierich, M.P. Gaub, S. Jakowlev, J. Jongstra, A. Krust, J-P. LePennec, P. Oudet and T. Reudelhuber (1984). Promoter elements of genes coding for proteins and modulation of transcription by estrogens and progesterone. in Recent Prog. Horm.Res., "The Proceedings of the Laurentian Hormone Conference", vol. 40, Ed. R.O. Greep, Academic Press, pp. 1-42.

Chandler, V.L., B.A. Maler and K.R. Yamamoto (1983). DNA sequences bound specifically by glucocorticoid receptor in vitro render a heterologous promoter hormone responsive in vivo. Cell 33: 489-499.

Dean, D.C., R. Gope, B.H. Knoll, M.E. Riser and B.W. O'Malley (1984). A similar 5'-flanking region is required for estrogen and progesterone induction of ovalbumin gene expression. J. Biol. Chem. 259: 9967-9971.

Heilig, R., R. Muraskowsky and J.L. Mandel (1982). The ovalbumin gene family. The 5' end region of the X and Y genes. J. Mol. Biol. 156: 1-19.

McKnight, G.S., and R.D. Palmiter (1979). Transcriptional regulation of the ovalbumin and conalbumin genes by steroid hormones in chick oviduct. J. Biol. Chem. 254: 9050-9058.

Moreau, P., R. Hen, B. Wasylyk, R. Everett, M.P. Gaub and P. Chambon (1981). The SV40 72 baise pair repeat has a striking effect on gene expression both in SV40 and other chimeric recombinants. Nucleic Acids Res. 9: 6339-6350.

Schrader, W.T. (1984). New model for steroid hormone receptors ? Nature 308: 17-18.

Shepherd, J.C.W., E.R. Mulvihill, P.S. Thomas and R.D. Palmiter (1980) Commitment of chick oviduct tubular gland cells to produce ovalbumin mRNA during hormonal withdrawal and restimulation. J. Cell Biol. 87: 142-151.

Wasylyk, B., C. Wasylyk, P. Augereau and P. Chambon (1983). The SV40 72 bp repeat preferentially potentiates transcription starting from proximal natural or substitute promoter elements. Cell 32: 503-514.

Structural Elements of Glucocorticoid Receptors

U. GEHRING
Institut für Biologische Chemie der Universität, Im Neuenheimer Feld 501,
D-6900 Heidelberg, FRG

Mouse lymphoma cells in culture have been of great advantage for the study of various aspects of glucocorticoide hormone action. These cells have several of the characteristics of small thymic lymphocytes, in particular they respond to glucocorticoids by growth inhibition and cell lysis (1-3). This type of cellular response can be used to select for variant cells which are fully or partially resistant to glucocorticoids. Based on this idea, a cell genetic approach to steroid hormone action has been initiated more than a decade ago with S49.1 mouse lymphoma cells as a model system (for reviews, see ref. 4-6). Interestingly these studies as well as related investigations with other lymphoid cell lines (7-9) turned out to be in fact centered on the specific glucocorticoid receptors since the majority of resistant cell variants characterized so far in some detail proved to have defects in their receptors. This observation strongly emphasizes that intact receptors are essential for the physiological hormone response to occur in target cells. The biochemical investigation of defective receptors has greatly helped to obtain a better understanding of the structure of the wild-type receptor and its mode of action. In addition, the quantitative role of receptors has become clear by the observation that independently isolated lymphomas of widely differing glucocorticoid sensitivity have a gradient in receptor levels (10).

Glucocorticoid Receptor Mutants

Resistant clones of S49.1 cells have been classified into different groups according to their hormone binding properties and the intracellular distribution of receptor-steroid complexes between cytoplasm and cell nucleus (4-6). The "receptorless" (r^-) phenotype is most abundant; it is characterized by greatly reduced or virtually undetectable steroid binding activity as measured either in whole cells or in cell extracts. In some clones of this phenotype a receptor related polypeptide of wild-type size has been identified by

Nucleocytoplasmic Transport
Edited by R. Peters and M. Trendelenburg

immunochemical techniques (11,12) suggesting a defect in the hormone binding site of the receptor. Two types of resistant cell variants have been obtained in which roughly normal hormone binding is seen, but the interaction of receptor-steroid complexes with nuclei, chromatin, or DNA is abnormal. In the "nuclear transfer deficient" (nt^-) type the receptors are defective in nuclear binding. In the phenotype of "increased nuclear transfer" (nt^i) the receptor-hormone complexes show increased nuclear binding and abnormally high affinity for DNA.

Hybrids between wild-type cells and variants of each of these resistance phenotypes as well as hybrids between different resistant variants have been constructed (for a review, see ref. 5). It turned out that the wild-type response was dominant over resistance and that complementation between the resistance phenotypes described above did not occur. This was interpreted to mean that these abnormal receptor states belong to the same genetic complementation group and that the active domains of the wild-type receptor for steroid binding and for nuclear interaction reside within the same polypeptide chain (5).

DNA-Cellulose Chromatography of Receptor-Steroid Complexes

Unfractionated DNA can easily be adsorbed onto cellulose (13). The use of this kind of affinity matrix has been of great advantage for the analysis of nt^- and nt^i mutant receptors in comparison with the wild-type (4, 5, 14, 15). As shown in Table 1, wild-type receptor-complexes of mouse lymphomas S49.1, WEHI-7 and P1798 eluted from DNA-cellulose with 170 - 190 mM salt while nt^i receptor complexes required significantly higher salt concentrations (210-230 mM). In the case of nt^- receptors a smaller proportion of complexes was retained by DNA-cellulose (4) and the bound fraction eluted with abnormally low salt (70-90 mM). These differences in chromatographic behaviour reflect increased and decreased affinities to DNA of nt^i and nt^- receptor-complexes, respectively, as compared to the wild-type (4, 16). They also mirror the distribution of steroid which one observes when intact cells are incubated with hormone at 37° and subsequently fractionated in crude cytoplasmic and nuclear fractions.

Photoaffinity Labelling of Receptors

Affinity labelling of steroid hormone receptors has in recent years become a particularly valuable analytical tool (for reviews, see ref.

Table 1: DNA Binding Properties and Molecular Weights of Receptor Types

Receptor type	M_r of steroid-labelled polypeptide native	M_r of steroid-labelled polypeptide after chymotrypsin	DNA-cellulose chromatography (mM KCl required for elution) native	DNA-cellulose chromatography (mM KCl required for elution) after chymotrypsin
S49.1 wild-type	94 700	38 000	175	236
S49.1 nt^- (clone 22R)	94 000	37 400	75	129
S49.1 nt^- (clone 83R)	94 000	37 400	86	87
S49.1 nt^i (clone 55R)	40 300	39 800	225	229
S49.1 nt^i (clone 143R)	40 800	41 000	210	209
WEHI-7 wild-type	94 000	38 000	186	233
P1798 wild-type	93 000	-	193	-
P1798 nt^i	38 500	-	212	-

(Data from ref. 14, 15)

17, 18) since it allows to determine the polypeptide molecular weights of receptors under denaturing conditions. The method has the eminent advantage that crude receptor preparations can be used because after removal of excess free ligand the hormone is only bound to specific receptor sites and consequently only these are covalently tagged. Steroids containing α,β-unsaturated ketones can be excited by long wavelength UV light and a covalent bond is then being formed with protein present in the immediate molecular vicinity (18). The high-affinity glucocorticoids dexamethasone and triamcinolone acetomide contain the appropriate unsaturated ketone structure and can easily be linked covalently to glucocorticoid receptors by illumination at $\lambda \geqq 320$ nm (14,15). These steroids were used in tritium-labelled form to analyze wild-type and mutant receptors of lymphoma cells. SDS gel electrophoresis followed by fluorography revealed a major labelled polypeptide band of molecular eight 94 000 $\pm$ 5000 for wild-type receptors of mouse lymphomas S49.1, WEHI-7 and P1798 and for receptors of glucocorticoid-sensitive human lymphoblastic leukemia cells CEM-C7 (14,15). This is shown for the wild-type S49.1 receptor in Figure 1, lane A.

When mutant receptors of the nt^- type were investigated by the same method the molecular weight was found to be identical to that of wild-type receptors (Figure 1 B, Table 1). In contrast, nt^i variant receptors revealed a major radiolabelled polypeptide band of 40 000 $\pm$ 2000 molecular weight and a minor band of about 37 000 (Figure 1 C). This was found for nt^i receptors of S49.1 and P1798 mouse lymphomas (15,19). The molecular weight data are summarized in Table 1.

Photoaffinity labelling was also used to investigate the question whether receptor activation involves a change in the molecular size of the receptor polypeptide which contains the hormone binding site. Whithin the resolution of the SDS gel electrophoresis polypeptides of the same size were found to be covalently labelled with radioactive hormone whether the receptor-complexes were subjected to photoreaction in the activated or non-activated states (14). This, of course, does not exclude the possibility that in the non-activated state the receptor polypeptide described here is complexed with other cellular components (see below).

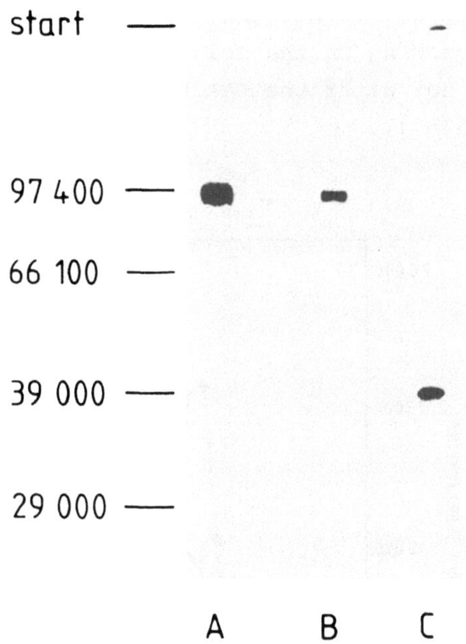

Figure 1: SDS gel electrophoresis of photoaffinity labelled receptors.

Receptor complexes with [^{3}H]triamcinolone acetonide were subjected to photolabelling and subsequent gel electrophoresis. Fluorography was used to detect radiolabelled bands. (A) S49.1 wild-type; (B) S49.1 nt$^-$ (clone 83R); (C) S49.1 nt^i (clone 55R). (Data from ref. 14, 15).

Partial Proteolysis of Receptor-Steroid Complexes

Abnormal DNA binding properties have previously been observed with wild-type receptors of rat liver (20) and mouse lymphomas (16, 21) treated with α-chymotrypsin. This can best be demonstrated by use of DNA-cellulose chromatography. Figure 2 shows an experiment in which the S49.1 wild-type receptor was exposed to α-chymotrypsin under mild conditions and subsequently chromatographed on DNA-cellulose. The hormone complexes required about 230 mM salt for elution as compared to 180 mM if untreated. This indicates a DNA affinity of the partially degraded receptor similar to that of native nt^i receptors (Table 1). Chymotrypsin treatment of nt$^-$ mutant receptors either did not change the affinity for DNA or produced a slight increase depending on the cell clone (Table 1). By contrast, α-chymotrypsin did not alter the DNA binding properties of nt^i mutant receptors (Table 1).

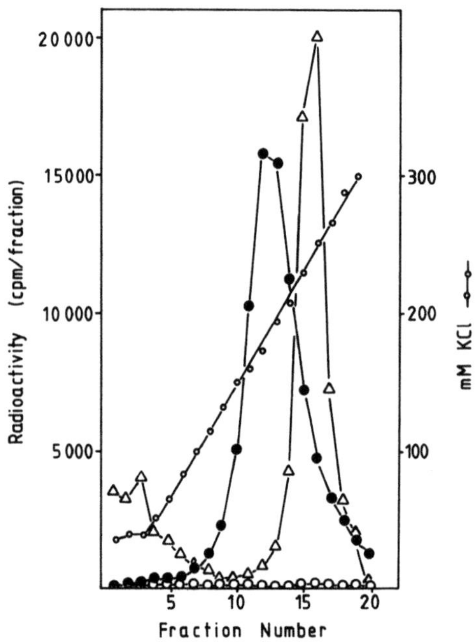

Figure 2: <u>DNA-cellulose chromatography</u>.
Unfractionated DNA from calf thymus was adsorbed onto cellulose (13). S49.1 wild-type receptor complexes with [^{3}H]triamcinolone acetonide were activated by a 30 min incubation at 20° and chromatographed on DNA-cellulose either in the native state (●) or following a 10 min treatment with 10 µg/ml α-chymotrypsin at 0° (Δ) or a 30 min treatment with 20 µg/ml trypsin (o). (Data from ref. 14, 15).

Chymotrypsin treated wild-type and mutant receptors were also subjected to photoaffinity labelling and subsequent SDS gel electrophoresis. With wild-type and nt^- receptors a major labelled polypeptide of molecular weight 38 000 was obtained (Table 1). Chymotrypsin, however, did not produce any change in the molecular size of nt^i receptors (Table 1).

Proteolysis of S49.1 wild-type, nt^-, and nt^i receptors with trypsin generated steroid labelled fragments of molecular weights of about 38 000, 29 000 and 27 000. The 38 000 fragment, however, was seen only under very mild conditions of digestion and was readily further degraded to the smaller receptor fragments. This is shown in Figure 3 for the wild-type. The tryptic receptor fragments of 27 000 to 29 000 which still carry the steroid label were not able to bind to DNA as was disclosed by DNA-cellulose chromatography (Figure 2). These fragments correspond to the so-called mero-receptors (22, 23).

Lymphoma cell receptors were also exposed to a bacterial enzyme, endoproteinase Lys-C, that specifically chaves polypeptides at lysine residues only. The patterns of steroid labelled receptor fragments detected by photoaffinity labelling and gel electrophoresis were very similar to those produced by trypsin (15). If, however, an arginine specific enzyme, endoproteinase Arg-C, was used there was no detectable cleavage of receptor polypeptides.

Domain Model for the Wild-Type Receptor Polypeptide

In order to be functionally active at the genomic level steroid hormone receptors need to contain at least two active domains within their molecular structure: one for hormone binding and another one for nuclear interaction. The experiments with glucocorticoid receptors described above, however, provide strong evidence for still another active domain existent within the wild-type receptor. This domain is involved in modulating nuclear interaction of receptor-steroid complexes or of DNA binding as in our model system of DNA-cellulose chromatography. The function of the modulating domain is to limit the affinity of receptor complexes to DNA such that the biologically relevant acceptor sites in chromatin can be recognized and specific genes are then being expressed.

The modulating domain has therefore also been called "specifier domain" (24). It may be clipped off from the wild-type receptor molecule by mild proteolysis or it may be missing due to some mutational event as is the case in nt^i mutants of S49.1 and P1798 cells. In nt^i type mutants the receptor-hormone complexes devoid of the modulating domain might bind too tightly to chromatin such that they would not have a chance to find the appropriate gene loci which need to be regulated. As to the cellular origin of nt^i receptors it is interesting to note that nt^i cells contain receptor mRNA of smaller size than the wild-type as detected with a specific cDNA probe (25). This suggests that the nt^i phenotype is either caused by a deletion mutation or by a defect in the splicing of pre-mRNA.

A domain model for the wild-type glucocorticoid receptor is presentid in Figure 4. The model implies that all three active domains of the receptor are contained within the same polypeptide chain of 94 000 molecular weight. The domains for steroid binding, for nuclear interaction, and for modulating nuclear binding are shown in the graphic presentation as blocks of about equal areas. In reality, however, the sizes of these domains appear to be quite different. According to the experiments discussed above the size of the modulating domain is expected to be about 55 000 daltons and the steroid and nuclear binding domains taken together have a molecular weight of about 40 000. Even though trypsin and other proteases cleave the 40 000 dalton polypeptide to steroid binding fragments of 27 000 - 29 000 it might not be justified to conclude that the DNA binding domain is strictly localized within a polypeptide region of about 10 000 daltons. The proteases might cleave off or destroy some of the essential parts of the DNA binding domain while leaving others intact. In other words, the three active domains of the receptor may partially overlap in ways which are not obvious from Figure 4. It should be emphasized that the sequential order of the three active domains along the 94 000 dalton polypeptide of the wild-type receptor is still unknown. What is clear is that the modulating domain is located distal to the cluster of steroid binding and nuclear interaction domains since chymotrypsin cleaves off this part of the molecule and leaves the other two domains linked.

The model of Figure 4 also indicates those areas of the receptor polypeptide which join the active domains. They are quite susceptible to proteolysis. The linker between the modulating domain and the DNA binding domain (indicated by closed arrow) appears to be some kind of a hinge region in the receptor molecule. It is especially sensitive to proteases and can easily be cleaved by chymotrypsin, trypsin, endoproteinase Lys-C and probably a great many other proteases including endogeneous cellular enzymes. The other linker, indicated by an open arrow, is probably somewhat less accessible to proteases but can be split, for example, by trypsin. In both linker regions tryptic cleavage occurs at lysine residues since a lysine specific protease, but not an arginine specific one, mimics the action of trypsin and produces the same receptor fragments.

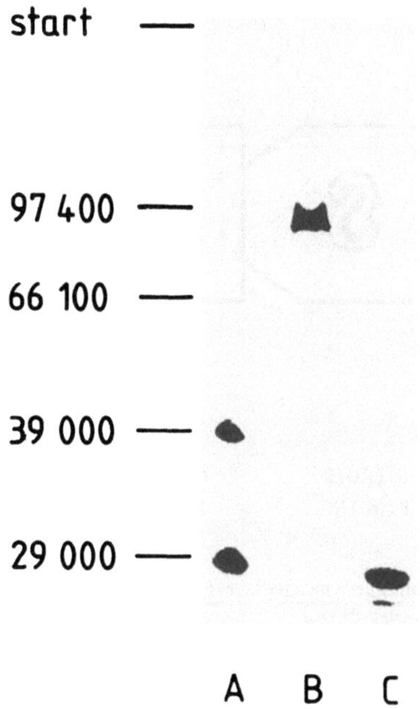

Figure 3: <u>SDS gel electrophoresis of trypsin-treated receptors</u>.

S49.1 wild-type receptor complexes with [^{3}H]triamcinolone acetonide were subjected to photoaffinity labelling followed by a 30 min treatment with trypsin in the cold. (A) Treatment with 10 µg/ml trypsin; (B) control without trypsin; (C) treatment with 20 µg/ml trypsin. (Data from ref. 15).

It is interesting to note that pyridoxal phosphate which supposedly reacts with ε-amino groups of lysine residues may protect receptors against the action of trypsin (26) such that at least the domains for steroid binding and nuclear interaction remain linked.

The receptor domain involved in modulating nuclear interaction appears to harbour the main antigenic determinants of wild-type glucocorticoid receptors. Antisera raised against the highly purified rat liver receptor did not react with the steroid labelled chymotrpytic receptor fragment (27, 28). They also did not bind nt^i receptors of P1798 lymphoma cells (29). Similar results were also obtained with monoclonal antibodies to the glucocorticoid receptor of rat liver (30).

While some of them bound the wild-type and nt^- mutant receptors of S49.1 lymphoma cells to a considerable extend they did not react with nt^i mutant receptors (11). In an interesting immunochemical study

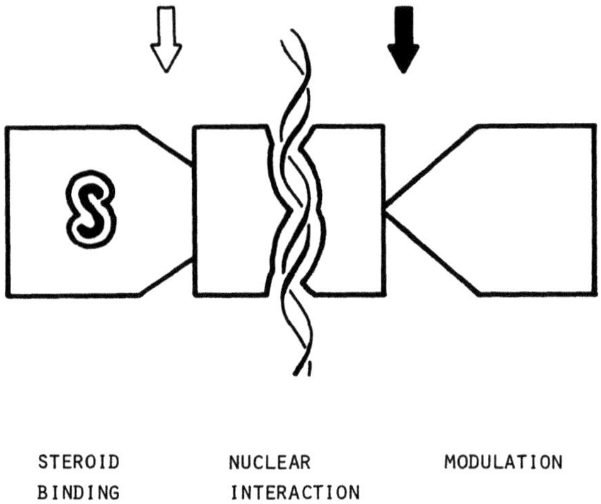

Figure 4: Domain model of the wild-type glucocorticoid receptor.

The domains for steroid binding, for nuclear interaction, and for modulating nuclear interaction are shown in this order. The arrows indicate linker regions which are highly susceptible to proteolytic cleavage.

with polyclonal antibodies it was shown (27) that the modulating domain cleaved off from the rat liver glucocorticoid receptor by chymotrypsin could be recovered by gel permeation chromatography as a separate entity. The modulating domain is thus not chopped into many fragments by chymotrypsin. This finding is in accordance with the model of Figure 4, in particular it supports the idea that a hinge region in the receptor structure links the modulating domain to the remainder of the molecule.

The tree active domains disclosed for the wild-type glucocorticoid receptor are probably not unique to this receptor type. It is assumed that the same kind of domain structure as outlined in Figure 4 is also present in receptors for other steroid hormones.

High Molecular Weight Complexes of Glucocorticoid Receptors

If receptor containing cell extracts are analyzed under non-denaturing conditions, complexes as large as 320 000 - 340 000 daltons can be detected (31-37). This raises the question as to how the hormone-carrying receptor polypeptide of molecular weight 94 000 is organized in such large structures. It has been postulated that these complexes are either homotetramers of steroid-binding subunits or contain two such subunits in association with two somehow related polypeptides (32-34,37,38). We addressed the problem of receptor subunit structure by comparing the hydrodynamic parameters of wild-type and nt^i mutant receptors (39) with the idea in mind that the high molecular weight form of the truncated nt^i receptor should be significantly different from that of the wild-type. If, for example, the nt^i receptor were a homotetramer it should exhibit a molecular weight of about 160 000 daltons; for a heteromeric structure containing only one hormone-binding polypeptide, however, a molecular weight of about 270 000 - 290 000 daltons would be expected.

Gel filtration and sedimentation analysis were used to determine Stokes radii (R_s) and sedimentation coefficients ($s_{2o,w}$), respectively. These were used to calculate the molecular weights of receptor complexes (40). The data, as summarized in Table 2, show a difference in molecular weights of about 40 000 daltons between wild-type and nt^i mutant receptors. This is in good agreement with the difference observed when the hormone-binding polypeptides were analyzed under denaturing conditions (Table 1). Moreover, if the wild-type complex was subjected to mild treatment with α-chymotrypsin a receptor form was recovered which is similar to the nt^i mutant (Table 2).

These studies suggest a structure for the large receptor complexes in which one hormone-binding polypeptide of molecular weight 94 000 or 40 000, respectively, is associated with other subunits. Recent experiments point to two types of molecules which may be components of these large receptor complexes: a heat shock protein of molecular weight 90 000 (41,42) and RNA (43). The biological functions of these and possibly other components in relation to the mechanism of glucocorticoid receptor action remain unknown at present.

Table 2: High Molecular Weight Receptor Complexes

Receptor type	Treatment	R_s (Å)	$s_{2o,w}$ (S)	M_r
S49.1 wild type	-	81	9.5	325 000
	chymotrypsin	69	10.0	291 000
S49.1 nt^i (clone 143R)	-	71	9.5	285 000

(Data from ref. 39)

References

1. Claman, H.N. (1972), N.Engl.J.Med. 287, 388-397.

2. Craddock, C.G. (1978) Ann.Intern.Med. 88, 564-566.

3. Harris, A.W., and Baxter, J.D. (1979) In: Glucocorticoid Hormone Action (eds. Baxter, J.D. and Rousseau, G.G.) Berlin, Springer-Verlag, 423-448.

4. Yamamoto, K.R., Gehring, U., Stampfer, M.R., Sibley, C.H. (1976) Recent.Progr.Horm.Res. 32, 3-32.

5. Gehring, U. (1980) In: Biochemical Actions of Hormones (ed. Litwack, G.) New York, Academic Press, 7, 205-232.

6. Gehring, U. (1984) In: Hormones and Cancer (eds. Bresciani, F., King, R.J.B., Lippman, M.E., Namer, M., and Raynaud, J.-P.) New York, Raven Press, 2, 245-254.

7. Harmon, J.M., and Thompson, E.B. (1981) Mol.Cell.Biol. 1, 512-521.

8. Huet-Minkowski, M., Gasson, J.C., and Bourgeois, S. (1981) Cancer Res. 41, 4540-4546.

9. Bourgeois, S., and Gasson, J.C. (1985) In: Biochemical Actions of Hormones (ed. Litwack, G.) New York, Academic Press, 12, 311-351.

10. Gehring, U., Mugele, K., and Ulrich, J. (1984) Mol.Cell.Endocrinol. 36, 107-113.

11. Westphal, H.M., Mugele, K., Beato, M., and Gehring, U. (1984) EMBO J. 3, 1493-1498.

12. Northrop, J.P., Gametchu, B., Harrison, R.W., and Ringold, G.M. (1985) J.Biol.Chem. 260, 6398-6403.

13. Alberts, B., and Herrick, G. (1971) Meth.Enzymol. 21, 198-217.

14. Dellweg, H.-G., Hotz, A., Mugele, K., and Gehring, U. (1982) EMBO J. 1, 285-289.

15. Gehring, U., and Hotz, A. (1983) Biochemistry 22, 4013-4018.

16. Andreasen, P.A., and Gehring, U. (1981) Eur.J.Biochem. 120, 443-449.

17. Simons, S.S., and Thompson, E.B. (1982) In: Biochemical Actions of Hormones (ed. Litwack, G.) New York, Academic Press, 9, 221-254.

18. Gronemeyer, H. (1985) Trends in Biochem.Sci. 10, 264-267.

19. Nordeen, S.K., Lan, N.C., Showers, M.O., and Baxter, J.D. (1981) J.Biol.Chem. 256, 10503-10508.

20. Wrange, Ö., and Gustafsson, J.-Å. (1978) J.Biol.Chem. 253, 856-865.

21. Stevens, J., and Stevens Y.-W. (1981) Cancer Res. 41, 125-133.

22. Sherman, M.R., Pickering, L.A., Rollwagen, F.M., and Miller, L. K. (1978) Fed.Proc. 37, 167-173.

23. Miller, L.K. (1980) In: Biochemical Actions of Hormones (ed. Litwack, G.) New York, Academic Press, 7, 233-243.

24. Vedeckis, W.V. (1983) Biochemistry 22, 1975-1983.

25. Miesfeld, R., Okret, S., Wikström, A.-C., Wrange, Ö., Gustafsson, J.-Å., and Yamamoto, K.R. (1984) Nature 312, 779-781.

26. Ninh, N.V., Arânyi, P., and Horvâth, I. (1982) J.Steroid.Biochem. 17, 599-601.

27. Carlstedt-Duke, J., Okret, S., Wrange, Ö., and Gustafsson, J.-Å. (1982) Proc.Natl.Acad.Sci.USA 79, 4260-4264.

28. Eisen, H.J. (1982) In: Biochemical Actions of Hormones (ed. Litwack, G.) New York, Academic Press, 9, 255-270.

29. Stevens, J., Eisen, H.J., Stevens, Y.-W., Haubenstock, H., Rosenthal, R., and Artishevsky, A. (1981) Cancer Res. 41, 134-137.

30. Westphal, H.M., Moldenhauer, G., and Beato, M. (1982) EMBO J. 1, 1467-1471.

31. Niu, E.-M., Neal, R.M., Pierce, V.K., Sherman, M.R., (1981) J.Steroid.Biochem. 15, 1-10.

32. Sherman, M.R., Moran, M.C., Tuazon, F.B., and Stevens, Y.-W. (1983) J.Biol.Chem. 258, 10366-10377.

33. Vedeckis, W.V. (1983) Biochemistry 22, 1983-1989.

34. Norris, J.S., and Kohler, P.O. (1983) J.Biol.Chem. 258, 2350-2356.

35. Stevens, J., Stevens, Y.-W., and Haubenstock, H. (1983) In: Biochemical Actions of Hormones (ed. Litwack, G.) New York, Academic Press, 10, 383-446.

36. Sherman, M.R., Stevens, Y.-W., and Tuazon, F.B. (1984) Cancer Res. 44, 3783-3796.

37. Sherman, M.R. and Stevens, J. (1984) Ann.Rev.Physiol. 46, 83-105.

38. Raaka, B.M., and Samuels, H.H. (1983) J.Biol.Chem. 258, 417-425.

39. Gehring, U., and Arndt, H. (1985) FEBS Letters 179, 138-142.

40. Sherman, M.R. (1975) Meth.Enzymol. 36, 211-234.

41. Joab, I., Radanyi, C., Renoir, M., Buchou, T., Catelli, M.-G., Binart, N., Mester, J., and Baulieu, E.E. (1984) Nature 308, 850-853.

42. Catelli, M.-G., Binart, N., Baulieu, E.E., Welch, W., Helfman, D., and Feramisco, J. (1985) Cold Spring Harbor Symposium on Heat Shock, Abstracts, 23.

43. Economidis, I.V., and Rousseau, G.G. (1985) FEBS Letters 181, 47-52.

The Role of RNA-Protein Interactions in Intracellular Targeting

I. W. MATTAJ[1]
Biocenter, Basel University, Klingelbergstrasse 70, 4056 Basel, Switzerland
[1]Present Address: EMBL, Meyerhofstraße 1, Postfach 10-2209,
D-6900 Heidelberg, FRG

Introduction

The eukaryotic cell is characterised by the possession of membrane bound subcellular organelles, of which the nucleus is generally regarded as the most important, being responsible for the control and regulation of most intra- and intercellular activities. In order to construct the nucleus it is necessary that nuclear components, especially proteins which are made in the cytoplasm, segregate properly into the correct subcellular compartment. The mechanisms by which molecules are targetted to the nucleus have been discussed in two recent reviews (De Robertis, 1983, Dingwall, 1985) and are the subject of several of the contributions to this volume. Other decisions to be taken by the nucleus during eg. development or differentiation are influenced by information exchange between the nucleus and cytoplasm, some probably involving the movement of regulatory molecules e.g. developmental determinants (Conklin, 1905, Gurdon, 1977) between the cytoplasm and nucleus. One mechanism by which such a change in intracellular location can be achieved is the subject of this paper. Two examples of macromolecules which change their intracellular location following RNA-protein interaction will be discussed. The first example is the exclusion from the nucleus of 5S RNA following its interaction with TFIIIA, a transcription factor necessary for 5S gene expression. The second is the translocation of the protein components of U snRNPs from the cytoplasm to the nucleus following their binding to U snRNA.

Nuclear Exclusion of 7S RNP

Previtellogenic *Xenopus laevis* oocytes have a very unusual RNA content. Roughly 75% of the total RNA at this stage is either 5S or tRNA, 5S RNA being 30-40% of total oocyte RNA, and only a small amount

Nucleocytoplasmic Transport
Edited by R. Peters and M. Trendelenburg

of the other ribosomal RNAs (5.8S,18S,28S) are present (Mairy and Denis, 1971, Ford, 1971). The 5S RNA is found in the form of two ribonucleoprotein particles, the 7S and 42S RNPs, (Ford, 1971, Picard and Wegnez, 1979), and it is the smaller of these two particles, the 7S RNP, which is of interest to us here. The 7S RNP consists of one molecule of 5S RNA and one molecule of a 39,000 Dalton protein. This protein has been shown by chemical and immunological studies to be TFIIIA (Pelham and Brown, 1980, Honda and Roeder, 1980), a transcription factor whose binding to 5S genes is necessary for their transcription by RNA polymerase III (reviewed by Brown, 1984). Unexpectedly however, when the localisation of TFIIIA in oocytes was studied by immunohistochemistry, it was found that all oocyte stages where TFIIIA was detectable showed a purely cytoplasmic localisation of the transcription factor (Mattaj et al., 1983). An example of this staining is shown in Figure 1a.

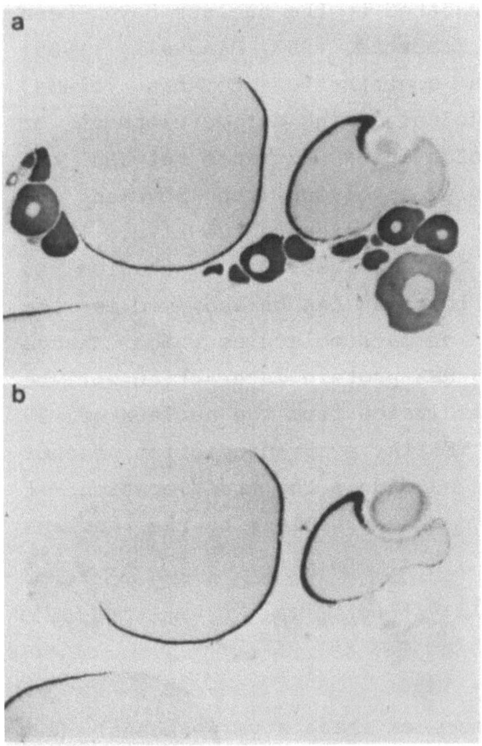

Figure 1. Immunolocalisation of TFIIIA on a section of Xenopus ovary.
(a) An ovary section incubated with a Rabbit polyclonal anti-TFIIIA antiserum before visualisation with protein A - peroxidase and 3-amino-9-ethyl carbazole.
(b) Non-immune antiserum control.
Reproduced from J. Cell Biol. 1983, 97, 1261-1265.

The apparent exclusion of TFIIIA from the nucleus is obvious in all the favourably sectioned stained oocytes. The size distribution of the stained oocytes indicates that this staining method detects TFIIIA only at its maximal concentration, which occurs in late stage I previtellogenic oocytes. At this stage the vast majority of the TFIIIA in the oocyte is complexed with 5S RNA (Picard and Wegnez, 1979, Dixon and Ford, 1982) and so what we are seeing on the sections is the distribution of the 7S RNP, containing the bulk of the TFIIIA. To prevent confusion it should be stated that some TFIIIA is clearly in the oocyte nucleus, complexed to the 100,000 or so 5S genes. Furthermore Birkenmeier et al. (1979) developed an in vitro transcription system from extracts of mature Xenopus oocyte nuclei which contains TFIIIA. However the fact remains that the bulk of the TFIIIA is cytoplasmic. Why should a DNA-binding transcription factor be largely excluded from the nucleus, and what is the mechanism which causes this distribution?

During studies on 5S gene transcription in vitro (Pelham and Brown, 1980) it was shown that 5S RNA in excess over 5S genes would compete for TFIIIA binding, and thus inhibit 5S gene expression, suggesting a mechanism of feedback regulation of 5S RNA synthesis. If 5S RNA binding results in the movement of TFIIIA to the cytoplasm, a much tighter control of nuclear TFIIIA concentration would be possible, which may be of relevance to control of 5S gene expression in general. In particular some hypotheses to explain the switch in expression from oocyte plus somatic type 5S genes to somatic type only, which takes place between the completion of oogenesis and gastrulation, invoke the need for a very low intranuclear concentration of free TFIIIA at this time (Brown, 1984).

Many mechanisms for the nuclear exclusion of the 7S RNP can be proposed, and since they are difficult to test experimentally, it has not been possible to decide which operates. One suggestion, that 7S RNPs are attached to some cytoplasmic structure or aggregate, and thus unable to diffuse freely, has been tested. 7S RNPs are known to be soluble in homogenates of immature oocytes (Ford, 1971). Such a homogenate was prepared and injected into a mature oocyte. The 7S RNP was excluded from the nucleus of the injected oocyte, although it diffused throughout the cytoplasm. Non-diffusibility does not therefore seem to be the mechanism.

Since 5S RNA can enter the oocyte nucleus in the absence of bound TFIIIA (De Robertis et al., 1982) and since it would seem that some TFIIIA in the free state must be able to enter the nucleus in order to bind to 5S genes, the most likely explanation of the nuclear exclusion seems to be that the RNA-protein binding prevents nuclear entry of the 7S RNP, while the two component parts of the RNP could separately enter the nucleus freely. Due to the insolubility of free TFIIIA it has been impossible to test this theory directly. However, in the next few pages I will describe a case where RNA-protein interactions have been clearly shown to be involved in the nuclear localisation of an RNP whose component parts individually have no apparent nuclear affinity.

U2 snRNPs

U2 small nuclear Ribo Nucleoprotein Particles consist of one RNA, U2 snRNA, which is 189 bases long and several proteins named A', B, B', B", D, E, F and G . It is not clear whether all U2 snRNP particles are identical in composition or what the molar ratios of the different snRNP components are (Hinterberger et al., 1983, Kinlaw et al., 1982; Mimori et al., 1984). Like the other U snRNPs U2 snRNPs are highly conserved nuclear components in the somatic cells of all higher eukaryotic species so far examined. In Xenopus oocytes the situation is rather different. The accumulation of the protein and RNA moieties of the U snRNPs is non-coordinate, resulting in the accumulation of an excess of free U snRNP proteins, which are stored in the oocyte cytoplasm (Zeller et al., 1983, Fritz et al, 1984). Previous to these discoveries it had been shown that naked U snRNAs, when microinjected into the oocyte cytoplasm, migrated to and became concentrated in the nucleus (De Robertis et al., 1982). Zeller et al. (1983) then showed that this movement of microinjected U snRNAs from the cytoplasm to the nucleus was accompanied by movement of the U snRNP proteins in the same direction. The conclusion which could be drawn from these experiments was that free U snRNP proteins had no intrinsic affiniy for the nucleus, and that the signal which caused accumulation of U snRNPs in the nucleus must either reside on the RNAs themselves or be formed as a result of RNA-protein interaction.

In order to distinguish between these two possibilities, a series of mutant U2 genes were constructed *in vitro*, which gave rise to mutant U snRNA transcripts on microinjection into Xenopus oocyte nuclei (Mattaj and De Robertis, 1985). These mutant RNAs were then tested for their ability to associate with various U snRNP protein components and for their ability to accumulate in the oocyte nucleus. The major conclusion of this work is summarised in figure 2.

In figure 2a the structure and sequence of wild-type Xenopus U2 is shown. One of the mutant U2 genes created, ΔC resulted in the exchange of the 12 underlined nucleotides by the 17 nucleotides in the lower part of the figure. Although this mutation had little or no effect on the transcription of ΔC (Fig. 2b, lane 3) the transcript of this gene was not immunoprecipitable by antisera of the Sm type, which recognise proteins B, B' and D on U2 snRNP particles. This means that the change in ΔC RNA resulted in deletion of the RNA sequences necessary for binding of these antigens. The effect of this on nuclear accumulation is shown in Figure 2c, where it is clear that while wild-type U2 accumulates strongly in the oocyte nucleus, ΔC seems to be unable to enter the nucleus at all.

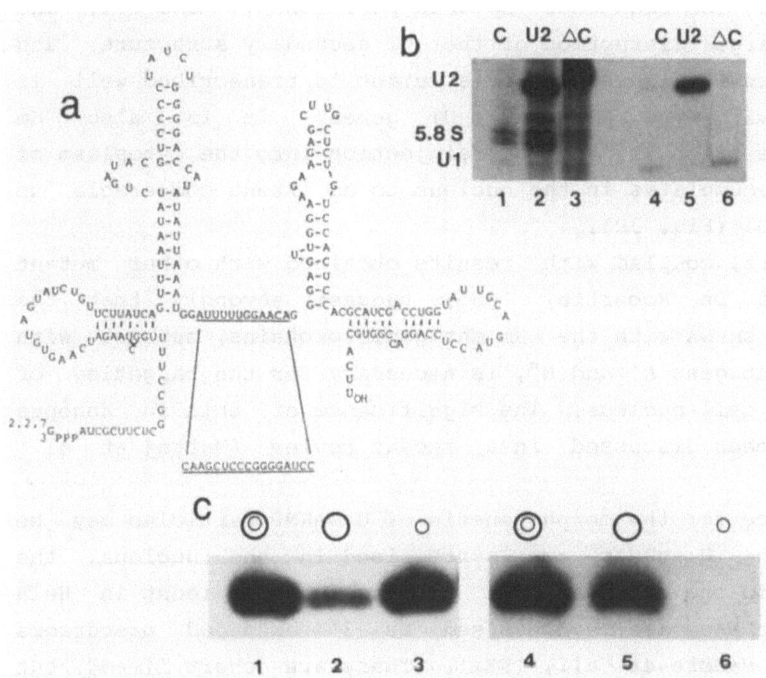

Figure 2: Analysis of ΔC, a mutant U2 that can no longer bind Sm antigen.

(a) The position and sequence of the ΔC deletion - substitution. The twelve underlined nucleotides in the proposed wild-type U2 secondary structure diagram have been removed and replaced by the seventeen underlined nucleotides below them.

(b) RNA transcripts from microinjected oocytes. Lane 1; Control oocytes. Lane 2; U2 injected. Lane 3; ΔC injected. Lane 4; Sm immunoprecipitate from control oocytes. Lane 5; Sm immunoprecipitate of U2 injected oocytes. Lane 6; Sm immunoprecipitate of ΔC injected oocytes.

(c) Intracellular location of U2 and ΔC transcripts 24hr after microinjection into the cytoplasm of Xenopus oocytes. Lanes 1-3; U2 transcript. Lanes 4-6; ΔC transcripts. The symbols indicate total oocytes or separated cytoplasmic and nuclear fractions.
Reproduced from Cell 1985, 40, 111-118.

A possible trivial explanation of these results, that mutations in the U2 RNA created new secondary structures incompatible with protein binding or nuclear migration, was tested in the experiment shown in Figure 3. A mutation was made in the U2 gene which resulted in the duplication of all the sequences between positions 71 and 134, and presumably in a large disruption of the U2 secondary structure, and called D1. As shown in Figure 3T, this mutant is transcribed well in comparison to wild-type U2 and U1 genes. It is also Sm immunoprecipitable (Fig. 3I) and, on reinjection into the cytoplasm of an oocyte, it reaccumulates in the nucleus to an extent comparable to wild-type U2 and U1 (Fig. 3R).

These results, coupled with results obtained with other mutant RNAs, (Mattaj and De Robertis, 1985) suggest strongly that the interaction of U2 snRNA with the Sm antigenic proteins, but not with the U2 specific antigens A' and B", is necessary for the targeting of U2 snRNPs to the cell nucleus. The significance of this in Xenopus development has been discussed in a recent review (Mattaj et al., 1985)

The importance for the morphogenesis of U snRNP particles may be the following. The U snRNAs are synthesised in the nucleus, the binding proteins in the cytoplasm. It appears that, at least in Hela cells, the U snRNAs are synthesised as 3' extended precursors (Elicieri, 1981, Madore et al., 1984). These are short lived, but appear to first enter the cytoplasm, then return to the nucleus during which time they are processed to the mature length. Probably the U snRNPs are formed during this short cytoplasmic sojourn of the RNAs, and return to the nucleus as complete particles. This might ensure an orderly assembly and entry of the U snRNPs, which would be more difficult to arrange if all the protein components had to have individual nuclear targeting signals and the assembly of the particles took place in the nucleus, where the RNA binding affinity of the abundant snRNP protein components might result in their interfering in transcription, processing or transport of other RNA species.

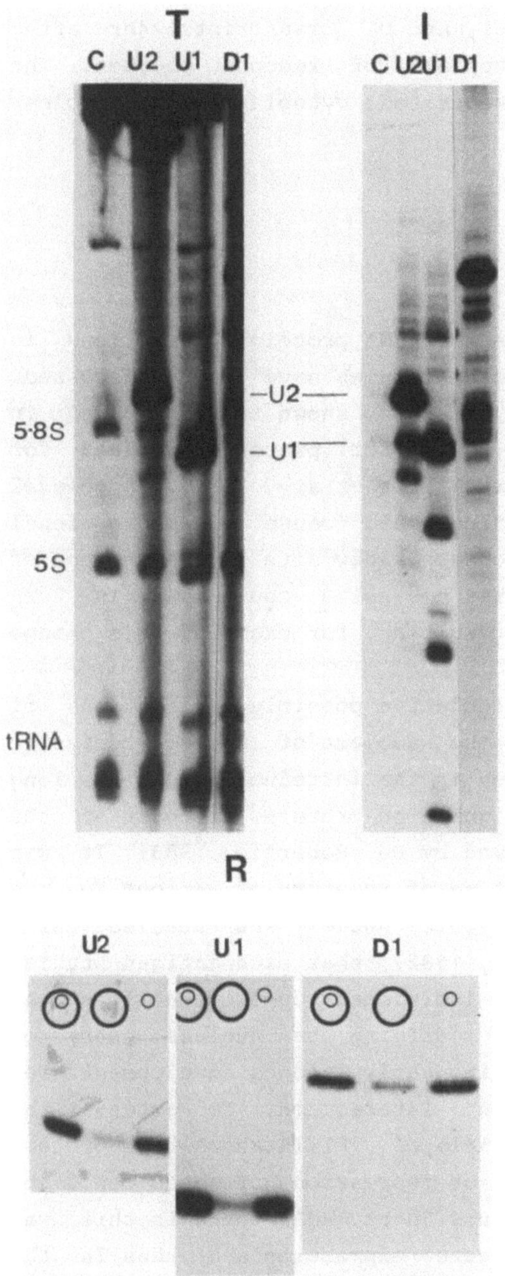

Figure 3 Analysis of D1, a mutant U2 containing a 64 base internal duplication.

(T) RNA transcripts from control oocytes (C) or from oocytes injected with either U2, U1, or D1 genes.

(I) RNA immunoprecipitated with Sm antiserum from control (C) or injected (U2,U1,D1) oocytes.

R) Intracellular location of U2, U1, and D1 transcripts 24hr after their microinjection into the cytoplasm of Xenopus oocytes. The symbols indicate total oocytes or separated cytoplasmic and nuclear fractions.

Conclusions and Discussion

Two examples of the involvement of RNA-protein interactions in the intracellular targeting of macromolecules have been discussed. Another RNP, the 7SL-containing SRP, has been shown to be involved in targeting nascent peptide chains, coding for proteins destined for export, to the endoplasmic reticulum (Walter et al., 1984). A general mechanism by which RNA-Protein binding would cause a conformational change in the protein to unmask or create a new site on the surface of an RNP is not hard to imagine.This new site could then in turn recognise new ligands or receptors resulting, for example, in a change in intracellular location.

In order to discuss more speculative possible applications of such a mechanism I will return to the subject of nucleocytoplasmic interactions in development mentioned in the introduction. It has long been clear that nuclear targeting signals on proteins are part of the mature, unmodified, protein (reviewed by De Robertis, 1983). It was therefore unexpected that when the time of re-entry of various Xenopus oocyte nuclear antigens into embryonic nuclei was studied using monoclonal antibodies (Dreyer et al., 1982), that each antigen studied entered the nucleus at a particular, individual, developmental stage. In other words something must be modifying the nuclear entry of potentially karyophilic proteins during early Xenopus development. One obvious possibility is RNA-protein interaction. To extend the discussion to gene control, the example of TFIIIA shows how protein factors required for the activation or repression of genes could be shuttled into, or out of, the nucleus. Short RNAs could in this way play a role in the modulation of gene expression and thus in the control of cell function.

Acknowledgements

I would like to thank Prof. E.M. De Robertis, in whose lab all of this work was carried out, and who played a primary role in the development of the experiments and ideas reported. I would also like to acknowledge the other members of the lab, particularly Susanne Lienhard and Rolf Zeller for their lively support. Finally I thank Dr. D.D. Brown for the gift of anti-TFIIIA antiserum and Christel Riedel for typing the manuscript.

References

Birkenmeier, E.H., Brown, D.D. and Jordan, E. (1978). A nuclear extract of Xenopus laevis oocytes that accurately transcribes 5S RNA genes. Cell 15, 1077-1086.

Brown, D.D. (1984). The role of stable complexes that repress and activate eucaryotic genes. Cell 37, 359-365.

Conklin, E.G. (1905). The organization and cell lineage of the ascidian egg. J. Nat. Acad. Sci. Philadelphia 13, 1-139.

De Robertis, E.M. (1983). Nucleocytoplasmic segregation of proteins and RNAs. Cell 32, 1021-1025.

De Robertis, E.M., Lienhard, S. and Parisot, R.F. (1982). Intracellular transport of microinjected 5S and small nuclear RNAs. Nature 295, 572-577.

Dingwall, C. (1985). The accumulation of proteins in the nucleus. Trends in Biochem. Sci. (TIBS) 10, 64-66.

Dixon, L.K. and Ford, P.J. (1982). Regulation of protein synthesis and accumulation during oogenesis in Xenopus laevis. Dev. Biol. 93, 478-497.

Dreyer, C., Scholz E. and Hausen, D. (1982). The fate of oocyte nuclear proteins during early development in Xenopus laevis. W. Roux's Arch. 191, 228-233.

Eliceiri, G.L. (1981). Maturation of low molecular weight RNA species. In The Cell Nucleus 8, H. Busch, ed. (New York: Academic Press), 307-330.

Ford, P.J. (1971). Non-coordinated accumulation and synthesis of 5S ribonucleic acid by ovaries of Xenopus laevis. Nature (Lond.) 233, 561-564.

Fritz, A., Parisot, R.F., Newmeyer, D., and De Robertis, E.M. (1984). Small nuclear U-RNPs in Xenopus laevis development: uncoupled accumulation of the protein and RNA components. J. Mol. Biol. 178, 273-285.

Gurdon, J.B. (1977). Egg cytoplasm and gene control in development. Proc. Royal Soc. London 198, 211-255.

Hinterberger, M., Pettersson, I. and Steitz, J.A. (1983). Isolation of small nuclear ribonucleoproteins containing U1, U2, U4, U5 and U6 RNAs. J. Biol. Chem. 258, 2604-2613.

Honda, B.M. and Roeder, R.G. (1980). Association of a 5S gene transcription factor with 5S RNA and altered levels of the factor during cell differentiation. Cell 22, 119-126.

Kinlaw, C.S., Dusing-Swartz, S.K. and Berget, S.M. (1982). Human U1 and U2 small nuclear ribonucleoproteins contain common and unique polypeptides. Mol. Cell. Biol. 2, 1159-1166.

Mairy, M. and Denis, H. (1971). Recherches biochimiques sur l'oogenèse. I. Synthèse et accumulation du RNA pendant l'oogenèse du crapaud sud-africain Xenopus laevis. Dev. Biol. 24, 143-165.

Madore, S.J., Wieben, E.D. and Pederson, T. (1984). Intracellular site of U1 small nuclear RNA processing and ribonucleoprotein assembly. J. Cell. Biol. 98, 188-192.

Mattaj, I.W. and De Robertis, E.M. (1985). Nuclear segregation of U2 snRNA requires binding of specific snRNP proteins. Cell 40, 111-118.

Mattaj, I.W., Lienhard, S., Zeller R. and De Robertis, E.M. (1983). Nuclear exclusion of TFIIIA and the 42S particle transfer RNA binding protein in Xenopus oocytes: A possible mechanism of gene control? J. Cell. Biol. 97, 1261-1265.

Mattaj, I.W., Zeller, R., Carrasco, A.E., Jamrich, M., Lienhard, S. and De Robertis, E.M. (1985). U snRNA gene families in Xenopus laevis in The Oxford Surveys of Eukaryotic Genes, in press.

Mimori, T., Hinterberger, M., Pettersson, T. and Steitz, J.A. (1984). Autoantibodies to the U2 small nuclear ribonucleoprotein in a patient with scleroderma-polymyositis overlap syndrome. J. Biol. Chem. 259, 560-565.

Pelham, H.R.B. and Brown, D.D. (1980). A specific transcription factor that can bind either the 5S RNA gene or 5S RNA. Proc. Natl. Acad. Sci. USA, 77, 4170-4174.

Picard, B. and Wegnez, M. (1979). Isolation of a 7S particle from Xenopus laevis oocytes: a 5S RNA-protein complex. Proc. Natl. Acad. Sci. USA. 76, 241-245.

Walter, P., Gilmore, R. and Blobel, G. (1984). Protein translocation across the endoplasmic reticulum. Cell 38, 5-8.

Zeller, R., Nyffenegger, T. and De Robertis, E.M. (1983). Nucleocytoplasmic distribution of snRNPs and stockpiled snRNA binding proteins during oogenesis and early development in Xenopus laevis. Cell 32, 425-434.

Synthesis and Structure of a Specific Premessenger RNP Particle

U. SKOGLUND, K. ANDERSSON, B. BJÖRKROTH, M. M. LAMB, and B. DANEHOLT
Department of Medical Cell Genetics, Karolinska Institute,
104 05 Stockholm, Sweden

In the nucleus premessenger RNA is associated with proteins into ribonucleoprotein (RNP) particles[1,2]; the RNA component of the particles is usually referred to as heterogeneous nuclear RNA (hnRNA). The functional significance of the protein constituent is still unclear but several suggestions have been made. It has, for example, been proposed that the proteins assist in the release of the RNA from the template during transcription, prevent RNA from being degraded, select RNA for transport, promote the translocation process and the passage through the nuclear pores. More recently, the particle proteins are believed to be implicated in RNA processing, notably in the splicing process[3].

Since the early studies during the 1960s much basic information has been collected on the structural features of the hnRNA-protein particles (hnRNP)[1,2]. The investigations have usually been carried out on the unfractionated population of hnRNP. A classic study by Georgiev and coworkers demonstrated the polyparticle nature of hnRNP[4]. Using RNase treatment they could show that the intact 30-200 S particles are fragmented into monoparticles sedimenting in a range of 30-50 S. They also reported that the protein composition is simple. It has later been shown that there is a small set of 5-10 abundant proteins, the core proteins[5-9], but there are also maybe 50 or more minor protein species[10]. Therefore, it is not surprising that the monomer particles are heterogeneous in size, shape and protein composition[10]. It is, however, important to stress that the hnRNP particles are organized in a non-random manner[11-13]. Furthermore, this specificity has been correlated to functional events. For example, Miller and coworkers could demonstrate that discrete structural features along growing RNP fibers form a gene-product specific pattern, and that processing of the transcript takes place close to particular granules[11]. In nuclease digestion experiments it has been shown that spliced regions of hnRNP particles have an aberrant sensitivity to nucleases[12,13]. The observed heterogeneity is potentially interesting from a functional point of view, and it is now an important task to search for correlations between specific protein components and possible functions associated with the hnRNP particles.

Nucleocytoplasmic Transport
Edited by R. Peters and M. Trendelenburg

In our laboratory we have focused our attention on the higher order structure of hnRNP. It is known that hnRNP particles in general are built from a 5-10 nm fiber that can form a 20-30 nm thick fiber and even higher order structures[14]. In order to study hnRNP at this higher order structure level we have chosen to analyze a specific hnRNP particle, the Balbiani ring transcription product in the salivary gland cells of the dipteran Chironomus tentans. We have attempted to characterize the higher order structure of the hnRNP particle, to follow the synthesis of the particle and to relate changes in the structure of the particle to phenomena like transport.

Balbiani rings

The salivary glands contain four polytene chromosomes with a large number of chromosomal puffs, i.e. transcriptionally active regions. Three puffs are of exceptional size and have been designated Balbiani rings. The two largest, BR1 and BR2, have been studied in considerable detail[15]. The BR1 and BR2 genes have been cloned and sequenced[16-19]. They display a similar hierarchic structure and belong to the same gene family. Giant transcripts (75 S RNA), corresponding to 37,000 bases[20], are being synthesized[21,22]. These RNA molecules are transferred into the cytoplasm without a major reduction in size[23]. The BR1 and BR2 RNA act as messenger RNA molecules and direct the synthesis of the salivary polypeptides Sp Ia and Sp Ib, respectively[24]. The large-sized polypeptides are used by the larva to spin a tube-like structure forming the dwelling of the larva. It can be concluded that not only the BR genes but also the primary BR transcripts and the corresponding messenger RNAs are of exceptional size. These properties of the system make it particularly suitable for ultrastructural work.

Structure of the nascent BR RNP particle

When salivary glands were studied in sectioned material by electron microscopy the polytene chromosomes as well as the Balbiani rings could be readily identified within the nucleus. The active genes appear as loops with growing RNP products attached to the chromatin axis[25-27]. In Fig. 1, a section through a Balbiani ring is displayed, and loop segments have been encircled. The chromatin axis can barely be seen, but the laterally extending RNP products are easily recognized. In order to establish the complete structure of such a segment, serial sectioning was performed, and a simple reconstruction could be done. Furthermore, it was possible to outline the structure of a complete putative gene from a large number of segments studied. The result is shown in Fig. 1 b. Specific portions of the gene

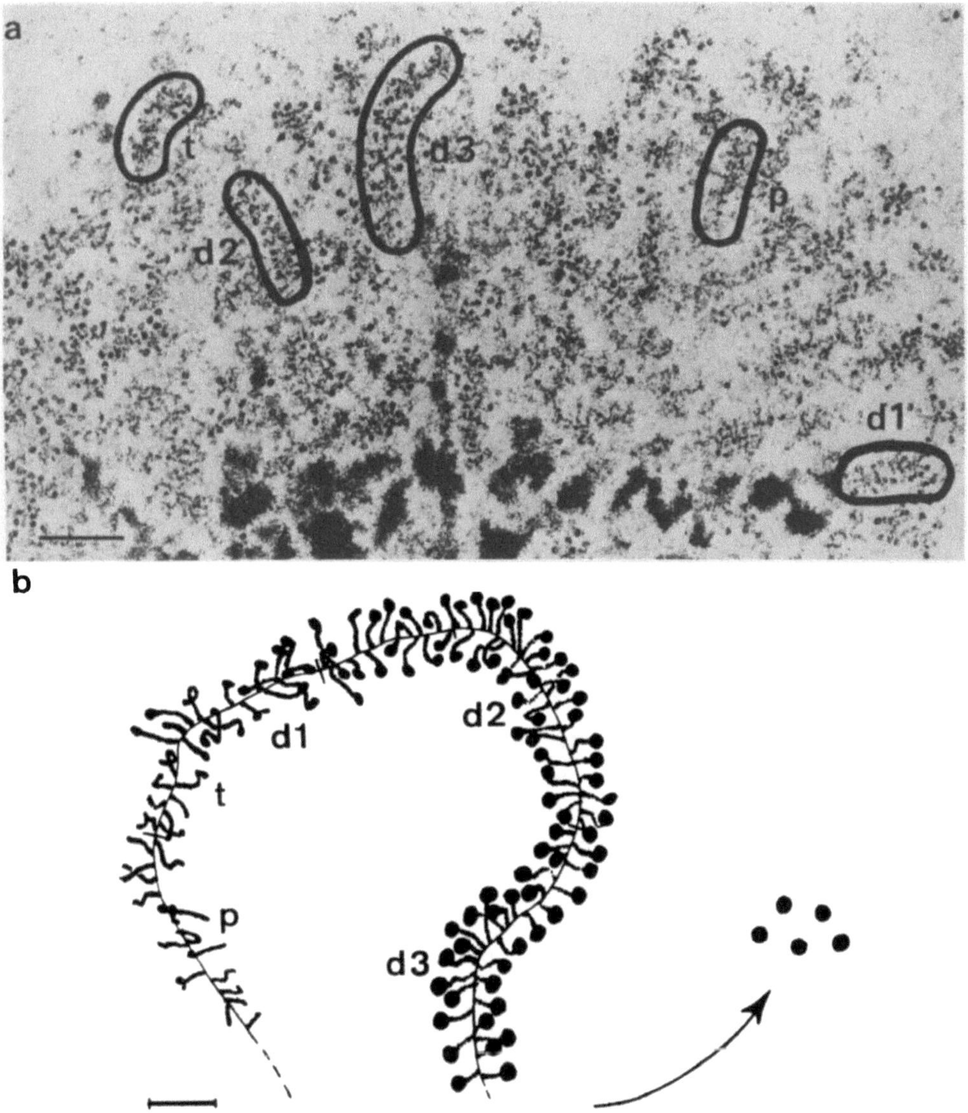

Fig. 1. The ultrastructure of a Balbiani ring gene. In a five representative segments of the gene have been encircled and designated: p = proximal, t = transitional, and d 1-3 = distal regions. Reconstruction of one complete Balbiani ring gene with the various segments indicated is displayed in b. The bar represents 500 nm in a and 200 nm in b.

has been given letter designations, and examples of the various regions are indicated in Fig. 1 a. The growth and packing of the transcription product can be followed in Fig. 1 b. The product appears as a 19 nm fiber and is subsequently packed into a dense globular structure; the growing particle with the globular portion can be described as a stalked granule. The globular part increases in size along the gene, and the completed product is released as a spherical particle.

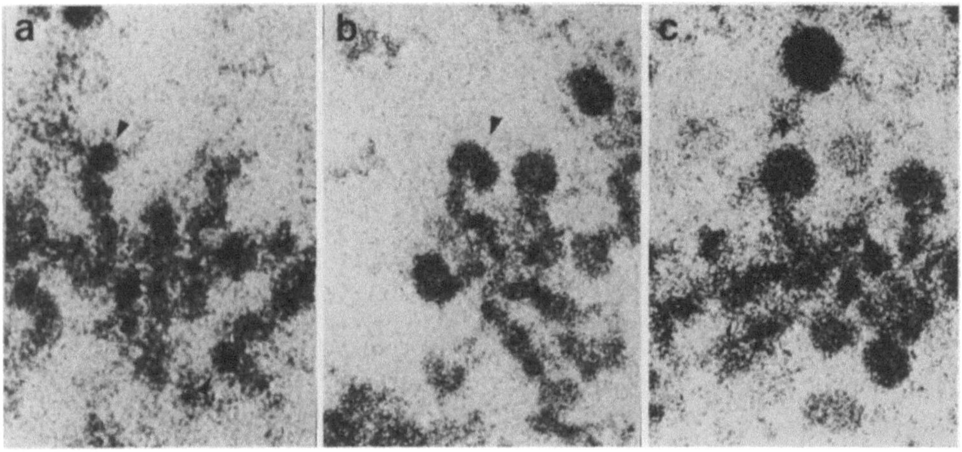

Fig. 2. Electron micrographs of growing RNP particles in various stages of maturation (a-c). In c a completed RNP particle can be recognized above the indicated stalked granule.

In a detailed study of the structure of the growing RNP products, we noted that the globular part is not just an expanding sphere but rather consists of a 26 nm thick RNP fiber, that is gradually increasing in length and moulded into the spherical shape of the particle[27]. This can be seen in the three examples of growing RNP particles displayed in Fig. 2. In Fig. 2 a the 19 nm fiber has just started to condense at the tip of the fiber. Further downstream the gene, the dense element can be seen as a thick fiber of 26 nm (Fig. 2 b), and towards the end of the gene the fiber curls up into a ring-like configuration (Fig. 2 c). In Fig. 2 c it is also possible to recognize a released particle lacking a stalk; the stalk material has most likely been included in the particle (see below).

The stalk of some growing RNP fibers is partly unfolded, and a constituent 10 nm fiber is then exhibited[27]. Miller spreads of active BR genes also suggested that the BR particle is composed of a 10 nm RNP fiber (Fig. 3)[28]. The spread genes showed the well-known "Christmas tree" structure, i.e. a thin chromatin axis is seen as well as a number of RNP fibers, sticking out perpendicularly to the axis and increasing gradually in length along the gene. Since the RNP fibers in the proximal portion as well as in the distal portion of the gene showed the same dimension and knobby morphology (Fig. 3), we conclude that the entire growing product, the 19 nm as well as the 26 nm component, is composed of the 10 nm RNP

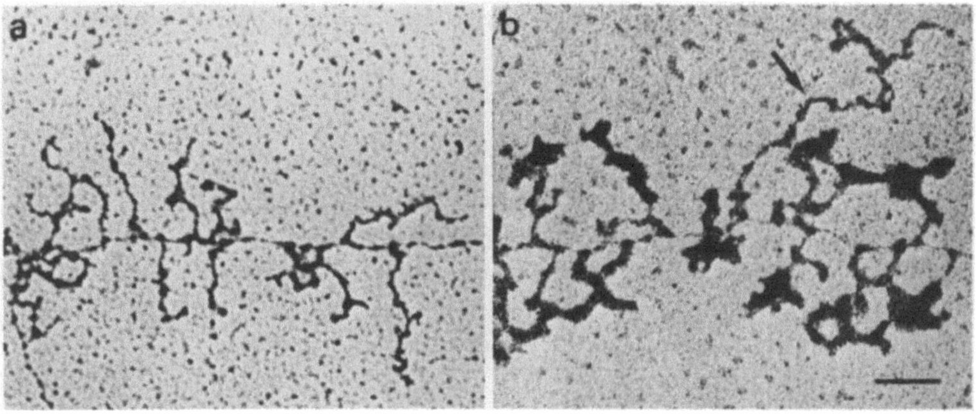

Fig. 3. Details from Miller spreads of the proximal (a) and distal (b) parts of the Balbiani ring gene. The growing RNP fibers appear in the proximal region as "knobby", 10 nm fibers. The RNP particles in the distal region do not spread as readily, but occasionally the entire RNP particle is unfolded and exhibits the 10 nm fiber element (arrow). The bar denotes 100 nm.

fiber. Evidently, the 10 nm element is immediately formed upon RNA synthesis and coiled into the 19 nm fiber. In a second step it is further packed into the dense 26 nm fiber. The packing mode in this second step cannot simply be a coil of the 10 nm fiber or a coil of the 19 nm fiber (for discussion, see ref. 27); a repacking of the 10 nm fiber into a more complex fold has to be assumed. The 26 nm fiber is finally bent into the ring-like structure, the BR granule.

3-D structure of the BR RNP particle

We have recently developed a method that allows three dimensional reconstruction of ultrastructural objects[29]. The method is referred to as electron microscopic computed tomography. It allows an objective and detailed analysis of ultra-structural objects at a molecular level.

In order to reconstruct a BR particle we started out from a section through a cell nucleus containing a number of released BR granules. Gold markers were added to the surface of the section to serve as precise position markers. The BR particle was photographed in the electron microscope at different angles of incidence, representing every ten degrees in the range $\pm 60^{o}$. The tilted series obtained represents a number of projections through the object. The electron micrographs were digitized, and due to the gold markers they could be properly aligned with an iterative least squares procedure[30]. The reconstruction was carried out by a back projection method[31] and was performed in a stepwise fashion completing one slice through the object after the other. The complete reconstruction is presented as a balsa wood model.

We have studied four individual particles and found that they display very similar gross morphology[29]. The calculated correlation coefficients turned out to be high (about 0.8), which permitted us to calculate the structure of an average particle. This structure is presented in Fig. 4. It can be seen in Fig. 4 a that the particle is asymmetric and has a hole in the center and a slit at 1-2 o'clock. The result confirms the earlier conclusion that the particle has a ring-like structure The RNP element is, however, a thick ribbon rather than a fiber.

Some details of the structure in Fig. 4 should be stressed. Four domains can be recognized and they are demarcated in Fig. 4. Domain 1 starts at the slit and extends downwards. There is a sharp transition to the broad domain 2, which comprises the bottom of the particle and reaches up on the other side of the particle. Domain 3 is a short and narrow portion of the ribbon between domain 2 and the broad and irregular domain 4. The ring-like configuration is somewhat skewed, which gives the structure a slight left-handedness.

The asymmetry of the particle made it possible to identify the 5' and 3' domains of the structure. This could be accomplished by detailed comparisons between the features of the growing RNP particles and the domain organization shown in Fig. 4. It was established that the smooth domain 1 is in the lead and that it is followed by the heavy domain 2 and the narrow domain 3; the alternative order, viz. 4, 3 and 2, could be excluded. We conclude that domain 1 corresponds to the 5' end and

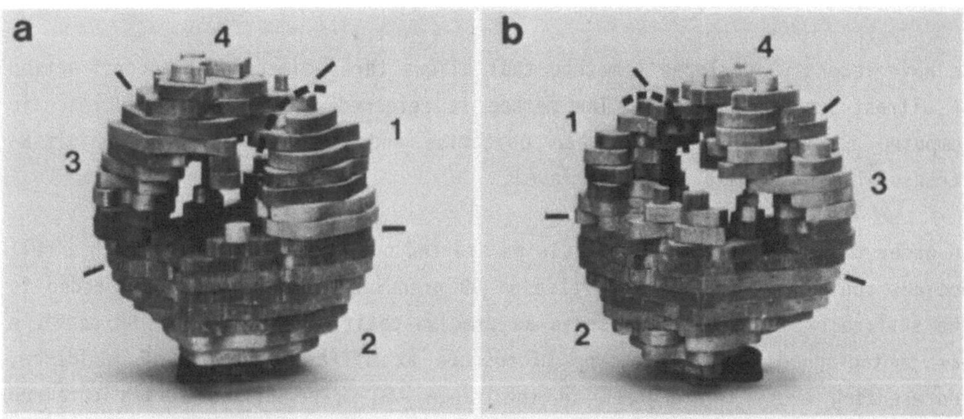

Fig. 4. The average structure of the reconstructed Balbiani ring RNP particle. The front view of the particle is shown in a and the back view in b. The four domains have been demarcated with numbers and division lines.

domain 4 to the 3' end of the transcription product. Towards the end of the gene it is possible to recognize domain 1, 2 and 3 but not 4 in the globular portion of the particle. Therefore, it seems reasonable to assume that the irregular domain 4 is generated from the stalk upon release of the particle.

At the present resolution of 80-90 Å we cannot discern the 100 Å fiber within the 3-D reconstruction. However, since the ribbon has a thickness of 100-150 Å it seems likely that the 100 Å fiber is repeatedly folded from one side of the ribbon to the other concomitant with the growth of the ribbon from domain 1 towards domain 4. We can improve the resolution in the reconstruction by including more tilted views and by using a higher magnification. Preliminary studies suggest that a resolution of 40-50 Å can be reached, which should be sufficient to follow the course of the fiber within the particle.

Functional implications of the higher order structure

The molecular mechanisms responsible for the folding of the thin RNP fiber into the specific higher order structure are unknown, but some observations on the nature of the packaging process in the Balbiani rings are of interest in this context. During transcription, the thin RNP fiber does not directly fold into the defined, higher order structure. Instead the RNP fiber first forms a coil and only after a delay period is it tightly packed into the ring-like RNP ribbon. It seems plausible that a monotonous, core structure of the RNP fiber is laid down initially (cf. Introduction), and that in a subsequent step the higher order structure becomes established, e.g. by specific modifications of the proteins of the fiber or by the addition of one or more packaging proteins.

The finished and released particles have all acquired the well-defined, higher order structure. Since in eukaryotic cells addition of the poly(A) tail as well as splicing take place after the termination of transcription[32], these processes should therefore occur on RNP particles in their mature 3-D configuration. Regarding the BR products it is known that poly(A) is added to the transcripts[33], but whether the transcripts are spliced has not been settled. We conclude that the higher order structure has to be taken into account when posttranscriptional events are being considered.

The higher order structure is likely to be closely related to the transport process. The structure of the particle is remarkably compact with the dense RNP ribbon moulded into the spherical shape of the particle, suggesting that such a packing mode is well suited for transport of a large transcript. The mechanisms involved in the actual transport of the particle from the gene to the nuclear pores are essentially unknown. It might be relevant that the BR granules appear closely associated with a fine fibrous network in the nucleus, but it is not clear whether these contacts are specific and do reflect an in vivo situation. A number of BR particles also appear positioned in front of the nuclear pores indicating a recognition of the pores.

During the translocation of the BR product through the nuclear pore, the particle changes its ring-like configuration to an extended structure with an approximate diameter of 250 Å and a length of about 1350 Å[34,35]. When the leading edge of the ribbon reaches cytoplasm, the relaxation of the structure continues and the unfolded, thin RNP fiber can be seen extending further into the cytoplasm. It is possible that the loss of the ring-like structure is a consequence of a forced passage through a narrow pore. It is, however, more plausible to assume that the translocation process consists of a sequence of events, the relaxation of the higher order structure of the particle being an early and regulated step in this process.

Conclusions

A specific premessenger RNP particle, the Balbiani ring RNP granule, has been studied with electron microscopy when it is being formed on the gene template as well as after it has been completed and released from the gene. Initially, a 10 nm RNP fiber is formed and coiled into a 19 nm fiber. The 10 nm fiber is subsequently repacked into a dense 26 nm fiber, which is bent into a slightly left-handed, ring-like structure, the BR granule.

The completed product has recently been studied by a new method, electron microscopic computed tomography. The ring-like configuration is confirmed, but the RNP element is better described as a thick ribbon than as a 26 nm fiber. The particle is asymmetric and consists of four domains; the 5' and 3' domains are located close to each other in the ring-like configuration. The higher order structure is discussed in relation to available information on the synthesis and processing of the particle, its attachment to the nuclear fiber network during transport and finally its translocation through the nuclear pore.

Acknowledgement

We thank Evy Vesterbäck for typing the manuscript. This research was supported by the Swedish Natural Science Research Council, the Swedish Cancer Society and

Gunvor and Josef Anērs Stiftelse. The electron microscopic computed tomography was made possible by generous support from the Wallenberg Foundation.

References

1. Pederson, T. (1983). J. Cell Biol. 97, 1321-1326.
2. Knowler, J.T. (1983). Int. Rev. Cytol. 84, 103-153.
3. Grabowski, P.I., Seiler, S.R. and Sharp, P.A. (1985). Cell 42, 345-353.
4. Samarina, O.P., Lukanidin, E.M., Molnar, J. and Georgiev, G.P. (1968). J. Mol. Biol. 33, 251-263.
5. Economides, I.V. and Pederson, T. (1983). Proc. Natl Acad. Sci. 80, 1599-1602.
6. Beyer, A.L., Christensen, M.E., Walker, B.E. and Le Stourgeon, W.M. (1977). Cell 11, 127-138.
7. Karn, J., Vidali, G., Boffa, L.C. and Allfrey, V.G. (1977). J. Biol. Chem. 252, 7307-7322.
8. Jones, R.E., Okamura, C.S. and Martin, T.E. (1980). J. Cell Biol. 86, 235-243.
9. Choi, Y.D. and Dreyfuss, G. (1984). Proc. Natl. Acad. Sci. 81, 7471-7475.
10. Jacob, M., Devilliers, G., Fuchs, J.-P., Gallinaro, H., Gattoni, R., Judes, C. and Stevenin, J. In The Cell Nucleus, vol. VIII (ed. Busch, H.), pp. 193-246. Academic Press, New York, 1981.
11. Beyer, A.L., Miller, O.L. Jr. and Mc Knight, S.L. (1980). Cell 20, 75-84.
12. Steitz, J.A. and Kamen, R. (1981). Mol. Cell Biol. 1, 21-34.
13. Ohlsson, R.I., van Eekelen, C. and Philipson, L. (1982). Nucleic Acid Res. 10, 3053-3068.
14. Sommerville, J. In The Cell Nucleus, vol. VIII (ed. Busch, H.), pp. 1-57. Academic Press, New York, 1981.
15. Daneholt, B. In Insect Ultrastructure, vol. 1 (ed. King, R. and Akai, M.), pp. 382-401. Plenum Publishing Corporation, New York, 1982.
16. Sümegi, J., Wieslander, L. and Daneholt, B. (1982). Cell 30, 579-587.
17. Wieslander, L., Sümegi, J. and Daneholt, B. (1983). Proc. Natl Acad. Sci. 79, 6956-6960.
18. Case, S.T. and Byers, M.R. (1983). J. Biol. Chem. 258, 7793-7799.
19. Case, S.T., Summers, R.L. and Jones, A.G. (1983). Cell 33, 555-562.
20. Case, S.T. and Daneholt, B. (1978). J. Mol. Biol. 124, 223-241.
21. Daneholt, B. (1972). Nature 240, 229-232.
22. Egyhazi, E. (1975). Proc. Natl Acad. Sci. 73, 947-950.
23. Daneholt, B. and Hosick, H. (1973). Proc. Natl Acad. Sci. 70, 442-446.
24. Edström, J.-E., Rydlander, L. and Francke, C. (1980). Chromosoma 81, 115-124.
25. Andersson, K., Björkroth, B. and Daneholt, B. (1980). Exptl. Cell Res. 130, 313-327.
26. Olins, A.L., Olins, D.E. and Franke, W.W. (1980). Eur. J. Cell Biol. 22, 714-723.
27. Skoglund, U., Andersson, K., Björkroth, B., Lamb, M.M. and Daneholt, B. (1983) Cell 34, 847-855.
28. Lamb, M.M. and Daneholt, B. (1979). Cell 17, 835-848.
29. Skoglund, U., Andersson, K., Strandberg, B. and Daneholt, B. (1985). Submitted
30. Bricogne, G. and Skoglund, U. In preparation.
31. Crowther, R.A., DeRosier, D.J. and Klug, A. (1970). Proc. Roy. Soc. A. 317, 319-340.
32. Darnell, J.E., Jr. (1982). Nature 297, 365-371.
33. Egyhazi, E. (1980). Eur. J. Biochem. 107, 315-322.
34. Beermann, W. In 13. Colloquium der Gesellschaft für physiologische Chemie am 3.-5. Mai 1962 in Mosbach/Baden, pp. 64-100. Springer Verlag, Berlin, 1962.
35. Stevens, B.J. and Swift, H. (1966). J.Cell Biol. 31, 55-77.

Subject Index

MIX
Papier aus verantwortungsvollen Quellen
Paper from responsible sources
FSC® C105338

If you have any concerns about our products,
you can contact us on
ProductSafety@springernature.com

In case Publisher is established outside the EU,
the EU authorized representative is:
Springer Nature Customer Service Center GmbH
Europaplatz 3, 69115 Heidelberg, Germany

Printed by Libri Plureos GmbH
in Hamburg, Germany